Following the Truth, Wherever It Leads

Following the Truth, Wherever It Leads

An Investigation of What Is Reality (and How It Affects Our Lives)

KENNETH G. REDDINGTON

WIPF & STOCK · Eugene, Oregon

FOLLOWING THE TRUTH WHEREVER IT LEADS
An Investigation of What Is Reality (and How It Affects Our Lives)

Copyright © 2015 Kenneth G. Reddington. All rights reserved. Except for brief quotations in critical publications or reviews, no part of this book may be reproduced in any manner without prior written permission from the publisher. Write: Permissions. Wipf and Stock Publishers, 199 W. 8th Ave., Suite 3, Eugene, OR 97401.

Wipf and Stock
An Imprint of Wipf and Stock Publishers
199 W. 8th Ave., Suite 3
Eugene, OR 97401

www.wipfandstock.com

ISBN 13: 978-1-62032-747-0

Manufactured in the U.S.A.   03/19/2015

Unless otherwise noted, Scripture quotations are taken from the New King James Version®. Copyright © 1982 by Thomas Nelson, Inc. All rights reserved.

Contents

Introduction: Following the Truth, Wherever It Leads | vii

PART 1: An Openness to Truth, Regardless of the Source

> Preface to Part 1 | 3

Chapter 1: What in Life Is Really Important? | 7

Chapter 2: Approaching Science with an Open Mind | 14

Chapter 3: Approaching the Bible with an Open Mind | 22

PART 2: What Do We Really Know about the World? The Science

Chapter 4: Charles Darwin and His Theory | 35

Chapter 5: What Is Modern Darwinism? | 49

Chapter 6: What Can We Learn from the Cosmos? | 56

Chapter 7: Evolution Is Real but Limited | 71

Chapter 8: Biology and the Complexity of the Cell | 80

Chapter 9: Geology and an Old Earth | 92

Chapter 10: Geology and a Young Earth | 103

Chapter 11: The Origin of the Human Race | 131

Chapter 12: Is Intelligent Design Science? | 146

Chapter 13: The Unwrapping of Evolution: Does Any Theory Satisfy Objective Analysis? | 172

Chapter 14: Summary of the Science: When All Is Said and Done What Do We *Really* Know about the Natural World? | 203

PART 3: What Do We Really Know about the World? The Bible

>Preface to Part 3 | 215

Chapter 15: A Philosophy of Science for Harmonizing Science and Scripture | 221

Chapter 16: The Traditional View of Creation | 229

Chapter 17: Ramm's View of Creation | 233

Chapter 18: Young Earth Creationism | 239

Chapter 19: Geology and Old Earth Creationism | 248

Chapter 20: The Cosmos and Old Earth Creationism | 255

Chapter 21: Other Views on Interpreting the Bible | 263

Chapter 22: Are Science and Faith Compatible? | 279

Chapter 23: An Evaluation of Evolutionism and Creationism | 290

Chapter 24: Response | 305

Chapter 25: When the Pursuit of Truth Hits a Speed Bump | 319

PART 4: How Significant Is This for Our Daily Lives?

>Preface to Part 4 | 337

Chapter 26: Where Does Truth Lead Us? | 339

Chapter 27: The Bible as Truth | 342

Chapter 28: Life's Questions | 345

Chapter 29: Is Truth a Destination or a Process? Experiencing Truth | 353

Chapter 30: Living Life to the Fullest: Can Clarifying One's Worldview Give New Meaning to Life? | 355

Appendix | 373

Endnotes | 390

Bibliography | 405

Indices

>Seeking the Truth by Scientific Discovery | 409

>Seeking the Truth by Revelation | 413

Introduction
Following the Truth, Wherever It Leads

What can we rely on as enduring truth in a world of constant change?

How open are you to truth in order to make the best out of your life in a rapidly changing world?

What does it mean to be human anyway?

How do we know that what we believe about ourselves and about our world corresponds to reality? As philosopher Francis Schaeffer expressed it, "How can we be sure that what we think we know of the world outside ourselves really corresponds to what is there?"[1]

My whole life has been a quest for knowledge about myself and about the world I live in. I grew up in a world of uncertainty and change. In school I liked mathematics best because I knew the answers were clearly either right or wrong. I also liked the natural sciences, especially physics, because the cause and effect theory of physics made sense to me. Concepts like the Doppler effect impressed me deeply—because they helped me make sense of the things I experienced. (The Doppler effect is the sound heard from a fast-moving train as it approaches with a higher pitch than it makes after it passes by. Albert Einstein used it as an illustration of the effects of his theory of relativity.)

In high school I majored in math and science (physics and chemistry, not biology). It was right after the atom bombs were dropped on Hiroshima and Nagasaki, and I was intrigued about nuclear physics and its power, and wanted to study it in college. I had a scholarship to a college of science, but was unable to go.

I think I was unconsciously seeking objective "truth" that could guide me in life and that I could build my life upon. As I observed so much change around me I think I was seeking principles and reality that do not change in any age or circumstance.

My wife and I spent over forty happy years in Japan. I have many happy memories of Japan, and a few not-so-happy ones.

Introduction

One happy memory was living for a year in a farming village in a narrow valley near the foot of Mt. Fuji. The villagers were very welcoming and it was a wonderful place to raise our small children. Everyone knew each other, and no one locked their doors at night. Our children were five and three. They were free to go anywhere and were treated so kindly they often didn't even come home for meals—they were fed wherever they happened to be at mealtime.

We learned to appreciate different ways to do things, and to evaluate more objectively the ways we learned to cope when growing up. In this village they didn't yet have electricity or even modern plumbing. We got our drinking water from a small stream right behind our home. It was very clean, and the whole area outside every home was clean. Cleanliness and sanitation were valued very highly in traditional Japan. Clean water was precious, and every night a farm family would fire up the bathtub (*ofuro*) in a shed outside of their home and every member of their family would wash outside the tub and then warm up and relax their muscles by soaking in the same hot water that every one used. In a community where water was scarce and firewood was prepared by much effort, this seemed to us a wise and practical custom.

We experienced many things like this that impressed on us that our American ways are not superior simply because that is what we are used to. Openness to other views and practices helped us to cope and live more effectively.

However, in observing and taking part in many new ways of doing things we also learned that the Japanese are just like us in so many ways. We share the same basic human nature.

Another happy memory was when we were invited as a family to live on the campus of a private middle school and high school in western Japan and to teach English there. Our children were of high school age and studied by correspondence. All the Japanese students in this school lived in dormitories on campus, and our three children spent much time together with them and grew up together. Some of the friendships developed there remain to this very day.

Our children and the others on campus were similar in so many ways. They no doubt had the same questions about the meaning of life: the same concerns about getting education and a good job, about eventually leaving home, getting married, and developing a career. They shared the same concerns about life and discovering how to live life to the fullest. They share the same human nature, with all its fears and doubts, encouragements and setbacks.

One unhappy memory was while teaching English composition at a private high school in Japan. I was teaching one afternoon and overheard a teacher in the next classroom tell his class that the Bible is a bunch of myths (fairy tales). I thought: myths? Had he studied it? Was he aware of the many archeological discoveries that confirm biblical events that happened thousands of years ago, and of cultures that existed then? Was he aware of the modern scientific discoveries that confirm what the Bible says about nature? Or was he just sharing an opinion that he had not honestly investigated, to students who would take his personal opinion as the well-thought-out conclusion of an "authority"?

Another unfortunate memory happened many years later. While watching national television I saw on the screen some kind of living thing crawl out of a dirty-looking body of water. As I continued to look, this thing gradually changed shape until it looked like an ape and finally like that of a human. I realized that this is a common perception of how humanity developed over millions or billions of years from very simple one-cell organisms. This is commonly taught in school textbooks in Japan, America, and probably all over the world. But where is the scientific evidence behind it?

I have remembered these two events in Japan ever since. They suggest popular images that may not be based on reality, and which I believe deserve close scrutiny. It will take someone with a broad knowledge of both the Bible and modern science to research and explain whether the teachings of the Bible are consistent with the discoveries of modern science. I had the broad knowledge of the Bible: I have an undergraduate degree and a graduate degree in Bible from Columbia International University. But I lacked the background in natural science. My second master's degree and PhD are in the social sciences (psychology, sociology, and education). Also, my life in Japan was so busy that I could not spend time needed to do the research.

After retiring in America, I have devoted a dozen years to a careful study of the scientific literature. I read Charles Darwin's *The Origin of Species by Means of Natural Selection or the Preservation of Favoured Races in the Struggle for Life* (when referring to this I will abbreviate it to *Origin*). I also read part of *The Descent of Man* by Darwin. I did much reading in cosmology, biology, and geology. Books on the philosophy of science also proved helpful.

I have found that recent discoveries in cosmology (astrophysics) and in biology are giving scientists increasing confidence in making some conclusions regarding the realities of nature. I also read the writings of leading scientists who consider themselves evolutionists but are now questioning much of what is still being taught as "evolution" and as "scientific fact" in public school textbooks, and even in many popular college textbooks.

The scientific establishment attempts to project an image of unity in beliefs about origins and evolutionary development. But even within the scientific community there are evolutionists who are saying that biological evolution like Darwin proposed is actually quite limited. Many scientists also now have doubts about the possibility that life could have developed on earth in the way evolutionary theory describes.

I believe a true scientific attitude makes one open to change his thinking when confronted with facts that have been adequately verified. Small changes in belief are easy to reconcile with one's general understanding of the world. But to adjust one's thinking to truth that challenges deep-seated beliefs is more challenging. Stephen Jay Gould was a well-known scientist who faced this dilemma. He remained an

Introduction

evolutionist while seeking a new theory that explains the facts better than the slow, incremental biological changes proposed by Darwin's theory. (We will report on Gould in chapter 12.)

Another scholar was the investigative journalist Lee Strobel.

A Change of Worldview

I did not study evolution in any detail when in public school. So here I would like to relate the experience of someone who did.

His name is Lee Strobel. He was an award-winning journalist at the *Chicago Tribune* for about twenty years. He has written three popular books that talk of the long journey he took that completely changed his thinking on science and religion. In *The Case for a Creator*, Strobel explains his journey from faith to skepticism and back to faith again.

Strobel says he was raised in a Christian home where faith in the existence of God as Creator was believed as something that no one questioned. But in school he heard his teachers explain how science has proven that the living things around us developed by evolution just like Darwin taught, and that faith in a Creator is old-fashioned. He saw marvelous illustrations of this in his biology textbooks. He became totally convinced that Darwinism was true and that belief in the Bible would gradually die out as scientific thinking triumphed.

In his life as an investigative journalist, Strobel was sent to gather facts and write newspaper stories about various events. Once he and a cameraman attended a meeting in a country town in West Virginia where a large crowd of Bible-believers were discussing whether to have their children boycott classes. They were angry because they felt some of the books that were being read in school taught things that were contrary to their beliefs. Strobel left that gathering wondering why people still believed in Christianity in spite of the scientific proofs that Darwin's theory of evolution had shown that such beliefs were outmoded.

Strobel had come to believe what prominent evolutionary biologist and historian William Provine of Cornell University later expressed in five concise statements. He decided that if Darwinism is true, then we can accept Provine's following conclusions:

- there's no evidence for God;
- there's no life after death;
- there's no absolute foundation for right and wrong;
- there's no ultimate meaning of life; [and]
- people don't really have free will.[2]

Introduction

One of the things that made a deep impression on Strobel in school was what he calls four "images of evolution." These were drawings and pictures in his biology textbook that left a deep and abiding impression on his mind. In his thinking, these illustrations were images that confirmed the truth of evolution.

The first image was the laboratory apparatus that Stanley Miller used in 1953 to artificially produce the building blocks of life. Miller tried to recreate the atmosphere of the primitive earth when life first began. Then he shot electric sparks through it to simulate lightning. Miller managed to produce a red goo containing amino acids. Because amino acids are essential for life, this shows that life could have developed on earth solely through natural processes. Strobel concluded that a Creator was not needed to produce life on earth. He called this "the most powerful picture of all."

Strobel relates the following experience related to the second image.

> The first time I read Charles Darwin's *The Origin of Species*, I was struck that there was only one illustration: a sketch in which he depicted the development of life as a tree, starting with an ancient ancestor at the bottom and then blossoming upward into limbs, branches, and twigs as life evolved with increasing diversity and complexity.
>
> As a recent textbook explained, Darwinism teaches that all life forms are "related through descent from some unknown prototype that lived in the remote past."[3]

This strengthened Strobel's conviction. Miller seemed to prove that life could develop on earth spontaneously, and Darwin explained how that primitive life could slowly and gradually develop into the biological complexity we observe today.

In the third image, he discussed the sketches of embryos made by German biologist Ernst Haeckel. Haeckel drew embryos of a fish, salamander, tortoise, chicken, hog, calf, rabbit, and a human. He placed these sketches side by side to show that they were almost identical. Haeckel showed how these embryos later became distinctly different. Strobel thought that this was dramatic evidence of universal common ancestry.

Strobel calls the fourth image the most famous fossil in the world. It is called the *archaeopteryx*, or "ancient wing." This fossil is considered to be about 150 million years old. As Strobel expressed it, "With the wings, feathers, and wishbone of a bird, but with a lizard-like tail and claws on its wings, it was hailed as the missing link between reptiles and modern birds."[4]

He says that the picture of this fossil convinced him that the fossil record supported Darwin's theory.

Strobel says these four illustrations, added together, totally convinced him that Darwin's theory was correct. First, Miller demonstrated it was possible for life to develop on earth spontaneously from non-life. Second, "Darwin's theory accounted for how so many millions of species of organisms could slowly and gradually develop over huge expanses of time." Third, Haeckel showed that, as embryos develop, they show dramatic similarities that demonstrate universal common ancestry. Fourth was a fossil of a half-bird, half-reptile that was clearly one of the missing links that Darwin predicted would confirm his theory.

Faced with this evidence of the validity of Darwin's theory, Strobel fully rejected all he had been taught about God. He believed that it showed that God was not needed for the creation and development of life. No supernatural deity was needed either to produce life or guide in its development.

He remarks that "one recent textbook was very clear about this: By coupling undirected, purposeless variation to the blind, uncaring process of natural selection, Darwin made theological or spiritual explanations of life processes superfluous."[5]

Strobel quotes two authorities on the evolution-creation controversy to support this view. The first is by Stephen C. Meyer, a Cambridge-educated philosopher of science. Meyer wrote that "many evolutionary biologists admit that science cannot categorically exclude the possibility that some kind of deity still might exist."[6] Meyer said it is possible that a divine designer might work creatively in such a way that scientists cannot tell that he was working. He could make his interventions in nature look like it was only natural processes at work. No one can deny that this is a possibility. Strobel concluded, "Yet for most scientific materialists such an undetectable entity hardly seems worthy of consideration."

Meyer added that "contemporary Darwinism does not envision a God-guided process of evolutionary change." Then Meyer mentions the often-quoted observation by evolutionary biologist George Gaylord Simpson: "Man is the result of a purposeless and natural process that did not have him in mind."[7]

The second authority Strobel quotes is Phillip Johnson, a law professor who has been at the forefront of the evolution-design controversy. Strobel quotes him as saying that "the whole point of Darwinism is to show that there is no need for a supernatural creator, because nature can do the creating by itself."[8]

How did Strobel respond to this "God is dead" emphasis by Darwinians? He says that he wasn't aware of these kinds of observations when he was a student. "I just knew intuitively that the theories of Darwin gave me an intellectual basis to reject the mythology of Christianity that my parents had tried to foist on me through my younger years."[9]

"The mythology of Christianity!" Where have I heard that thinking before? It took me back decades to when I taught in Japan and overheard a teacher tell his students that "the Bible is a bunch of myths."

Strobel as a teenager read in the *World Book Encyclopedia* the following: "In the Bible, God is held to be the Creator, the Sustainer, and the Ultimate End of all things."[10] It went on to say that many Christians find it impossible to reconcile this conviction with the idea of evolution like Darwin depicted.

Strobel's says that when he read that everything fell into place.

> My assessment was that you didn't need a Creator if life can emerge unassisted from the primordial slime of primitive earth, and you don't need God to create human beings in His image if we are merely the product of the impersonal forces of natural selection. In short, you don't need the Bible if you've got *The Origin of Species*.[11]

At this point Strobel quotes the British atheist philosopher Bertrand Russell, who wrote about science having revealed the world as "purposeless" and "void of meaning." Russell laments that, because mankind is a product of natural causes, death ends all of an individual's hopes, aspirations, and human genius, and they are destined to extinction. "Only on the firm foundation of unyielding despair, can the soul's habitation henceforth be safely built."[12]

Strobel says that Russell seems to be lamenting the purposelessness and lack of meaning in a life with no hope beyond the grave. But he says he responded very differently.

> Rather than facing this "unyielding despair" that's implicit in a world without God, I reveled in my newly achieved freedom from God's moral strictures. For me, living without God meant living one hundred percent for myself. Freed from someday being held accountable for my actions, I felt unleashed to pursue personal happiness and pleasure at all costs.
>
> Who cared if scientific materialism taught that there is nothing other than matter and therefore no person could possibly survive the grave? I was too young to trifle with the implications of that; instead, I pursued the kind of immortality I could attain by leaving my mark as a successful journalist, whose investigations and articles would spur new legislation and social reform.[13]

But a few years later Strobel's wife became a Christian. He reacted negatively. "I simply couldn't comprehend how such a rational person could buy into an irrational religious concoction of wishful thinking, make-believe, mythology, and legend."

He continues, speaking of the change in his wife:

> In the ensuing months, however, as Leslie's character began to change, as her values underwent a transformation, as she became a more loving and caring and authentic person, I began asking . . . in a softer and more sincere tone of genuine wonderment, *"What has gotten into you?"* Something—or as she would claim, Someone—was undeniably changing her for the better.[14]

Strobel could see a positive change in his wife's life and attitude that I believe is often evident when people believe in God in their hearts and are "born again"—as the Bible terms this transformation. When this change of heart is accompanied by a sincere desire to obey the Bible in daily living, the resulting change soon becomes evident. Strobel's response is interesting.

Introduction

> Clearly, I needed to investigate what was going on. And so I began asking questions—a lot of them—about faith, God, and the Bible. I was determined to go wherever the answers would take me—even though, frankly, I wasn't quite prepared back then for where I would ultimately end up.[15]

After seeing the positive change in his wife, Strobel spent the next few years investigating the historical evidence for Jesus' life on earth and seeking answers to the many questions he said troubled him in his youth. Then some years passed, and during much of this time he was busy as the legal affairs editor for the *Chicago Tribune*. He called it "the most powerful newspaper in the Midwest." (His education to prepare for this work include a journalism degree from the University of Missouri and a master of studies degree from Yale Law School. This training proved helpful in his later investigations in the fields of science and theology. He says he learned how to question people like a lawyer does. This approach often leads to a better understanding of the real facts of the matter under discussion.)

At the end of this time with the *Chicago Tribune* he returned to his pursuit of answers to questions of science and faith. He used his training as an investigative reporter to interview experts in various fields. As fruit of these investigations, and to document his journey from atheism to faith, he wrote the award-winning book *The Case for a Creator* (2004). It is subtitled *A Journalist Investigates Scientific Evidence That Points toward God*.

Because he was a well-known journalist, Strobel had the opportunity to interview many leading thinkers in America. He says, "My approach would be to cross-examine authorities in various scientific disciplines about the current findings in their fields. I sought doctorate-level professors who have unquestioned expertise, are able to communicate in accessible language, and who refuse to limit themselves only to the politically correct world of naturalism and materialism."[16]

In his interviews he determined not only to ask people questions, "but to go wherever the answers would take me." He quotes Linus Pauling, the two-time Nobel Prize winner: "Science is the search for truth." His books show that he confronted each expert he interviewed with direct questions to see if they have the scientific or historical facts to support each of their conclusions about scientific truth and spiritual truth.

He says that with an open (unprejudiced) mind he sought answers to questions such as the following: Will science and faith always be at war with each other? To be science-minded, must an individual avoid matters of religious belief? Should spiritual matters and scientific issues be viewed in a fundamentally different way?

Strobel encourages others to be willing to challenge things they had been taught in the classroom, in the same way he did. "Scientists themselves will tell you that this is entirely appropriate." Then he quotes no less an authority than the National Academy of Sciences as saying that "all scientific knowledge is, in principle, subject to change as new evidence becomes available."[17]

Introduction

Investigating the Truth

Considerable time has been spent here explaining Strobel's quest for the science that supports Darwinism, and the importance of investigating the truth. This is because he has read some of the same scientific books I have. After that he interviewed some of these authors to challenge their arguments the way a lawyer might do in court. He says of these interviews: "I would stand in the shoes of the skeptic, reading all sides of each topic and posing the toughest objections that have been raised. More importantly, I would ask the experts the kind of questions that personally plagued me when I was an atheist."[18]

I will refer to some of Strobel's interviews as I present the findings of modern-day science, especially in the fields of cosmology and biology. This may prove invaluable at times, because not many of us have the opportunity to meet these experts and question them as he did.

A Journey toward Truth

Will you join me on a journey to discover what is reality: both in the natural and in the spiritual world? I think such a journey should start by considering how important it is, and how we can profit from such a journey. I feel it should also consider how we can prepare to think about what we discover with an "open mind." The subject matter is quite broad. It seeks to embrace the solid observations of modern science. It attempts to explain in simple terms what the Bible says about these scientific facts. And it tries to put the known facts of modern science and the clearly revealed truths about nature in the Bible together. As such, it attempts to shed the light of science on biblical truth and the light of Bible revelation on scientific truth. It attempts to do so in a way readers who have no specialized knowledge in either field can understand.

The Importance of Discovering Truth

I believe the most important question for us to consider is one that goes beyond the realm of this natural world. Science is considered a study of what is true—what is reality—in the natural world. I believe it can be demonstrated that the Bible also declares truth about the natural world. If this is true, does the Bible also declare truth about a world beyond nature? Is life on earth, and being happy, comfortable, and relatively successful in this life all there is to existence? Or is our short period of living on earth a preparation for a longer, more important existence—something like a school curriculum that is not an end in itself? It seems to me that answering this broad question will help us to live more wisely and confidently from day to day.

Strobel read what William Provine and Bertrand Russell wrote about there being no absolute foundation for right and wrong and no ultimate meaning of life, and

Introduction

ended up trying to find meaning for his life by seeking fame as an investigative reporter for the Chicago Tribune. He succeeded in his efforts, but after the dramatic change in his wife he began a new quest. He began to question whether his rejection of his former faith in God was really based on a solid foundation. I believe that such willingness to test assumed facts and be open to change when the evidence requires it is a good example for us all.

PART 1

An Openness to Truth,
Regardless of the Source

Preface to Part 1

The Cambridge University–educated philosopher of science Stephen Meyer reminds us that both religions and sciences claim to investigate and explain truth.

> Philosophers have noted that religions as well as sciences make truth claims... For example, both make claims about the origin and nature of the cosmos, the origin of life, and the origin of man; both make claims about the nature of human beings, the history of certain human cultures, and the nature of religious experience.[19]

Some people wonder, "Can science be trusted? Can we accept the discoveries of science today as accurately describing the realities of nature?" Others may wonder, "Can the Bible be trusted? Can we accept the statements in it as objectively describing what 'reality is'—what is really true? Does it accurately describe events that happened long ago? When it speaks about the natural world as it is today, can we rely on it to be fully accurate?"

To many people science has proven itself. The technical advances have produced marvelous inventions. We hear of the ability of NASA to understand the laws of nature and calculate mathematically so accurately they can send a rocket to Mars and place men on the moon. Surely few who have investigated such scientific claims doubt that science has a clear understanding of at least some of the realities of nature.

To many, the claims and promises of the Bible are as trustworthy as the claims of science. The famous scientist Galileo argued that where science cannot investigate objectively, the Bible should be accepted as truth. Galileo put his trust in both science and the Bible. Numerous people today think that faith is needed to trust what the Bible says but is not needed when it comes to science. But another scientist tries to correct this thinking. Charles Townes, a Nobel Prize winner in physics, points out that faith is needed to accept the declarations of science as well as the Bible. He says, "Many people don't realize that science basically involves assumptions and faith. But nothing is absolutely proved."[20]

Preface to Part 1

Scientists today remark at how dramatically in the last few decades science has *confirmed* ancient pronouncements of the Bible on the realities of nature. Take for example the biblical teaching that the universe has a beginning—that God created it out of nothing. Philosopher of science Stephen Meyer tells us that

> during the twentieth century a quiet but remarkable shift has occurred in science. Evidence from cosmology, physics and biology now tells a very different story than did the science of the late nineteenth century. Evidence from cosmology now supports a finite universe, not an infinite one, while evidence from physics and biology has reopened the question of design.[21]

Even such fundamental laws of physics like Einstein's theories support what the Bible teaches.

> Thus general relativity now stands as one of the best-confirmed theories of modern science. Yet its philosophical implications and those of the big bang theory are staggering. Taken jointly, general relativity and the big bang theory provide a scientific description of what Christian theologians have long described in doctrinal terms as *creation ex nihilo*–creation out of nothing.[22]

This is but one example of how modern science has developed to where the ancient declarations of the Bible have been found to be scientifically accurate. The first verse of the Bible says that "in the beginning God created the heavens and the Earth." As another verse in the Bible puts it, "By faith we understand that the worlds were prepared by the word of God, so that what is seen was not made out of things which are visible" (Heb 11:3).

"By faith" we know these things. We can know certain realities "by faith" because the Bible declares them. Many hearing this will put the teachings of the Bible and the discoveries of science in different categories because they think faith is not needed to accept the claims of science. But, as we noted above, Charles Townes informs us that faith is equally needed to accept the findings of science.

My fellow teacher at the high school in Japan told his students that "the Bible is full of myths." This was clearly a statement of his "assumptions and faith." Scientists are in the business of testing their assumptions to ascertain their validity. I suggest that you and I explore both the claims of modern science and the claims of the Bible to see for ourselves whether or not they are factual in their assertions about the natural world. I trust that, if you do so, you will discover that much of the evidence gained from scientific discoveries today can be trusted to declare what is real in the world of nature, and that the words of the Bible can also be relied on to do the same.

These matters cannot be developed adequately in only a few pages. The next three chapters are devoted to exploring several important issues. Chapter 1 considers fundamental matters that go far beyond science. It also explains some facts important to our ability to have an openness to learn new and important things. Chapter

2 discusses the need to consider the discoveries of science with objective and open minds. This is followed by a consideration of the basis for looking to the Bible as a source of objective truth.

Does all this have any relationship to how we lead our daily lives? I think that as you read through the book you will discover that it does. I have read of scientists who admit that evolution has not been proved scientifically, and yet believe it gives meaning to their research day by day. There are also Christians all over the world who say that it is their faith in the teachings of the Bible that gives meaning to their daily lives.

The world has gained much through the scientific study of the laws of nature and how to utilize them to better our daily lives. A quest for understanding of spiritual laws may likewise be of benefit to us. Let us seek an openness to new realities when there is sufficient evidence to support it.

I encourage the reader to follow me on the path of *following the truth—wherever it leads*.

1

What in Life Is Really Important?

It seems clear that science, philosophy and religion have all had a huge impact on the modern world. All three claim to investigate what "reality" is. All three claim to seek "truth" and to declare it. Science professes to study the world of nature. Philosophy professes to study the world of ideas. Religion professes to study a world of reality that transcends both nature and human reasoning.

In this book we will explore modern findings in science. We will seek to reconcile these findings with what the Bible teaches about the origins of the universe, life on earth, and the origin of the human race. My approach to the issue is different from Strobel's. This is because the purpose of my book is different. It is to investigate whether or not there is a basic conflict between the facts of science and the words of the Bible—when both are properly understood. Science claims to search for *reality* in the natural world. The Bible claims to proclaim *reality* in the spiritual world. But it also speaks about the natural world. If its statements about the natural world and human nature are true, then where the Bible and science speak about the same things, there should essentially be agreement on the facts.

Yet their spheres of study are different. Stephen Meyer points out that science and religion represent two distinct types of human activity and require different activities to pursue their own goals. Meyer also observed that increasingly "philosophers of science have realized that the real issue is not whether a theory is scientific, but whether a theory is true or warranted by the evidence."[23]

Science is not qualified to judge or evaluate reality in the spiritual world. The Bible is not a textbook on science. But *when the two proclaim reality in the natural world*, their voices should agree. God's purpose is primarily spiritual in revealing in the Bible how this universe and life on it originated. It is to communicate to humanity that *He* is the Creator, and that He is the sustainer of the universe. It is also to reveal truth about His nature, and His desire to have a personal relationship to humanity—whom He created "in His image."

The perception of many is that science studies and explains objective reality, and religion subjective reality. This is not completely valid. However, there is an element of truth in it. You will experience the effects of the law of gravity whether or not you believe in it. But there are spiritual truths that must be believed to be experienced. A certain condition of heart is needed to experience much spiritual truth that is not needed to experience scientific truth. (Yet, it must also be said the Bible speaks about spiritual laws that affect all people, whether or not they know and believe them.)

I, and also a great many scientists, believe that what the Bible says about creation and about past events is objective reality, apart from what the individual believes.

Both natural reality and spiritual reality are important. A famous philosopher who was not a Christian put it this way:

> When we consider what religion is for mankind and what science is, it is no exaggeration to say that the future course of history depends upon the decision of this generation as to the relations between them.[24]

Can the Two Be Reconciled?

There are scientists like Richard Dawkins who write books from their philosophical position that there is no spiritual world and the realm of the natural is all there is. Their writings are filled with "facts," but only those that support their view. There are also scientists like Henry Morris who write from their religious position. They believe that the universe is no more than a few tens of thousands of years old. They argue that the evidence of life on earth reveals an "appearance of age" even though they believe that fossils can be only a few thousand years old. Their books are also filled with scientific facts, but mostly those that support their position.

Other scientists who are Bible believers are willing to accept the findings of modern science that this universe appears to be billions of years old and life on earth may also be very old.

What might be accomplished by seeking to discover what modern science has actually discovered about the origin of the earth and life on it, and comparing this with what the Bible clearly teaches? Might it result in one or more ways to reconcile the two? I began the research reported in this book from the belief that they can be reconciled, and invite the reader to consider this issue carefully so he can arrive at his own conclusions.

Here some important questions come to mind. Is such a study really worth the time and effort? Does it deal with any of the issues that affect how you and I live, work, and find satisfaction and meaning in life? What is "an open mind," and what commonly hinders having one?

Another need is to consider that we also need to be aware of our presuppositions and convictions and those of the experts we rely on. This is because these affect how we evaluate scientific reality, and also how we can accurately interpret the Bible. These will be considered in the next two chapters.

Finally, we should consider the fields of science where the two must be reconciled. These are primarily cosmology, biology and geology. These will be explored in later chapters.

Is Such a Study Worth the Time and Effort?

To answer this question adequately we should ask if the answer is important to us. Does it help us understand ourselves, our world, and gain any insights into how to live better? I believe the answer is clearly "yes."

First, What Are the Big Questions That Affect Our Lives?

What are the big questions in life? How do they affect us personally? Someone has suggested that these are the following: What or who am I? Where did I come from? Is there any real meaning to my life? Is this life all there is to my existence? Does it really matter how I live my life? Where can I look for the answers to life's important questions?

These questions seem important in a quest to find meaning to life beyond our daily routine—beyond the coping and problem solving sequences that make up daily living for most of us. Other broader but equally basic questions are also important. What is true and what is false? What is the basis of morality? What is the meaning of human history? Why is the world the way it is, and where does it come from? What is our future as a race, and how can we improve it? Is this life all there is, or is this life only a preparation for a more permanent existence?

These important questions about life and existence are commonly called a person's "worldview." A worldview is a way of looking at reality, and includes our assumptions about reality that we often are not consciously aware of.

Our worldview is said to do four important things for us. It seeks to *explain* why things are the way they are; *evaluate* our experiences and choices; *reinforce* our decisions, and *integrate* new information into a coherent whole.

We need such a framework because it ties everything together and helps us to understand our social and natural world and our place in it. Such a framework or worldview helps us understand reality and cope with it. It also aids in making the critical decisions that shape our futures and help us survive and succeed in life.

The Bible teaches that there is a spiritual world as well as a natural world. It also states that the spiritual world is enduring. This is a major teaching of the Bible. First John 2:17 is one verse that illustrates this: "And the world is passing away, and the lust of it; but he who does the will of God abides forever." I believe a worldview that includes both the natural and the spiritual world is important for all who desire to live wisely on this earth.

Next, What Is an "Open Mind"? How Can We Be Open to Truth That Is New to Us?

The philosopher Francis Schaeffer wrote the following about the influence of an individual's worldview on his thinking:

> People have presuppositions, and they will live more consistently on the basis of these presuppositions than even they themselves may realize. By *presuppositions* we mean the basic way an individual looks at life, his basic world view, the grid through which he sees the world. Presuppositions rest upon that which a person considers to be the truth of what exists. People's presuppositions lay a grid for all they bring forth into the external world. Their presuppositions also provide the basis for their values and therefore the basis for their decisions.[25]

Adlerian psychologists tell us that all people have a tendency toward what is called a "biased perspective." What we see and remember is influenced by our psychological needs. A common illustration of this occurrence would be when adult siblings get together and discuss their childhood memories. It is quite common for them to have very different memories of the same shared experience. When growing up their perceptions of events were "biased." This is because their thinking was influenced by the dynamics of their interpersonal relationships and their psychological and social needs at the time.

We all, consciously or unconsciously, evaluate all we see and hear and tend to sift through the conflicting claims for what is true and what is important to us. So if we desire to be truly open-minded we should be willing to examine our own worldview and those worldviews of the "experts" we believe and rely on. Only then can we become *conscious* of the effects of our own presuppositions on how we see reality. But this is not easy. There has been an explosion of information and ideas throughout our modern world. Many of these ideas conflict with each other. Open-minded people naturally want to know what of all this information is true and factual. Then they want to know how it affects their personal lives if it does.

How Can We Know What Is Real?

We live in a world of suspicion and doubt. Sometimes we hear different accounts of a traffic accident—and know both accounts can't be true. When we hear of court case after court case where two versions or more of the "truth" are presented we are prone to wonder, "Which one describes what *really* happened?"

The mass media feeds on scenes like this. In politics and in religion differing views of "reality" are spotlighted and even encouraged. Interestingly though, it seems like this is not common in the arena of science. Science news that supports the prevailing view of the scientific establishment is often featured, even when its factual basis is weak.

Subjective and Objective Truth

If something is "true," is it true for all people, or only for me? Is "truth" objective and universal, or subjective and different for different people? Or are there both objective and subjective truths in the world of reality?

Is there such a thing as objective reality—truth that is reality at all times and to all people in all places? How can we know what is real from what is not? Do we look to science for the answer? Or to religion? Or somewhere else?

As the philosopher Francis Schaeffer put it,

> We need absolutes if our existence is to have any *meaning*—my existence, your existence, Man's existence. Even more profoundly, we must have absolutes if we are to have a solid epistemology (a theory of knowing—how we know, or how we know we know). How can we be sure that what we think we know of the world outside ourselves really corresponds to what is there?[26]

This suggests, at the very least, that the issue of whether or not absolute truth exists has an important relationship to our subject.

Many philosophers of science have pointed out that modern science has its roots in the Jewish and Christian worldview, and developed in an era when superstition abounded. It was born out of the Christian worldview. Schaeffer observed that "the rise of modern science did not conflict with what the Bible teaches; indeed, at a crucial point the Scientific Revolution rested upon what the Bible teaches."[27] This has even been reported by famous men who were not Christians, like Alfred North Whitehead (1861–1947) and J. Robert Oppenheimer (1904–1967). Schaeffer quotes Whitehead as saying that Christianity is the mother of science because of "the medieval insistence on the rationality of God." Whitehead also said that Christian thinking gave the early scientists "faith in the possibility of science."[28]

Both science and theology believe that objective truth exists, and advocates of each freely admit that man cannot know it perfectly. Science looks to discovered truth by looking for evidences of objective reality, and religion looks to revealed truth. Thus there are two primary ways that people accept information that is beyond the realm of their personal experience.

PART 1—An Oppenness to Truth, Regardless of the Source

What Is Ultimate Reality?

Is there such a thing as "ultimate reality? Is everything relative and changeable, or do some things remain unchanged throughout time and location?

What is ultimate reality? And how can we know what it is? Theologians and scientists who are Christians say that we can know it only by revelation.

Science is a knowing by seeking. Religion is a knowing by revelation. Philosophy and metaphysics are a knowing by reasoning. But that which is beyond our experience and beyond our knowing by seeking and testing, and by reason, must be revealed. The Bible claims to be a revelation from God. It claims to reveal eternal truth based on an unchanging God.

The Bible maintains that God is eternal and immutable in His existence and nature. It also teaches that God created the universe out of nothing. If God is eternal and unchangeable in His existence and His nature, then *God* is the ultimate reality. Such a claim cannot be made of this natural universe. Rather, "the only unchanging thing is change" in the world around us!

The Bible also says that God is a spiritual Being, and His existence is in a separate realm from the space-time continuum that forms this universe. If so He cannot be discovered or known by scientific investigation. He can be known only by self-revelation.

Whether or not the above statements about the Bible are accepted as representative of reality or not depends on one's "faith commitments." But it is also important to realize that statements made by the scientific community are also dependent on "faith commitments." Many of those who adhere to very different belief systems do agree on much of what modern science has discovered. But it is important to differentiate between objective scientific discoveries and the broad theories that are derived from their interpretations. Scientific facts and theory are often mingled in reports on science. It is always important in scientific reports to differentiate the two because the interpretations can and often are influenced by the desire to make the facts support a particular theory.

Scholars have pointed out that, although science and Christianity deal with different realms, at the beginning of modern science they shared three essential beliefs. These are that the natural world and God are real, they are both rational, and both can be understood (but not perfectly). The Bible speaks about both natural and spiritual reality. Christianity is a complete worldview, and as such embraces both the natural and the spiritual world.

Conclusion

Philosophers have said that religion as well as science claim to explain the truth. Both seem to make clear and authoritative statements about some of the same subjects. As Meyer explains it, "Both make claims about the origin and nature of the cosmos, the origin of life, and the origin of man; both make claims about the nature of human beings, the history of certain human cultures, and the nature of religious experience."[29]

How this universe came into being, and how mankind began on earth has profound implications for every human being. It clearly relates to the meaning of our individual existence, and to how we understand our world, our daily experiences, and even ourselves. It relates to our values, to what we seek in life, and also to ethics and morals. This is because how we evaluate all these things is dependent on our understanding of the realities of nature and of the spiritual world.

The origin of the universe and life would seem a very important issue to those who seek understanding and meaning for their lives beyond the routine and coping with problems of daily life. It is vitally related to science, philosophy and religion and to what we embrace as "true" and what we reject as "false." It affects our decisions, although we are probably unaware of this influence most of the time.

To study how we and this world came to exist seems clearly worthy of a careful and open-minded consideration. But it should be acknowledged that approaching anything from a truly open mind is difficult for us humans. We approach everything in life with presuppositions. Even so, many of us are seemingly unaware that we do have presuppositions, and that they do color what we accept as true and reject as untrue.

The arguments we hear from scientists about the origins of life on earth and what we hear from some churches seem irreconcilable. But is it possible to look at the issue in a new way? Can we focus only on the verifiable discoveries of science on the one hand, and the clear statements of Scripture on the other, and see if these can be reconciled? Can they fit into a worldview that is a more accurate reflection of what we can trust as "truth" in both the spheres of science and religion?

The next two chapters are meant to present a background for understanding this issue in an open-minded way. In chapter 2 we will look at the presuppositions of science and seek an understanding of how these affect the issue. In chapter 3 we will consider the presuppositions of biblical interpretation and seek an understanding of how these relate to the issue. After that we should be more ready to consider the findings of science and how they can be reconciled with the clear teachings of Scripture.

2

Approaching Science with an Open Mind

A striking thing I often notice in America today is that there are great numbers of news articles about science which present findings based on the assumption that humans evolved from a very primitive form to modern man. This is done in spite of the statements of some prominent paleontologists and biologists that there is no clear proof for this linkage. A recent article referred to the evolutionary development in empathy in humans. The article described research showing scans of the brain that light up when the individual experiences pain, and contrasting scans of the same individual observing the reaction to pain in their partner. Then it moved to observations of chimpanzees comforting each other. Explanation was given about how evolution has led chimpanzees and humans on different paths, and that human brains are much larger and very different from chimpanzee's brains. The article stated, "Our brains today are quite unlike a chimp's in size, organization and cellular complexity." All this with no mention of the fact that this explanation is based on a theory that is lacking scientific proof that chimps and humans had a common ancestor, or noting that there are other non-evolutionary explanations for the scientific information presented.

How do we approach science with an open mind? How do we try to make sense out of what we hear and read without it being distorted unconsciously because of our personal biases?

This article is one of many that present recent scientific findings in a biased way. How do we discern facts from interpretation when we hear of a new discovery in science? Often these are presented in the mass media in a dramatic way to try to grip the attention of the listener or the reader. Adding to this problem is the media's tendency to present science from a philosophical position that mixes observation and unproven theory. This illustrates the need to separate observed facts from interpretation as we try to fit new knowledge into our personal worldview.

Charles Hummel points out two errors that he says cloud any discussion regarding science and religion. The first has to do with the nature of modern science and the status of its laws.

> Through the influence of naturalism, a philosophy that considers the natural world to be the whole of reality, the scientific method has come to be widely accepted as the only valid approach to understanding reality. Scientists are thought to be objective in their search for facts, as opposed to others (especially theologians) who are considered biased by their faith.[30]

But the reality is often quite different. True objectivity is a virtue that is not easy for us to achieve. Many scientists admit that the Darwinian theory that guides their research has not been proved scientifically. But then they add that it provides a theoretical framework that gives meaning to their work. In fact, a theoretical base seems essential to all or most scientific research. New research is almost always based on a *theoretical framework* and is intended to build on previous research. However, quite often this framework has not been adequately tested scientifically. It is important for scientists to test theories with their research and not simply seek only an interpretation that fits the theory.

How can we know that the theoretical framework of the scientist—or of the reporter—represents objective reality? In order to do that, another "reality" needs to be understood. Scientists of different theoretical persuasions sometimes look at the same scientific data and come up with very diverse conclusions. This is true in geology, in biology and sometimes in other natural sciences and also in the social sciences. Some things that are observed in both the natural and social sciences are interpreted quite differently by "experts." How scientists understand and interpret nature is clearly affected by their theories and their worldview. This is important to understand when we consider "scientific fact."

The second point Hummel makes has to do with proper interpretation of the Bible.

> Many think that one can prove almost anything from Scripture. They point to a bewildering spectrum of popular doctrines to support their views. Yet the biblical texts cannot legitimately be made to teach whatever a person wants to find.[31]

Models have been developed through the ages to make sense of what we observe in nature. Many have become accepted as reliable, but then are sometimes discarded This has happened far too often to defend a closed-minded position that now we have the final answer. It is unwise to take a position that we can now rest assured that science will never again discard the present paradigms and theories. It may do so when the present paradigms and theories are challenged by newly discovered facts. At the same time it appears that science has discovered some truths about natural laws that will stand the test of time.

What Is Science?

John Collins points out that "a science is a discipline in which one studies features of the world around us, and tries to describe his observations systematically and critically."

Science makes observations, and then develops and tests theories. It also makes speculative theories in order to seek solutions for observations that cannot be explained by existing theories. Scientists Wayne Frair and Gary Patterson give an example. They tell us that the orbit of Mercury made small deviations from the mathematical formula of Newtonian mechanics. These remained unexplained until Einstein developed a new theoretical framework to answer the questions, What is mass? What is space? What is time? What is gravity? Einstein's new theories helped to explain what astronomers were observing that couldn't be explained before. In this way the progress of science is advanced by formulating new theories to account for observations that cannot be explained by current theories.[32] It is also advanced by revising existing theories to take into account new knowledge.

Science can be defined as the branch of knowledge that includes four things:

1. It observes and accumulates facts related to the natural world.
2. It measures and systemizes them.
3. It proposes theories, and then tests them.
4. It seeks to discover and verify general laws of nature.

The word "science" refers to a system of acquiring knowledge in the pursuit of truth in nature. This pursuit is based on empiricism, experimentation, and methodological naturalism. The basic unit of knowledge is the theory, which is a hypothesis that is predictive. "Science" also refers to the knowledge that has been collected and organized from this research.

Definitions of science commonly include the word "empirical" meaning "to depend upon experience or observation alone." This includes information gained from a direct study of nature and by controlled experiments. Results can be supported or contradicted by other experiments. In this way scientific theories can be tested. When scientists look at events in nature they seek to formulate a theory that explains their observations. They may feel confidence in their theory, but they see their explanation as subject to change or modification. They see it as their best available view of reality, until further research produces data that conflicts with their theory—or find a better or simpler rationale for what they see in nature.

Hummel points out that a popular conception of a scientific researcher is someone who is detached and objective, who solves scientific problems through cold logic and observation alone. But Hummel says the reality is "radically different." He quotes Michael Polanyi, a well-known professor of physical chemistry and philosopher of science, as saying that all human knowledge takes place within a framework of unprovable commitments (a faith structure) that motivates and guides the knower in acquiring knowledge. Hummel explains it this way:

> A person's faith structure includes a wide range of beliefs, from ultimate presuppositions (the universe is orderly) to a mundane confidence (the sun will rise tomorrow). The former must be assumed; the latter is based on sense perception... Whether ultimate or mundane, faith is not blind; it arises from and is embedded in evidence assimilated from our experience. The main point is that for everyone, in all fields of study—including science—faith is a motivating and unifying component in knowing.[33]

Hummel goes on to explain that every scientist's faith structure affects his research from beginning to end. Usually the scientist unquestioningly accepts as valid both the presuppositions and the methods of his discipline. Science, as a discipline, is practiced by many of different religious persuasions, including atheism. It is not the *practice* that divides them, but the *interpretations* they place on the results of scientific inquiry. It is important not to forget that a person's worldview has great influence on the paradigms and theories that are used to make sense of raw scientific data.

The Limits of Science

Science and the scientific method have their limitations. One reason is that past historical events cannot be observed or tested in the present. Natural phenomena can be observed using only the five senses or enhanced by using telescopes, microscopes, or other aids to observation. Some phenomena cannot be observed in this way. Since they cannot, our understanding of them is no better than the assumptions we make about them based on observable data. An example of this is the theory of dark energy, which is assumed to exist because it helps astrophysicists to make sense of what they can observe.

We should always keep this in mind. Natural phenomena that can be observed and tested today are suitable subjects for the scientific method. A high level of confidence can be placed on conclusions that are often observed and tested about natural states and processes in the present. But this does not mean we can place a high level of confidence about scientific conclusions concerning events in the distant past.

Much care is necessary in reaching conclusions about natural events in the past. Biological experiments today can, at best, determine if certain biological transformations are possible. Geology can support evolutionary theory when fossils are discovered that seem to illustrate biological transformations that are found to be possible today. In this sense, attempts to reconstruct past events are not without meaning. We should be careful to avoid claiming a theory has been proved in relation to past events based on fragmentary and inconsistent geologic records, even when biological possibilities are known.

Scientific investigation, then, has limits in its attempt to reconstruct events in the distant past and in its attempt to study phenomena that cannot be directly observed and measured. But the limits of science do not extend only to the distant past. Another caution is needed. Nobel Prize winner Charles Townes cautioned that all conclusions of scientific research should be considered as tentative because science involves assumptions and faith, and "nothing is absolutely proved."

Scientific Naturalism

A dominant worldview of many secular scientists is called "scientific naturalism." Scientific naturalism first claims that nature is all there is and therefore there is nothing that is supernatural. Second is the belief that all the realities of nature fall within the scope of the scientific method. Two principle alternatives are presented to the public. One is the naturalistic view, where a self-existent universe is considered to be all there is and there is no need for a creator. The other is called "intelligent design." Intelligent design includes the theistic view where God is presented as Creator.

With few exceptions, a strict methodological naturalism is taught in schools of science everywhere and it is said to be the principle that guides all of modern science. Methodological naturalism assumes that what is observable in nature can be explained only by natural causes without a need to consider the existence or nonexistence of the supernatural. This is a largely unspoken rule in both the sciences and humanities today. This view is different from previous times, when the existence of an immaterial soul and a transcendent ethical and supernatural order was assumed. Methodological naturalism does not deny the possibility of supernatural forces on natural phenomena, but does not take them into account. Philosophical naturalism, on the other hand, insists that all phenomena can and should be explained in terms of natural causes and laws. Philosopher of science Stephen Meyer points out that Darwinism, as it is embraced today, insists on "an exclusively naturalistic mechanism of creation"—one that rules out both creationism and intelligent design.

There are several reasons that naturalism has become so dominant in scientific thinking today. First is the success of the physical sciences to explain events and processes once thought of as mysterious or supernatural origin, such as diseases, earthquakes, and floods. Another reason is that naturalists have equated all other beliefs with superstition, a rejection of science. A third reason is the difficulty of relating such things as transcendent ethical principles, an immaterial soul, and God to the natural order.

Most members of the scientific community have embraced at least some of the assumptions of scientific naturalism. This is also true of a number who accept theism. Many scientists accept "methodological naturalism" in their research while rejecting "philosophical naturalism" as a belief system. From what we read in newspapers and current journals it is clear that scientific naturalism is the predominate worldview in the American Academy of Sciences. But there are scientists within the academy and outside it that are increasingly challenging that worldview. Some philosophers of science like Edward Davis and Robin Collins are convinced the naturalism accepted today will have to change radically in order to explain the realities discovered in nature itself and in human experience.

The Faith Commitments of Science

Most early scientists, like Galileo, held very different assumptions from those of philosophical naturalism. As Galileo expressed it,

> I have no doubt that where human reasoning cannot reach—and where consequently we can have no science but only opinion and faith—it is necessary in piety to comply absolutely with the strict sense of Scripture.[34]

Galileo openly claimed to have a religious faith. Is "philosophical naturalism" also a religion? This is hotly debated in some circles. If it is possible to have a religion that believes neither in a deity nor in the supernatural, then it seems to be one. Carl Sagan declared, "The cosmos is all that is or ever was or ever will be." Physicist Howard Van Till expresses the core belief of naturalism as follows:

> The universe is self-existent and needs no transcendent Creator to give it being or to sustain it in being. There is no Creator of whom creatures could be aware and hence no possibility of a Creator-creature relationship that could serve as a source of meaning and purpose.[35]

Van Till calls naturalism a religion that is presented simply as science. He questions whether it can be called a true science when he says, "Answers to ultimate questions regarding purpose, meaning, value, significance or source of being are, when offered in this context, treated as if they were either self-evident or easily derivable from simple and logically unassailable extensions of scientific reasoning alone."[36] In other words, naturalism seeks to answer important philosophical questions, but many of the answers given are not supported by scientific experiment or data. They are promoting a specific worldview in the name of science without presenting the factual evidence to support it.

Many scientists and philosophers argue that science really has no answers to the following important questions: Is the cosmos all there is? Is nature the ultimate reality? Is the existence of the universe self-caused, as numerous proponents of naturalism have argued? It seems clear that naturalism is a faith commitment in place of, and opposed to, theism. Many scientists who *practice* naturalism in their approach to science actually do believe in God. They do so in the belief that faith in the supernatural has no place in scientific investigation. They are naturalists in their scientific methods only.

However, other scientists who accept naturalism do so in a more absolute and philosophical sense. One example is that of a leading evolutionist, Richard Lewontin, who, when writing against the account of creation in Genesis, acknowledged many of the weaknesses of modern science. He admitted the obvious "absurdity of some of its constructs." He acknowledged how the scientific community tolerates reports that support materialism even though they are not supported by the scientific facts. He says they do so "because we have a prior commitment, a commitment to materialism." He goes on to say that

> we are forced by our *a priori* adherence to material causes to create an apparatus of investigation and a set of concepts that produce material explanations, no matter how counter-intuitive, no matter how mystifying to the uninitiated. Moreover, that materialism is absolute, for we cannot allow a Divine Foot in the door.[37]

Evolutionary biologists such as Richard Dickerson and E. O. Wilson have written similar thoughts, as well as well-known advocates for Darwinism such as Stephen Jay Gould and Richard Dawkins.

Conclusion

All sciences, as well as all religions, are based on assumptions that are unproved and possibly unprovable in the natural world. Leading scientists are no different from theologians in their acknowledgement that they hold faith commitments that help them organize their thinking. The prominent faith commitment behind evolutionary theory is naturalism. From what we have discussed it is clear that a naturalistic philosophy governs the theorizing of most scientists. It supports and is supported by Darwin's general theory of evolution. Many scientific theories are developed and guided by presuppositions that rule out God and the supernatural from consideration. Evolutionary scientists explain natural events primarily on the basis of scientific naturalism. This is true even when alternative explanations are scientifically possible. Philosophical naturalism rules out even the possibility that anything exists outside the cosmos. Because of this, it precludes considering the possibility of any supernatural influence on or within the universe.

Evolutionary scientists and those who believe in creation both admit that the conclusions of science are at best tentative and subject to revision at any time. They acknowledge that theories are vital for guiding fruitful research. Scientists from both perspectives further contend that no one can be sure whether or not future discoveries may contradict and disprove them. Science is limited by its inability to reconstruct historical events that can no longer be observed or replicated in every detail. There is also the problem that many phenomena cannot be observed directly—either by the five senses or by aids such as telescopes and microscopes. Such phenomena can only be known indirectly. Being experienced indirectly, they are meaningful only when interpreted, and the knowledge gained is dependent on the accuracy of the interpretation. An obvious example here is the concept in astronomy that there exist in the universe realities that cannot be seen such as black holes, dark energy, and dark matter.

However, let us not take the admission that the findings of science are always tentative too far. New scientific discoveries will no doubt necessitate the modification of current theories in the future. Some theories may have to be abandoned altogether. Some conclusions of science, such as the general laws of thermodynamics, the theories of relativity propounded by Einstein and Newton's law of gravity appear to be substantiated by enough evidence that I would not expect them to ever be overturned completely. I feel confident we truly know *some* things about the realities of the natural world.

Materialism is based on an absolute presupposition that is not required by the nature of science itself. The presupposition is that any supernatural explanation for the existence of the solar system is ruled out. Lewontin and other prominent evolutionists fully acknowledge this. It is clear that other scientific theories, those that are not based on naturalism, are scientifically possible.

"Can we trust what science tells us?" The simplest answer is, "Yes, by taking a scientific attitude toward science!" This is accomplished by separating the clear discoveries of science from the interpretations of these discoveries and seeking to understand the *assumptions* on which these interpretations rest. Scientific theories are important to help us make sense out of observations of nature that result in scientific knowledge. It is important to test theories to determine how adequately they are supported by the hard facts of nature.

Many scientists have recently observed that in the late twentieth century and now into the twenty-first century, we are much more able to do test current theories with verifiable scientific discoveries than before, particularly in the fields of cosmology and biology.

3
Approaching the Bible with an Open Mind

> Science cannot be used to justify discounting the great monotheistic religions of the world, which rest upon centuries of history, moral philosophy, and the powerful evidence provided by human altruism. It is the height of scientific hubris to claim otherwise. But that leaves us with a challenge: if the existence of God is true (not just tradition, but actually true), and if certain scientific conclusions about the natural world are also true (not just in fashion, but objectively true), then they cannot contradict each other. A fully harmonious synthesis must be possible.[38]
>
> —Francis Collins

We have now given some thought about how we can have an open mind as we "follow the truth wherever it leads." We have also thought about science, and how to notice the distinction between the discovered facts of nature in a given news report and the interpretations put on them. In some cases there seems to be ample evidence that these interpretations have a solid scientific foundation. In other cases they seem to reflect more the philosophies of the reporters. For this reason I think it is important for us to discern the difference between the actual scientific findings and the interpretations.

Now we come to a discussion that relates to religion. Some of us may be atheists. Others of us are Buddhists. Others of another worldview. Some of us may have some knowledge of the Bible already. Can we each open our minds and hearts to consider what the Bible says about nature, and about its beginning, and together "follow the truth wherever it leads"?

When we consider what the Bible says about creation we are, of course, entering what is commonly thought of as the area of religion. Religion is a word that is often used differently by different people. It is commonly defined as belief concerning the supernatural, sacred, or divine, and the moral codes, practices, values, institutions, and rituals associated with such belief. It is also considered broadly to be the sum total of answers given to explain mankind's relationship with the universe.

As defined above, religion includes answering questions about the human race and the universe. We will say little about religions in general. This is because our interest is confined here to the biblical account of creation and whether or not it is trustworthy. But the astrophysicist Hugh Ross made an interesting observation on religion and the Bible. He said, "The Bible alone describes God as a personal Creator Who can act entirely independent of the cosmos and its four space-time dimensions. The God of the Bible is not subject to length, width, height, and time. He is the One Who brought them all into existence."[39]

The God that the Bible talks about is eternal in His existence. He is not a part of nature. He is the one Who created the universe and all things in it, as the very first sentence of the Bible declares. The Bible begins with these words: "In the beginning God created the heavens and the earth."

Ross, speaking as a Christian astrophysicist, makes an interesting observation.

> Pantheism claims there is no existence beyond the universe, that the universe is all there is, and that the universe always has existed. Atheism claims that the universe was not created and no entity exists independent of the matter, energy, and space-time dimensions of the universe. But all the data accumulated in the twentieth century tells us that a transcendent Creator *must* exist. For all the matter, energy, length, width, height, and even time, each suddenly and simultaneously came into being from some source beyond itself . . . Not only does science lead us to these conclusions, but so does the Bible, and it is the only holy book to do so.[40]

It should be acknowledged that this belief in creation by God is shared by Jews as well as Christians. It is shared by Muslims also. It is common to all religions that recognize the Old Testament as God's Word. But Ross points out that the Bible is unique in its explanation of creation and the Creator.

What Is the Bible?

The Bible claims that it is a revelation from God, and that this revelation is without factual error. The theologian Henry Thiessen makes an important point regarding this.

> If we can prove the genuineness of the books of the Bible and the truthfulness of the things they report on other subjects, then we are justified in also accepting their testimony in their own behalf.[41]

Thiessen is saying that one proof that the teachings of the Bible are truly a revelation from God is in their agreement with known and verified facts. These are not just the facts that were known to ancient wise men when the Bible was written, but also to realities that have only been discovered recently. We shall see later in this book what some of these realities are in the natural world. One of these realities is that recent discoveries in the cosmos have clearly shown that the universe is not eternal, but came into existence suddenly at some point.

There are three important points to note when we consider why so many take the Bible seriously, even in scientific matters: (1) It claims to be a revelation from God. (2) It claims to be without error. (3) It claims to be authoritative.

1. It Claims to Be a Revelation from God

The Bible claims for itself that it is both revealed in its contents and inspired in its expression in human language. There are literally hundreds of places in both the Old and New Testaments where the claim appears that what is written came from God and was inspired by God. *Revelation* means the communication of truth that cannot be discovered otherwise. *Inspiration* means that God's Spirit guided the writers to accurately record the truth. So here the Bible claims two things. First, God showed the writers of the Bible what to write. Second, God helped them so the words they wrote were without factual error or omission. Many things written by the authors of the Bible were unknown to people at that time, as the Bible talks about things that happened before humans were on the earth, and also about events that will happen in the future.

Jesus said that the whole Old Testament is verbally inspired, down to the smallest details (Matt 5:18). He also said that the prophesies in the Old Testament will be literally fulfilled.

2. It Claims to Be Without Error

The Bible claims to be without error in the original texts. One proof that the Bible is true is that scientists are discovering things today that God revealed long ago. This is especially true regarding facts about the creation of the cosmos and how biological life originated.

3. It Claims to Be Authoritative

The Bible, from cover to cover, clearly proclaims that it has authority and is to be obeyed. All the Christian theologians in the first eight centuries after Christ (with one known exception) accepted the Bible as the sole and final authority in matters of faith and conduct. Jesus also acknowledged that the Old Testament is God's Word and has authority as such.

Authority in matters of faith and conduct includes the statements of the Bible about historical and scientific facts. Science investigates many things that are not touched upon in the Bible. But where the Bible does speak clearly about matters regarding nature it is considered authoritative and trustworthy. This authority also covers matters of morals and ethics, and matters of religious belief and observance.

In this way the whole Bible claims that it is a revelation from the God depicted in the Bible, and is inspired, reliable, authentic and authoritative.

Interpreting the Bible

The Bible is written in human language and in many literary forms. This requires that it be understood and interpreted by following the laws that govern language in general.

Not all people of faith interpret some verses in the Bible in the same way. These differences seem exaggerated because of the tendency of the mass media to report new and different interpretations—even when these are clearly not based on adequate biblical scholarship. People may get the impression that biblical interpretation does not follow strict rules like scientific interpretation is assumed to follow. But Charles Hummel reminds us that each field of study has its own proper methodology, and these need to be acknowledged and respected to obtain valid results. It is as important to follow correct procedure in interpreting the Bible as it is in interpreting the findings of science.

I have visited churches of many denominations and fellowshipped with their pastors. Most of these churches accept the Bible as the verbally inspired Word of God. It has been my experience that almost all of the differences among us are on the interpretation of minor passages in the Bible that in no way affect the major teachings of Christianity. There is far more agreement among Bible-believing Christians than what the mass media would lead us to believe.

To interpret the Bible properly one must consider carefully the form of language used in a given passage. Various literary forms are used in places: for example, parables and poems have their own guidelines for interpreting them. Understanding the meaning to people in the situation at the time of writing is also important in many places.

Many declarations in the Bible are so clear that no interpretation is needed, especially those that relate to salvation and a person's relationship to God. But passages like those on nature and creation in Genesis are not written in scientific language, nor are they exhaustive statements. They are very brief, and leave much out that is of scientific interest. In some cases they are in the form of poetry. For this reason special care is needed when seeking to know how to reconcile them with the findings of modern science.

PART 1—An Oppenness to Truth, Regardless of the Source

Most of Genesis is written as history, and the rules for understanding history are fairly straightforward. Historical accounts are to be understood literally and in context.

The Bible contains many literary forms, and some of these forms require special linguistic tools to properly interpret such texts. But the creation account in Genesis is recognized by biblical scholars as a historical record. It was revealed by God to Moses as history, and because it is a historical record a literal interpretation is required.

The Bible and Creation

What is the main thing we need to understand in this chapter? It is whether or not the Bible can be trusted when it speaks about creation and the realm of nature The Bible teaches about the existence of a God Who tells the truth and Whose words can be relied on. It claims to be a revelation from God that was given to many writers over many hundreds of years. In fact, scholars marvel at the consistency and unity of the Bible in spite of the fact that the backgrounds and circumstances and time periods of these writers were so different.

It claims to be inspired in such a way that its writings are without error. It claims to be a revelation from the God Who created the heavens and the earth and Who continues to uphold and sustain them. If these claims are valid, then what the Bible says about creation and nature are trustworthy.

If what the Bible says about nature is in agreement with the findings of modern science this is indeed amazing! It clearly points to a supernatural source for this knowledge. There were no telescopes in those days. There were no microscopes. The ability of mankind to thoroughly study the heavenly bodies did not exist then. Unless the Being Who planned and created the universe communicated these things to humans there is no way in which they could be known until modern times.

There is another major thing to consider. Human languages develop as a way to communicate thoughts and experiences. The technical and scientific vocabulary that we are familiar with did not exist before the technical and scientific revolution. The words used in the Bible to describe nature are often not as precise as those used by scientists today.

The theologian Henry Thiessen gives an important clarification. When considering the validity of the Bible's pronouncements on nature and creation, he emphasizes the following:

> The Bible is not a textbook on either science or history; but if it is verbally inspired, then we expect it to speak truthfully whenever it touches on either of these subjects. It will help us to note that, just as our scientists still speak of the rising and setting of the sun, the four corners of the earth, etc., so the Bible often uses the language of appearance. The account of creation can be

harmonized with the assured facts of geology. Scientists today admit that light is earlier than the sun. Geologists and anthropologists are too much divided as to the age of man to consider their estimates an objection to the representations of Scripture.[42]

The Bible and Science

A great many historians and philosophers of science have emphasized that modern science developed out of a culture of educated people who believed in the Bible. Most of the founders of science were devout Christians who had a worldview that gave modern science its basis. These include Copernicus, Kepler, Galileo, Newton, Boyle, and Pascal. At the heart of their belief system was faith in an infinite, eternal, personal God Who created the universe from nothing. They also believed that this God established the laws of nature.

Kenneth Samples, a scholar educated in both science and theology, notes that "the Christian worldview supported the underlying principles that made scientific inquiry possible and desirable."[43] The careful scholarship used to rightly understand the Bible throughout history is the same as the scientific attitude essential for scientific investigation.

Samples continues his argument as follows. He says that science is based on certain philosophical beliefs and that these are rooted in Christian theism. He includes the following five beliefs:

1. That a real objective world exists
2. This world can be understood
3. We can trust our senses and rational thinking in understanding it
4. Nature is orderly and uniform, and
5. Mathematics and logic are valid as means of understanding it

Samples also refers to the Christian theistic answer to the oft-asked question, "Why is there something rather than nothing?" Why does the universe exist? If everything that exists has a cause, what caused the universe? Samples asks which is more reasonable to believe: that the universe came into being "from nothing by nothing" or that "in the beginning God created the heavens and the earth" like the Bible says? He clearly sees this as a powerful argument for the biblical account of creation. He further argues that current thinking on big bang cosmology seems uniquely compatible with Christian theism.

Many Christians who are scientists are convinced that theologians and scientists as a whole are sincerely seeking truth. Howard Van Till explains,

PART 1—An Openness to Truth, Regardless of the Source

> To the best of my knowledge, both science and theology are honestly seeking growth in authentic human knowledge about ourselves and about the universe in which we reside. Furthermore, I shall presume that each of these two enterprises is committed to maintaining the highest standards of both professional competence and intellectual integrity.[44]

Let us fully recognize that we can learn much from embracing the verifiable truths of modern science and the clear teachings of the Bible.

The philosopher Francis Schaeffer put it this way:

> There may be a difference between the methodology by which we gain knowledge from what God tells us in the Bible and the methodology by which we gain it from scientific study, but this does not lead to a dichotomy as to the facts. In practice it may not always be possible to correlate the two studies because of the special situation involved, yet if both studies can be adequately pursued, there will be no final conflict.[45]

The Bible as History

Stephen Meyer explains the historical nature of the Bible as follows. "The Bible . . . makes specific factual claims, chiefly about the history of the Jewish people, the life of Jesus and the early church, but also about other factual matters including the nature and origin of the natural world."[46]

Bible scholars from ancient times, both in the eras before and after Christ's coming, have interpreted the Bible as accurate history. The Jews throughout the Old Testament times kept careful records of genealogies and important events, attesting to the high value they placed on preserving accurate historical accounts. Genesis was written long after the events recorded in it, but it has been accepted as factual history throughout the ages. However, as we shall see in part 3, it has been acknowledged by many scholars that the age of the universe and life on earth cannot be known from Scripture.

Conclusion

Any attempt to reconcile "truth" from science with "truth" from the Bible may be considered by some trying to compare oranges and apples. Or worse—like comparing oranges with kangaroos! Are they really so different that an attempt to compare them is a waste of time?

Why even make the attempt? For one important reason: We all live in a world of physical realities and spiritual realities. What we believe about both the physical world and the spiritual world does affect how we live day by day. It also affects the future we are building for ourselves. And, according to the Bible, it greatly affects our eternal future. To live life to the fullest we need to be aware of both natural and spiritual realities.

If what the Bible says about nature, written before humans became able to discover it with modern telescopes, is found to be accurate this attests to the divine origin of the Bible. This gives the seeker of truth good reason to study the Bible to discover other truths that can affect his life.

Science seeks to understand reality by observation and measurement of a physical world that can be seen. The Bible claims to be a revelation from the God Who created the universe and all that exists in it. Can observed truth and revealed truth be compared, or should we conclude as some do that they speak about different domains and should not be put together?

But our task is really not one of comparisons. This is because science and the Bible are primarily concerned with different things. The primary task of science is to measure a reality that exists objectively in the natural world. The primary purpose of the Bible is to declare spiritual truth that exists beyond the natural world. Comparison between the spiritual realm and the natural realm is not the issue. Science is in no position to speak with authority about spiritual things.

Our area of concern is limited to the realm of nature. The issue is more of discovering the things in which both science and the Bible agree about nature, and where they might possibly disagree. It is an investigation of the verified "truths" of science and the clearly declared "truths" of the Bible, to see if they do indeed agree with and support each other—when they speak about the realm of nature. *It can be considered a test of what is true. If there is essential agreement about what both reveal about the natural world this suggests that both should be taken seriously. It is a clear indication that the Bible is from God. This suggests that the Bible contains spiritual truth that can enrich our lives on earth and prepare us for the life beyond.*

Where the two appear to conflict is not in the observation and description of nature as we experience it today. It is in the explanation of how the realities we observe in nature *came to be the way they are today*. The Bible claims to include a revelation of how the objective realities we observe in nature came into being. Science seeks to reconstruct the past—from the beginning of the universe and of life on earth—to develop theories of how these realities came into being.

The central question I would like to consider in this book is this: *Can the clearly known facts of the universe and life on earth be reconciled with the clear teaching of the Bible?*

This refers only to reconciling the facts of the universe and life on earth as discovered by science and supported by adequate observations of nature. It does not refer to the many theories that are lacking in scientific support. As to the Bible, it refers to what the Bible says unambiguously.

PART 1—An Oppenness to Truth, Regardless of the Source

It is often said that the Bible is not a textbook on science. This is true. The Bible claims to declare the truth whenever it speaks about the facts of nature. However, it was written primarily to deal with spiritual rather than scientific reality. The Bible claims to be the revealed word of a God Who knows all things and speaks truth truthfully. If this is true, then it can be relied on to speak truth at all times. However, for the Bible to be properly understood, it must be interpreted by following the laws of language that are appropriate. And much that we modern humans would like to know about natural things is not revealed in the Bible at all.

I have confidence that *some* scientific conclusions are supported by sufficient evidence that they can be depended upon to not later be proved in error. And I also have confidence that the biblical account of creation, when carefully studied following the laws of Bible interpretation, can also be depended on. How to test this and seek to harmonize the two is the subject of the rest of the book.

Myths?

Is the Bible really "a bunch of myths" as the teacher in Japan stated those many years ago? Or is it a reliable record of events that can be trusted—even on the strictest historical and scientific grounds?

Hundreds of thousands of Jewish scholars have poured over the Old Testament for several thousands of years, believing that the Bible is absolutely true in its history and teachings. Hundreds of thousands of Christian scholars have studied the Hebrew and Greek texts of the Bible for almost two thousand years, and remain convinced that the Bible contains no factual errors.

Is the Bible a reliable record that can be trusted even on historical and scientific grounds? We have in our day fresh knowledge by which to test the veracity of the Bible. Let us test the creation story of Genesis with the known facts of nature, as we "follow the truth—wherever it leads." In doing so I hope that we will find new meaning to our own personal lives.

The Task

How do we keep an open mind while sifting through the competing claims about what is the truth about the origins of the universe and life on earth? Can we open our minds to accept new information that might conflict with things we have learned in the past—when they seem to be supported by adequate factual information? Strobel was able to do this. I believe we can also.

Can we accept from science what has been demonstrated as appearing to have adequate proof? Can we also accept the claims of the Bible to be the inspired revelation from an eternal God—a God Who created it and so is the only "eyewitness" to what actually happened? If we do, will we come up with one or more ways of harmonizing the two? This book does indeed attempt to do just that.

Wayne Frair and Gary D. Patterson wrote this, when speaking of scientists who believe the Bible:

> We believe that a Christian does not need to abandon a biblical perspective in order to carry out effective science. Accurate exegesis and reliable interpretation of the Bible along with valid scientific conclusions are the goal of all scientists who are Christians."[47]

I trust this same spirit will help us understand better the contents of this book. This is a spirit of openness to new truth when it is adequately supported by the facts, and a desire to "follow the truth wherever it leads."

As Frair and Patterson phrased it, "It is our conviction that a proper understanding of the Bible will be consistent with all valid scientific observations of the world in which we live."[48] I believe and trust that we will find that there are several ways to harmonize Scripture and science without sacrificing the integrity of either.

This should be the focal point of our investigation. Do the observed realities of our cosmos support the Bible revelation of the cosmos's beginning? Do the processes scientists observe in the biological world support what the Bible says about the creation of life on earth? Does what we know about the human race seem consistent with the biblical account of the creation of mankind?

Or are there "irreconcilable differences"? The mass media today clearly presents a picture of science that is very different from the Bible account of the origin of life on earth. But do the *confirmed* discoveries of science today do so? This is an important issue to consider as we *follow the truth—wherever it leads*.

PART 2

What Do We Really Know about the World?

The Science

4

Charles Darwin and His Theory

Now let's think about Darwin and how he changed his thinking about the natural world. To understand this it is helpful to consider the views of his day, and how they were changing.

Charles Darwin proposed the theoretical framework that has dominated much of scientific thinking ever since his day. His ideas directly impacted the study of biology. They also affected the development of scientific thinking in general. They even influenced developing theories in the social sciences. I don't think his influence on modern thinking can be overstated. I'm sure all of us have been influenced by his ideas.

Darwin the Man

Charles Darwin was born into a well-to-do family in Shrewsbury, England, on February 16, 1809. He died in 1882. Charles attended Christian services with his mother as a child. His father was a highly successful physician who sent him to study medicine at Edinburgh University while still in his teens. He studied there for two years, but neglected his medical studies in order to study animals in nature. Even from childhood Darwin had fantasies of making fabulous discoveries in natural history, and it seems this interest kept interfering with his studies. Then, at his father's encouragement, he enrolled in Christ's College, Cambridge, in 1827 to study for the ministry. But his attention was soon diverted to collecting beetles and writing about the rare species he discovered. In 1831 he was invited to fill a position aboard HMS *Beagle* as captain's companion, working as a naturalist. When Darwin left England he had a belief in the Bible and some scholars believe he took the Genesis account of creation literally.

The World of Darwin's Day

It is helpful to consider the world of Darwin's day and what people believed then. The Church of England had a narrow view of some Bible passages. The Church had a strong influence on the political scene but there was a reaction to its political influence and also resistance because of certain developments in natural science.

At that time many people in Europe believed that the earth was only about six thousand years old. This was based on the assumption that the genealogies in the Old Testament are complete. A calculation based on this idea by Archbishop James Ussher in 1701 put the date of creation at 4004 BC. This idea of such a young earth was generally accepted in Europe in Darwin's day, even though it was not universally accepted by all the early church fathers. In fact, no one before Ussher seemed to have assumed what he did—that a person could determine the age of creation in that way.

In that day, species of plants and animals were thought to breed true to type, generation after generation, without any significant change. If there was variation, it was only trivial and within clearly defined limits of the species or type. But farmers were beginning to breed what seemed to be new species of plants and animals.

Early Ideas of Evolution

There were developments in science at that time that challenged these traditional beliefs. One was the findings of the eighteenth-century scientist James Hutton (1726–1797). Hutton looked at river canyons formed by erosion, and at fossils. These fossils appeared to have been formed slowly by stony material replacing the original organic material of plants and animals. Hutton concluded the earth must actually be millions of years old. Another development was the observations of the French anatomist George Leopold Cuvier (1769–1832). Cuvier discovered what he thought were evidences of evolution from simple life forms to more complex forms in fossils in the geological strata he studied.

By the end of the eighteenth century, "materialist thinkers" like Comte de Buffon (1707–1788) were suggesting that life could have been created on earth by spontaneous generation. They, like Jean Baptiste Lamarck (1744–1829), were also suggesting comprehensive theories about life evolving from primitive to more advanced forms. Darwin grew up in a society where scholars were debating and advocating these ideas—which conflicted with the dominant view that God had created life, even advanced forms of life, only a few thousand years before.

Darwin was also familiar with Charles Lyell's (1797–1875) many researches into changes in geological structures brought about by natural processes. Later he and Lyell became close friends.

Darwin's Voyage

Darwin boarded the HMS *Beagle* in December, 1831, and visited Patagonia, South America, and some islands in the Pacific Ocean. One object of the voyage was to survey Patagonia; and as biochemist Michael Denton put it, "Its result was to shake the foundations of Western thought."[49]

The trip took over four years. He studied the geological structure of the Cape Verde Islands, and examined the coral reefs extensively. He compared animals on the mainland with those on nearby islands, and also compared living animals to the fossil remains of the same species. Darwin's task as naturalist on the voyage was to study the animals and vegetation, and to bring back samples of plant life. He observed that in the Galapagos the same species on one island had physical features somewhat different than did those on a nearby island. What Darwin saw led him to think that new species were produced when populations became separate in isolated locations and subject to new conditions. He concluded that the differences were a result of different environmental conditions.

Later Darwin made many other observations that, to him, confirmed the idea of plants and animals undergoing change over time to better adapt to their environments. An individual bird or animal did not itself change; change came in subsequent generations of their offspring.

His study of the Galapagos finches is famous. He observed that the finches on one island were clearly different from those on another island. He assumed these finches all came from the same ancestors, but changed or "evolved" because of adaptation to different conditions in their environments. As Michael Denton explains it,

> The *Beagle* revealed to Darwin a new world, one that bore no trace of the supernatural drama that Genesis implied, and one which seemed impossible to reconcile with the miraculous biblical framework he himself had accepted when he left England. All the new evidence seemed to point to an immensely long geological past.[50]

Darwin observed geological features on Patagonia and in South America that he believed were clearly formed during a much longer period of time than most Christians believed then. As Denton puts it, "Nothing Darwin had witnessed on the *Beagle* implied that evolution on a grand scale had occurred, that the major divisions of nature had been crossed by an evolutionary process." But what Darwin did observe seemed to him clearly inconsistent with his previous belief in a supernatural and recent creationism. Denton continues,

> In his book Darwin is actually presenting two related but quite distinct theories. The first, which has sometimes been called the "special theory," is relatively conservative and restricted in scope and merely proposes that new races and species arise in nature by the agency of natural selection, thus the

PART 2—*What Do We Really Know about the World?* The Science

> complete title of his book: *The Origin of Species by Means of Natural Selection or the Preservation of Favoured Races in the Struggle for Life*. The second theory, which is often called the "general theory," is far more radical. It makes the claim that the "special theory" applies universally... This "general theory" is what most people think of when they refer to evolution theory.[51]

This distinction is very important, because there is much scientific evidence today to support Darwin's special theory. This evidence is often used to argue for the scientific validity of his general theory. However, much greater evidence in nature is required to demonstrate that this general theory is true. Many research reports in biology have been published that ask serious questions about how far evolution can proceed. Biochemist Michael Behe is one of many scientists who have found clear evidence for Darwin's special theory in the organic world but who have also discovered that the scope of evolution is severely limited.

After his *Beagle* trip, Darwin continued to study and develop his ideas. In the research published on paleontology and geology, he found support for his theory that evolution had occurred. By 1838 he was convinced evolution had occurred on a wide scale, but did not publish his theory for another twenty years—in an essay in 1858 and in book form in November 1859. Darwin credits a reading of "Malthus on Populations" for giving him the idea of the struggle for existence, and he wrote that this gave him the theory he was searching for.

> Darwin himself saw no limit to the extent of evolutionary change or to the power of natural selection to mould even the most complex of adaptations. At the end of the *Origin* he does not shrink from the ultimate implication that all life had evolved from a common source.[52]

This is clear from the way Darwin ends *Origin* with these words:

> There is grandeur in this view of life, with its several powers, having been originally breathed by the Creator into a few forms or into one; and that, whilst this planet has gone cycling on according to the fixed law of gravity, from so simple a beginning endless forms most beautiful and most wonderful have been, and are being evolved.[53]

Darwin wrote four books on his scientific discoveries and theories, the most famous being *The Origin of Species* and *The Descent of Man*, written in 1871. The biologist E. O. Wilson called Darwin's *The Origin of Species* "the greatest scientific book of all time." This is because Darwin proposed a perceptual *framework* for understanding how to interpret the new scientific facts that were being discovered in the nineteenth and later in the twentieth century. Darwin and Galileo are considered the great pioneers of modern science. Galileo is because he invented the telescope and his discoveries began the revolution of astronomy and our knowledge of our place in this vast universe. Darwin also is because his theory has guided the development of modern scientific thinking.

The impact of his ideas shocked the England of his day. The religious world attacked his ideas, and younger scientists claimed the right to defend Darwin. They claimed the right to promote their findings in biology and geology without regard for the prevailing belief in a young earth and direct creation. It was this direct challenge of the new science to long-held beliefs in the Bible and a young earth that Denton says set modern man adrift from his worldview and made such an impact on all spheres of learning. No other intellectual revolution in modern times (with the possible exception of the Copernican) so profoundly affected the way people view themselves and their place in the universe.

Darwin's Theory of the Origin of Species

Darwin titled his book *The Origin of Species*, but did not attempt to explain how life originated. Darwin called his theory "the theory of descent with modification through variation and natural selection" and expected a reaction to it.

He describes his theory as consisting of "the accumulation of innumerable slight variations, each good for the individual possessor," and that "there is a struggle for existence leading to the preservation of profitable deviations of structure or instinct . . . The truth of these propositions cannot, I think, be disputed."[54]

Put simply, species of plants and animals mutate and change over time. They improve in their ability to survive in their specific environments. This is because plants don't always produce identical plants every time; sometimes the descendant plant will be slightly different. This slight change may help the plant survive better in its environment. Then, simply because it survived longer, it will produce a greater number of new plants. He called this change *random* in the sense that he believed that no law of nature or supernatural power guided it. These "daughter plants" will inherit the beneficial change and pass it on to their descendants. Later one of these descendants will be slightly different in a beneficial way, and will pass on this beneficial change to its descendants. This accumulation of beneficial slight changes helps these plants survive, while other plants not as well adapted to the environment die out. The plant species gradually "evolves" in this way. The same is true of animals. Darwin called this process *natural selection*. This process, happening over long periods of time, is how Darwin understood the evolution of a species to occur.

Darwin was convinced that the great varieties of living things that exist today are the result of random chance and natural selection. One scholar described this principle as pure chance, absolutely free but blind, at the very root of evolution. Darwin's theory, that when reduced to its core concept, is actually quite elegant in its clarity and simplicity. It is based on the idea of random change by mutation and by natural selection. This is in the context of what Herbert Spencer called "the survival of the fittest." A characteristic developed randomly by natural mutation that bestows greater chance of survival will generally be passed on to descendants. To Darwin, "descent with modification," "natural selection," and "struggle for existence" define his theory. To really grasp the theory and how it has since been tested, we need to fully comprehend these three concepts.

Descent with Modification

Darwin was aware that farmers can modify species of plants and domesticated animals, and felt it was easy to understand how this occurs in nature. He said that "man can and does select the variations given to him by nature, and thus accumulates them in any desired manner. He thus adapts animals and plants for his own benefit and pleasure."[55]

> There is no reason why the principles which have acted so efficiently under domestication should not have acted under nature. In the survival of favoured individuals and races, during the constantly-recurrent Struggle for Existence, we see a powerful and ever-acting form of Selection.[56]

Darwin argued that, if man can patiently select variations useful to him and bring about modifications, why can't this be done in nature through natural selection or "the survival of the fittest"? He said, "I can see no limit to this power, in slowly and beautifully adapting each form to the most complex relations of life."[57] No supernatural "guiding hand" is needed to accomplish this; it is determined solely on suitability for survival.

In the last chapter of *The Origin of Species*, Darwin presented a lengthy argument. This is that the biological similarities of various types in the animal and bird worlds strongly supports "descent with modification" from a common ancestor as a better explanation for these facts than does the creation of separate species. He believed that he had "distinct evidence" that "within each kingdom all the members are descendant from a single progenitor." He thought he found this evidence in certain similarities in the structures of embryos even though these similar structures may perform different functions in different species.

He summarized his ideas this way:

> I have now recapitulated the facts and considerations which have thoroughly convinced me that species have been modified, during a long course of descent. This has been effected chiefly through the natural selection of numerous successive, slight, favourable variations; aided in an important manner by the inherited effects of the use and disuse of parts; . . . and by variations which seem to us in our ignorance to arise spontaneously.[58]

In reading *Origin*, I have a strong sense that Darwin had difficulty deciding how widely "the doctrine of modification of species" should be applied. He acknowledged that the wider he tried to apply it, the weaker the scientific support for it appeared to be. Yet he clearly wanted to believe that it had very wide application.

> I cannot doubt that the theory of descent with modifications embraces all the members of the same great class or kingdom. I believe that animals are descended from at most only four or five progenitors, and plants from an equal or lesser number.
>
> Analogy would lead me one step farther, namely, to the belief that all animals and plants are descended from some one prototype. But analogy may be a deceitful guide. Nevertheless all living things have much in common, in their chemical composition, their cellular structure, their laws of growth, and their liability to injurious influences.[59]

Natural Selection

The terms "selection" and "natural selection" appear hundreds of times in *Origin*. "Methodical and unconscious selection" is another expression Darwin used. The term simply means that beneficial changes in an organism give it a survival advantage. Changes that cause an organism to be weak or adjust poorly to the environment hinder its chances of surviving long enough to produce viable offspring. This is considered to be nature's way of selecting the stronger and better adapted individuals of a species to reproduce, and this eventually improves the whole species.

> As all the living forms of life are the lineal descendants of those which lived long before the Cambrian epoch, we may feel certain that the ordinary succession by generation has never once been broken, and that no cataclysm has desolated the whole world. Hence we may look with some confidence to a secure future of great length. And as natural selection works solely by and for the good of each being, all corporeal and mental endowments will tend to progress towards perfection.[60]

This was Darwin's conclusion on the last page of *Origin*. He saw natural selection as beneficial for the organism, and as continuing from the very beginning on into the distant future.

Struggle for Existence

> Darwin deduced that there must be a "struggle for existence," in which any slight advantage would be crucial. Those individuals with variant characteristics conferred such an advantage would survive and reproduce, passing

> the character on to their offspring. Those with harmful characters would be eliminated. This process of natural selection would, thus, gradually adapt the species to any changes in its environment.[61]

It should be pointed out that Darwin believed that the *appearance* of a given variation was by random chance, but that its survival in future generations of descendants often was not, as is explained above. To Darwin, evolution was random and without a guiding purpose, being governed solely by advantage bestowed in the struggle for existence and survival. It might be helpful to illustrate this last point; that mutations that bestow advantages tend to be passed on to descendants.

> The finches of the Galapagos Islands are a good example. The islands varied in the kinds of food available to the birds. On one island, nuts were plentiful, and a wide and thick beak was best for cracking them. In such an environment, a finch with a beak even a little wider or thicker than normal would have an advantage, and would therefore tend to be more successful in obtaining food. This bird would therefore survive to pass on his characteristic beak to his descendants, while other finches, not so well endowed, would perish. On another island, where insects, not nuts, formed the most plentiful supply, a differently shaped beak would confer an advantage, and so the finches would develop in a different direction.[62]

Darwin's research in the Galapagos Islands has inspired much interest that even today scientists are studying the changes in beak sizes of the finch populations there. Darwin called this "a struggle for existence leading to the preservation of profitable deviations of structure or instinct." Then he added, "The truth of these propositions cannot, I think, be disputed."[63] Darwin also said, "In the survival of favoured individuals and races, during the constantly-recurrent Struggle for Existence, we see a powerful and ever-acting form of Selection."[64]

The above arguments about "descent with modification," "natural selection," and "struggle for existence" that make up the core of his theory were primarily arguments from the science of biology. However, Darwin also wrote much about the fossil record, which he believed gave some support to his ideas.

Darwin on Geology

Darwin was interested in knowing how plants and animals changed through time, and sought to study this in the fossils he found during his trip aboard the *Beagle*. As he continued this study, he came to believe that the fossil record in some ways supported his theory of biological change over time and in other ways challenged it.

Darwin went into this problem in considerable detail. He insisted that there should be fossilized remains that show very gradual change over long periods of time as simple and primitive organisms develop into more complex and modern organisms.

And he frankly admitted that he could give no satisfactory answer to why these fossils had not been discovered during his lifetime. He was particularly concerned that complex fossils have not been discovered from before the Cambrian period. The Cambrian period is also called the "Cambrian Explosion." The fossil record shows that nearly all known types of complex animals appeared suddenly over a brief period of time. It is estimated to have occurred around 540 million years ago. All fossils found that have been dated prior to this time seem to be of very simple biological organisms. Darwin estimated this period as having existed some sixty million years ago. He figured that for evolution to have occurred as he theorized, at least two hundred million years would be needed. However, present-day scientists believe a much longer period was necessary.

Darwin began his two chapters on geology in *Origin* with these words:

> Why then is not every geological formation and every stratum full of such intermediate links? Geology assuredly does not reveal any such finely-graduated organic chain; and this, perhaps, is the most obvious and serious objection which can be urged against the theory. The explanation lies, as I believe, in the extreme imperfection of the geological record.[65]

Darwin admitted that in the geological record "we do not find interminable varieties, connecting together all extinct and existing forms by the finest graduated steps."[66] But he argued that this was because of migration, and primarily because the transitional stages were probably quite short and the time between them quite long. Darwin conclude that "he who rejects this view of the imperfection of the geological record, will rightly reject the whole theory."[67]

Darwin maintained a belief that the fossil record supported his theory and admitted he could only resort to the argument for an extremely poor fossil record to claim geological support. He repeats this argument near the end of the book: "If we admit that the geological record is imperfect to an extreme degree, then the facts, which the record does give, strongly support the theory of descent with modification."[68]

The way Darwin reasoned is interesting. He repeatedly detailed the geological evidence that showed there was a lack of transitional forms in the fossil record, acknowledged this threw his whole theory into doubt—and then repeatedly said that the lack of fossils itself "greatly diminishes or even causes to disappear" the difficulties he frankly acknowledged! Wishful thinking? Surely not "science"!

Darwin's Dilemma

Darwin insisted that natural selection could only produce change in a species gradually and over an extended period of time.

> According to the theory of natural selection an interminable number of intermediate forms must have existed, linking together all the species in each group by gradations as fine as are our existing varieties.[69]

"As natural selection acts solely by accumulating slight, successive, favorable variations, it can produce no great or sudden modifications; it can act only by short and slow steps."[70] Yet he was deeply troubled because he did not see any "intermediate forms" in the fossil record.

> The several difficulties here discussed, namely—that, though we find in our geological formations many links between the species which now exist and which formerly existed, we do not find infinitely numerous fine transitional forms closely joining them all together;—the sudden manner in which several groups of species first appear in our European formations;—the almost entire absence, as at present known, of formations rich in fossils beneath the Cambrian strata,—are all undoubtedly of the most serious nature.[71]

On the one hand, Darwin cautions:

> The abrupt manner in which whole groups of species suddenly appear in certain formations, has been urged by several palaeontologists . . . as a fatal objection to the belief in the transmutation of species. If numerous species, belonging to the same genera or families, have really started into life at once, the fact would be fatal to the theory of evolution through natural selection.[72]

Here Darwin acknowledged that "the abrupt manner in which whole groups of species suddenly appear" in the fossil beds suggests his theory is not based on solid scientific evidence. Then he added that there is an ever more serious problem, and this is how species belonging to several of the main divisions of the animal kingdom suddenly appear in the lowest known fossiliferous rocks.

Yet, on the other hand, Darwin concluded the chapters on geology with these words:

> If then the geological record be as imperfect as many believe, and it may at least be asserted that the record cannot be proved to be much more perfect, the main objections to the theory of natural selection are greatly diminished or disappear. On the other hand, all the chief laws of palaeontology plainly proclaim, as it seems to me, that species have been produced by ordinary generation; old forms having been supplanted by new and improved forms of life, the products of Variation and the Survival of the Fittest.[73]

Darwin was basically saying two things in his concluding words on the geological record. He first said that, since there is no proof the fossil record is very "perfect," the lack of a fossil record to support his theory cannot be used as an argument against the theory. This is certainly an interesting argument from a scientist who took great pains to explain and support his ideas. Then he said that the fossil record supports his assertion that "new and improved forms of life" have taken the place of older forms.

His main argument to explain why the fossil record does not support his theory is the extreme imperfections of the fossil record.

> The noble science of Geology loses glory from the extreme imperfection of the record. The crust of the earth with its imbedded remains must not be looked at as a well-filled museum, but as a poor collection made at hazard and random intervals.[74]

Darwin was convinced that the fossil record supported the theory of evolution, frankly admitted that few fossils of "transitional creatures" have ever been found, and that the fossil record suggests that new species seemed to appear suddenly.

This was clearly "Darwin's Dilemma." On the one hand he acknowledged that his grand theory of macroevolution would stand or fall on the basis of whether or not living forms changed slowly—in the accumulation of a large number of tiny changes over a very long period of time. But, on the other hand, the fossil record revealed no such gradual progression. Even worse, it showed whole groups of new species appearing suddenly without any "links" between them and earlier forms of life.

He could only resort to arguments like the following: "I can answer these questions and objections only on the supposition that the geological record is far more imperfect than most geologists believe."[75] He seemed to say he expected the gaps to be filled eventually. "Only a small portion of the world has been geologically explored."[76]

Darwin died with this dilemma unresolved. And, as modern geologists like to point out today, the proof of gradual, incremental change continuing over a long period of time has never been discovered in the fossil record anywhere in the world.

Darwin's Theory of the Descent of Man

In *The Descent of Man*, Darwin explained his theories about the origin and development of the human race. He did not write much on this subject in *Origin*, probably because he knew it would invite arouse much criticism. He seemingly waited until after the public responded to *Origin*. However, he did include man in an interesting comment in *Origin* about similarities among large classes of living things.

> The similar framework of bones in the hand of a man, wing of a bat, fin of the porpoise, and leg of the horse,—the same number of vertebrae forming the neck of the giraffe and of the elephant,—and innumerable other such facts, at once explain themselves on the theory of descent with slow and slight successive modifications.[77]

As Darwin anticipated, comments like the following in *The Descent of Man* alarmed many who believed in a literal interpretation of the creation story in Genesis

1 and 2: "We thus learn that man is descended from a hairy, tailed quadruped, probably arboreal in its habits, and an inhabitant of the Old World."[78] He offered no scientific proof for so bold a conclusion.

The moral implications of Darwin's theory that all living things, including human beings, "evolved" from more primitive organisms upset many. As *The History of Science and Religion in the Western Tradition* explains it,

> Darwin had been aware from the start of his theorizing that evolutionism would affect our ideas about human nature in a way that would undermine the traditional concept of the soul. His mature views on this issue were eventually presented in his *Descent of Man* (1871). He believed that many aspects of human behavior are controlled by instincts that have been shaped by natural selection. Our moral values are merely rationalizations of social instincts built into us because our ape ancestors lived in groups.[79]

Darwin was always careful to anticipate that there would be negative reaction to his ideas. In *Descent* he argued that even if his theory is wrong it would not be harmful to science.

> False facts are highly injurious to the progress of science, for they often endure long; but false views, if supported by some evidence, do little harm, for every one takes a salutary pleasure in proving their falseness; and when this is done, one path towards error is closed and the road to truth is often at the same time opened.[80]

Yet, for all his caution, he maintained the following: "He who is not content to look, like a savage, at the phenomena of nature as disconnected, cannot any longer believe that man is the work of a separate act of creation."[81] To support this claim Darwin lists several biological commonalities between humans and other mammals, and concludes that "all point to the conclusion that man is the co-descendant with other mammals of a common progenitor."[82]

> ... All higher mammals are probably derived from an ancient marsupial animal, and this through a long series of diversified forms, from some amphibian-like creature, and this again from some fish-like animal.[83]

Yet Darwin did acknowledge that man is different from other animals in his moral sense.

> A moral being is one who is capable of reflecting on his past actions and their motives—of approving of some and disapproving of others; and the fact that man is the one being who certainly deserves this designation, is the greatest of all distinctions between him and the lower animals.[84]

But he rejected the idea that man's religious sense set him apart.

> The belief in God has often been advanced as not only the greatest, but the most complete of all distinctions between man and the lower animals. It is however impossible . . . to maintain that this belief is innate or instinctive in man. On the other hand a belief in all-pervading spiritual agencies seems to be universal, and apparently follows from a considerable advance in man's reason, and from a still greater advance in his faculties of imagination, curiosity and wonder. I am aware that the assumed instinctive belief in God has been used by many persons as an argument for His existence.
>
> But this is a rash argument, as we should thus be compelled to believe in the existence of many cruel and malignant spirits, only a little more powerful than man; for the belief in them is far more general than in a beneficent Deity.[85]

He also commented as follows:

> There is no evidence that man was aboriginally endowed with the ennobling belief in the existence of an Omnipotent God . . .
>
> If, however, we conclude under the term "religion" the belief in unseen or spiritual agencies, the case is wholly different; for this belief seems to be universal with the less civilized races.[86]

In discussing how man developed from some lower form, Darwin concluded as follows:

> And natural selection arising from the competition of tribe with tribe, . . . together with the inherited effects of habit, would, under favourable conditions, suffice to raise man to his present high position in the organic scale.[87]

Darwin saw competition, the "struggle for existence" as the basis for the highly developed state of human beings today. He believed man's moral sense arose from his social instincts and the power of language. Thus, in *The Descent of Man*, Darwin clearly included the human race, along with all other life, as a product of evolution. The source of the human race, he was convinced, was the same as that of all of the higher animals.

Conclusion

Although Darwin was troubled about the lack of support for his theory in the fossil record, his arguments and illustrations that natural selection does produce new variations and species were clear and persuasive. These arguments from biology are accepted by most natural scientists today, even by some who also believe in the Genesis account of creation.

Since these arguments are accepted, why is Darwin's theory rejected by so many scientists today? It is because Darwin didn't stop with the conclusions that were based on his research. He began to theorize that this evolution on a limited scale that seems evident in nature actually has no limit, and can account, ultimately, for the existence today of all the various life-forms we observe. It is this extension of his observation of a limited change in plant and animal varieties to a much broader, in fact unlimited, change or "evolution" of living organisms that is being challenged by scientists from many specialties today.

Darwin himself drew attention to some of the issues that would test the validity of his broad theory. These concerns are to what extent natural selection can produce biological change, the lack of "transitional creatures" in the fossil record, and how evolution could account for the "noble qualities" of a human being.

Darwin was constantly evaluating his own conclusions. He frankly admitted that if he were found to be wrong at a few strategic points his general theory would be disproved. One point is regarding how small adaptive changes could eventually result in the development of the complex organs that exist today.

> If it could be demonstrated that any complex organ existed which could not possibly have been formed by numerous, successive, slight modifications, my theory would absolutely break down. But I can find out no such case.[88]

Another is how well future discoveries of the fossil record would demonstrate macroevolution.

> Geology assuredly does not reveal any such finely-graduated organic chain; and this, perhaps, is the most obvious and serious objection which can be urged against the theory.[89]

Darwin acknowledged this was a serious problem that he had no answer for, although he did think it would be solved in the future.

Darwin alluded to a third strategic issue in *Origin*. How could life have developed on earth—or anywhere else for that matter—"spontaneously" by natural means? Darwin left this problem to future generations to explore.

> It is no valid objection that science as yet throws no light on the far higher problem of the essence or origin of life. Who can explain what is the essence of the attraction of gravity?[90]

However, the advance of science has shown this to be a bigger problem than Darwin seemed to realize. This will become clear when we consider recent discoveries, especially those in the disciplines of biology and geology.

Darwin was a very perceptive and careful scientist who developed a broad theory that set the direction of much of the scientific research in the field of biology since his day.

5

What Is Modern Darwinism?

> Darwin provides a solution, the only feasible one so far suggested, to the deep problem of our existence.[91]
>
> —Richard Dawkins

We have now reviewed some of the experiences that led Darwin to forsake a belief in the Bible and develop a theory to explain the complexities he observed in nature by natural processes. Darwin was not the first to conclude that the young earth interpretation of Genesis that was prevalent in Europe in his day did not seem to explain what scientists were observing in nature. But his careful research and ideas provided a theoretical basis for the rapidly developing science of biology.

Harvard biology professor Edward O. Wilson called Darwin's *The Origin of Species* "the greatest scientific book of all time." This is because Darwin proposed a theoretical *framework* for understanding and evaluating the new scientific facts that were being discovered in the nineteenth century and later.

Darwin's theory soon came to dominate the world of natural science in the nineteenth century. By the twentieth century most scientists were attempting to understand and interpret their research findings to fit into a Darwinian framework.

Neo-Darwinism or Modern Darwinism

The term "Darwinism" refers commonly to Darwin's theory in contrast with "creationism." It is based on Darwin's ideas of random mutation and natural selection, and, usually, on the belief in a common ancestry for all living organisms on earth. Scientists often avoid using the term *Darwinism* and use the term *evolution* for these ideas. But *evolution* is often used in a much broader sense. It is used to refer to ideas of evolution that do not assume all three of Darwin's ideas that are mentioned previously above. It is also used for the small changes in species that are actually observed in nature. *Darwinism* refers to evolution on a much greater scale.

Darwinism is in its essence the idea that evolution in organisms occurs as the environment gradually changes. Individuals within a population of living things that happen to adapt better to these changes are more likely to survive and have more offspring. In this way those with characteristics that are more adapted to the environment survive and those less adapted tend to die out. Over a period of many generations, the characteristics of the population change and eventually new species are formed.

The term *neo-Darwinism* describes the modern theory of evolution based on Charles Darwin's idea of natural selection. The theory has developed since the 1930s by combining Darwin's theory of evolution by natural selection with Austrian biologist Gregor Mendel's theory of genetic inheritance.

Neo-Darwinism extends the scope of Darwin's concept of natural selection. It included new scientific discoveries and concepts such as DNA and genetics. (Genetics is the branch of biology that deals with the heredity of characteristics.) Genetics has been also applied to the social sciences. This modern synthesis includes the idea that "there is no purpose in nature and no goal toward which evolution is striving." Or, in the words of Thomas Henry Huxley, "We are the products of a cosmic accident."

As evolutionist James Rachels expressed it,

> There is no "more evolved" or "less evolved" in Darwinian theory; there are only the different paths taken by different species, largely, but not entirely, in response to environmental pressures. Natural selection is a process that, in principle, goes on forever, moving in no particular direction; it moves this way and that, eliminating some species and altering others, as environmental conditions change.[92]

Science historian Peter Bowler points out that "the rise of scientific Darwinism in the last half of the twentieth century has, however, been matched by a continued reluctance on the part of outsiders to admit that the theory can offer a complete explanation of the development of life."[93]

But evolution gives Richard Dawkins, and many others, a complete explanation for the development of life. Dawkins said he wrote his famous book *The Selfish Gene* to present one central argument. This is that we, and all other animals, are machines created by our genes. He also wrote that "we animals are the most complicated and perfectly-designed pieces of machinery in the known universe."[94]

James Rachels wrote that,

> as Darwin clearly recognized, we are not entitled—not on evolutionary grounds, at any rate—to regard our own adaptive behaviour as "better" or "higher" than that of the cockroach, who, after all, is adapted equally well to

live in its own environmental niche. Natural selection favours creatures whose conduct enables them to win the competition to reproduce. Not only human behaviour, but the behaviour of countless other species, has this result.[95]

It is of course natural for evolutionists to apply the theory of neo-Darwinism in the biological world to that of human society. In 1975 Edward Wilson coined the word "sociobiology" for a new science he called the systematic study of the biological basis of all social behavior. Rachels refers to Wilson's ideas as follows:

> Sociobiology, he proclaimed, will take over the territory previously occupied by moral philosophy. In the opening sentences of Sociobiology, he chides "ethical philosophers who wish to intuit the standards of good and evil," but do not realize that their moral feelings really spring from the hypothalamus and the limbic system. Biology, not philosophy, he says, explains ethics "at all depths."[96]

Rachels, Wilson, and other evolutionists deny that humans are "morally special." This is, of course, a denial that religion has any real place in moral and ethical issues. There has been considerable reaction even outside religious circles against the idea there is no real progress or purpose in nature, or any objective basis of morality.

The Presuppositions for Evolutionism

The general theory of evolution has been defined as the theory that all the living forms in the world have arisen from a single source that was originally inorganic. This theory is often called "evolutionism" to express its nature as an opposing theory to "creationism." This is because Darwinian evolution is considered by so many to be a naturalistic alternative to belief in creation.

Professor D. M. S. Watson, a leading biologist of the early twentieth century, wrote, "Evolution [is] a theory universally accepted not because it can be proven by logical coherent evidence to be true, but because the only alternative, special creation, is clearly incredible."[97] Evolutionary biologists such as Richard Dickerson and E. O. Wilson have written similar thoughts. So have well-known advocates for evolution such as Stephen Jay Gould and Richard Dawkins.

Woven into almost every public discussion of evolution are the assumptions of Darwinism. Darwinism is presented as being completely supported by scientific research, even though the reality is far different. We hear this on television, over the radio, and in newsprint. Is this clever propaganda? In most cases I don't think it is. It seems to be the natural thinking of people who see all science in the philosophical framework of evolution. They do not separate the assumptions of evolution from the verifiable scientific data that support the theory. The assumptions and the science are simply woven into one cohesive whole in their thinking. This can be seen in its most

PART 2—*What Do We Really Know about the World? The Science*

entertaining and eloquent form in Richard Dawkin's popular book of 2009 *The Greatest Show on Earth: The Evidence for Evolution*.

The Case for Darwinism

Darwin's general theory that all life-forms on earth evolved from one or a very few sources is clearly the dominant paradigm that most scientists use today. Some say they use this paradigm to give meaning to their research. There is good reason for this. A form of "evolution" has been demonstrated as something that happens now in our day.

Michael Denton is a biochemist and a researcher in molecular genetics. In the following quotation Denton uses the term "microevolutionary phenomena" to mean small biological changes that improve the survival of a species.

> Since 1859, a vast amount of evidence has accumulated which has thoroughly substantiated Darwin's views as far as microevolutionary phenomena are concerned. Evolution by natural selection has been directly observed in nature, and it is beyond any reasonable doubt that new reproductively isolated populations—species—do in fact arise from pre-existing species.[98]

The following arguments are often made to support Darwin's theory:

1. It is believed by many scientists that all forms of life existing today could have evolved from one purely natural source. It is considered scientifically possible that living things could have evolved from nonliving matter, and because of this, no other explanation of the existence of life need be considered.

2. Random natural selection over a large period of time can explain the diversity of living things that exist today. Great changes in plant and animal species have been observed and documented by many scientists, and it is assumed that this process of change can continue indefinitely.

   Denton notes that there is considerable evidence in the natural world to support Darwin's "special theory" that new species develop naturally (microevolution) and this theory appears to be "largely correct." He observes that "the fact that organisms can undergo a considerable degree of evolution under perfectly natural conditions has always been one of the most persuasive facts conducive to an overall evolutionary view of nature."[99]

3. Sufficient time has elapsed in the history of planet earth to make Darwin's theory scientifically possible.

It seems that the argument for Darwin's general theory stands or falls on the ability of science to support all three of these arguments. To throw any one of these into doubt on the basis of scientific discovery would seriously undermine the whole edifice. Philosopher of science Stephen Meyer emphasizes the importance that leading evolutionists place on natural selection. He wrote,

> Evolutionary biologist Ernst Mayr observes, "The real core of Darwinism... is the theory of natural selection. This theory is so important for the Darwinian because it permits the explanation of adaptation, the 'design' of the natural theologian, by natural means, instead of by divine intervention." And Francisco Ayala, evolutionary biologist and 1994 president of the American Association for the Advancement of Science, explains:
>
> "It was Darwin's greatest accomplishment to show that the directive organization of living beings can be explained as the result of a natural process, natural selection, without any need to resort to a Creator or other external agent ... (Darwin's) mechanism, natural selection, excluded God as the explanation accounting for the obvious design of organisms."[100]

The Case against Darwinism

Denton points out that there are "two fundamental axioms of Darwin's macroevolutionary theory." He explains that "macroevolutionary theory" means the following: all life forms are connected because they all come from the same source, and they have all developed by adapting to their environments by "a blind random process." Then he remarks that neither of these two fundamental axioms "have been validated by one single empirical discovery or scientific advance since 1859."[101]

Then Denton puts his finger on what he considers the two most important weaknesses of Darwin's theory. First there is no proof whatsoever that life could have arisen from a single source. And second, not only is macroevolution by "a blind random process" counterintuitive, nothing in natural science suggests it has ever happened or even could possibly happen.

This second statement seems very bold, even for a molecular biologist with an MD degree and a PhD degree in developmental biology. Whether or not it is supported by current developments in modern science will be explored in the next few chapters. Denton remarks further that evolutionary theory still does not tell us anything at all about how new forms of life arise. He believes that the problem of random chance developing the cell structures and functions he found in humans is enough by itself to throw Darwin's theory into doubt. We will look at Denton's arguments more closely in chapter 7.

John Collins, the author of the book *Science and Faith*, points out that even though Darwin envisioned that natural selection would cause organisms to "tend to progress toward perfection," the reality is far different. He quotes J. B. S. Haldane, one of the originators of the neo-Darwinian movement, as follows: "We are therefore inclined to regard progress as the rule in evolution. Actually it is the exception, and for every case of it there are ten of degeneration."[102]

A third weakness of Darwinism is often mentioned. It states that the estimates biologists suggest as the minimum time required for macroevolution are much longer than the evidence suggests this earth has actually existed. Philosopher of science Stephen Meyer, in expressing what he considers a strong objection to Darwinism, has concluded that "the theory of common descent, arguably the central thesis of Darwin's *Origin of Species* (1859), does not explain [the facts] by natural law." He argues it attempts to explain them by presenting "a hypothetical pattern of historical events that, if actual, would account for a variety of currently observed data."[103]

Here Meyer is acknowledging that the laws of nature we observe today cannot be used to show that common descent actually happened. It is only a theory that seems to explain *some* of the things scientists observe, though it admittedly is a very popular theory. The closest evolutionary science can come is to demonstrate that, based on our verified knowledge of scientific laws and possibilities today, it *could* have happened. Even Michael Ruse, a Darwinist philosopher of science, eventually concluded that Darwinism, like creationism, "depends upon certain unprovable metaphysical assumptions."[104]

Conclusion

Evolutionary scientists for the most part argue that Darwinism is adequately supported by scientific discoveries. But no research to date has demonstrated how organic life could have developed from inorganic sources by purely natural means, nor how random mutations and natural selection could have produced the biological complexity that is observed around the world. Astrophysicist Hugh Ross asks whether *any* theory of evolution or origins is adequately supported by the facts of science.

Darwin was not the first to conclude that the young earth interpretation of Genesis did not seem to explain what scientists were observing in nature. Starting before the nineteenth century, scientists were studying geological formations and fossils. They were concluding that the earth and life on it were much older than just a few thousand years. Farmers were developing new strains of plants that produced better fruit. They were crossbreeding animals to produce better products. An ever-widening difference was developing between those who looked to the Bible for the answers to life and natural events and those who looked only to science. Darwin's great contribution was to provide a theory that seemed to explain what scientists were actually observing in nature. This included conjecture on how life could have developed on earth as a result of natural processes. It also provided a seemingly plausible explanation of how the great variety of plant and animal life could have developed naturally.

As biochemist Michael Denton reminds us, Darwin presented two related but quite distinct theories. His special theory is that new species develop naturally through "natural selection." There is nothing especially radical about this. It simply means that new strains and species develop in natural environments without a "guiding hand" such as farmers use to get the results they want. The fact that such changes do happen in nature is almost never challenged today. The cause of these changes is being debated however.

Darwin's general theory is that this process of natural development is unlimited. He even theorized that all forms of life that exist today developed in a natural, unplanned way, from one or only a very few ancestors.

Scientists and philosophers of science have tested and developed Darwin's ideas for 150 years. The results have been decidedly mixed. Scientists of many specialties have considered Darwin's idea that life developed spontaneously in "some warm little pond" on earth and have concluded that such a simplistic scene is contrary to the laws of chemistry. Some are seeking the origin of life in a source outside our planet.

The influence of Darwinism spread far wider than natural science. It suggested that human development, like that of animals, might also be the result of "survival of the fittest" and environmental factors. This had great impact on the social sciences as well as concepts of ethics and morality. James Rachels has even gone so far as to declare that the moral behavior of humans is no better than that of cockroaches. In other words, there is no objective, unchanging standard of right and wrong—just more effective coping behavior.

It is not our task in this chapter to conclude whether or not Darwin's general theory that all life evolved from a single nonliving source is scientifically defensible or not. Before we can do this we need to consider current discoveries and interpretations in the fields of biology and geology, and to some degree astrophysics. Here we will content ourselves with the foregoing presentation of what neo-Darwinism is and some of the current common arguments both for and against it. These arguments help to clarify some of the issues science must address in order to support the theory.

The next chapters will consider the development of science since Darwin's day. They will also consider the scientific evidence for and against Darwin's general theory that all life developed on earth from one or only a few common ancestors. The important question is expressed well by Hugh Ross when he asked whether any scientific position is consistent with accumulating discoveries and increasing knowledge?

Darwin himself expected his theories to be tested in the future by scientists. As to the fossil record, he acknowledged, "Geology assuredly does not reveal any such finely-graduated organic chain; and this, perhaps, is the most obvious and serious objection which can be urged against the theory." But he expected that continuing research would demonstrate the truth of his ideas.

Let us consider the fields of cosmology, biology, and geology, as we "follow the truth, wherever it leads." We will find that there is more "verifiable truth" available than there was in Darwin's day. This should answer some, but not all, of the important issues regarding evolution and creation.

6

What Can We Learn from the Cosmos?

Can Naturalism Explain the Origin of the Universe?

> Probably no discipline of science has seen such remarkable recent advances in both observation and theory as cosmology. At the beginning of the 20th century, astronomers had seen only a tiny fraction—less than one-millionth—of the potentially observable universe. Today, they can see all of it, all the way to the very limits imposed by the laws of physics... Today, they have mapped out virtually all of its visible matter and most of its dark matter... Today, astronomers can say with confidence they have witnessed all of cosmic history.[105]

The universe is expanding! It will go on expanding forever—at an increasingly fast pace! Space is curved! Most of the universe is invisible to us! These are some of the voices we hear today from cosmologists and astrophysicists.

The pronouncements of astrophysicists at the beginning of the twenty-first century sound so dramatic. Countless numbers of scientists have looked through their telescopes and poured over the observations made by the Hubble Telescope in space and by other powerful mechanical eyes.

These are truly exciting times for scientists who study the heavens and for the many people who eagerly read their books. In many ways cosmology appears to be the "growing edge of science today."

Yet the question must be asked whether theoretical cosmology has progressed faster than the objective science needed to support it. Cosmologists speak of "dark matter" and "dark energy" and explain that most of the universe is invisible to even the best of telescopes. America's National Public Radio reported on a meeting of astrophysicists in April 2009. They admitted at the meeting that they still don't know what makes up 96 percent of the universe. These scientists said that dark energy is the biggest mystery in cosmology. Comments at that meeting include the following: "Empty space appears to have energy." "Every prediction in this area has failed up until now. We need a new revolution in physics." It seems the "mysterious universe" gets more mysterious as the years go by!

For millennia people have looked up at the night sky and wondered at the vast number of specks of light that they could see. This is expressed poetically in the Bible in verses like Psalm 19:1–3:

> The heavens declare the glory of God; And the firmament shows His handiwork.
>
> Day unto day utters speech, and night unto night reveals knowledge.
>
> There is no speech nor language where their voice is not heard.

The renowned astronomer John A. O'Keefe wrote the following in 1992:

> We are, by astronomical standards, a pampered . . . cherished group of creatures; our Darwinian claim to have done it all by ourselves is as ridiculous and as charming as a baby's brave efforts to stand on its own feet and refuse his mother's hand. If the universe had not been made with the most exacting precision we could never have come into existence. *It is my view that these circumstances indicate the universe was created for man to live in.*[106]

Research in cosmology and astrophysics have progressed rapidly in the 150 years since Darwin wrote *The Origin of Species*. Famous names like Albert Einstein, Fred Hoyle, Edwin Hubble, Stephen Hawking, Fang Li Zhi, and William Lane Craig immediately come to mind. Craig is mentioned here because Lee Strobel interviewed him, and there is a very helpful report of that interview in chapter 5 of *The Case for a Creator*.

The big bang, black holes, even dark energy and dark matter are concepts that have revolutionized the thinking of scientists in recent decades. Black holes are described as stars that do not produce visible light, but can be detected because they give out light as X-rays. Scientists like those with NASA are constantly seeking to find other planets out there like our own that can accommodate advanced life, and are looking actively for primitive forms of life somewhere else in the universe. New theories are being devised all the time to attempt to explain how our universe came into being seemingly without a cause, and to explain how life can exist without a Creator.

In this chapter I will primarily explain the research findings of Hugh Ross, a cosmologist and astrophysicist. Much of the technical detail Ross uses to substantiate his conclusions will not be mentioned here. The reader is encouraged to read Ross's scientific explanations in his books. These include *The Creator and the Cosmos*, *Creation and Time*, and *Creation as Science*.

Who Is Hugh Ross?

Hugh Ross (1945–) is a cosmologist and astrophysicist, and also an active Christian. He says his interest in cosmology began when he was still in elementary school. He earned a bachelor of science degree in physics from the University of British Columbia in Canada, and a master of science and PhD in astronomy from the University of Toronto. Then he did postgraduate research on quasars and galaxies at the California Institute of Technology.

When around nineteen, he got a Bible and began to read it, beginning at Genesis 1. He was very impressed that Genesis, written thousands of years ago, described a creation process that is in remarkable agreement with current discoveries about the cosmos that he was studying. He was convinced that Moses could not have known these things 3,400 years ago, and concluded that there is no other explanation for this besides that Moses was inspired by a Creator-God to write as he did.

Ross says of this experience, "I came to Christ through a two-year personal study of the Bible. By the end of that study, I was convinced the Bible was free of contradiction and error doctrinally, historically, and scientifically."[107]

Ross explains that

> science is an attempt to interpret the facts of nature. Christian theology is an attempt to interpret the words of the Bible. Since, according to that theology, God created the universe and is also responsible for the words of the Bible, and since He does not lie or deceive, there can be no contradiction between the words of the Bible and the facts of nature. Any conflict between science and Christian theology must be attributed to human misunderstanding.[108]

Ross believes that after the scientific discoveries of the nineteenth century, many people lost faith in the conviction that both the Bible and natural science were reliable and could be trusted. This is particularly true in the areas of biology and geology. Forced to make a choice, some rejected the reliability of the Bible. Others rejected the reliability of secular science. He feels that rejecting either is a big mistake. He found that in his own area of specialty—cosmology—recent discoveries *confirm* rather than contradict the words of the Bible.

Ross was struck with the fact that, for life to exist on earth as we experience it today, a "just right universe" was necessary. Not only that, but the right galaxy, the right star and the right planet were needed. These were conditions so special that mathematically there is no possibility that they all could come together by chance— no matter what the length of time.

What Can We Learn from the Cosmos?

He encourages scientists, theologians, and all others, to consider what science has recently discovered about our universe, and compare this with what the Bible tells us about creation. He believes that when they do this with an open mind, reflecting carefully on what the Bible actually says, they will see a marvelous agreement between them.

Ross has authored at least nine books that argue this position, and coauthored several more. They give detailed explanations from astrophysics, geology and what the Bible reveals about the creation. These books include *The Fingerprint of God* (1989), *The Creator and the Cosmos* (1993), and *Creation and Time* (1994).

Later we will consider what he reports in more recent books.

It seems to me, from reading these three books, that there are five main ideas in his theory. These are the *big bang*; the *transcendent Creator*; a *just-right universe*; *time*; and *long days of creation*. I will report them in this order. The first four will be considered in this chapter. The last will be covered in part 3. Under *time* we will review Ross's arguments for an *expanding universe*.

How a Christian Astrophysicists Explains the Universe Today

1. The "Hot Big Bang"

Ross explains that the universe began about the size of a period on a piece of paper, and has already expanded to its present size. This discovery was anticipated by Einstein, with his general theory of relativity (including the ideas of a curved space and $E=MC^2$, etc.).

Ross is convinced that the universe was caused by an extremely hot, extremely compact explosive creation event because only such an event could bring about the enormous amount of entropy the universe is observed to possess. As this explosion was sudden and hot, it is called the "hot big bang." He explains that entropy describes the degree energy dissipates or radiates as heat in a closed system, and so isn't available to perform work.

He refers as follows to discoveries made in 1992 on background radiation temperatures by the Cosmic Background Explorer satellite:

> These new results do more than just prove that the universe began with a hot big bang. They tell us which kind of hot big bang. The 1990 results left room for the possibility that the big bang could have been a tightly spaced succession of "little" bangs. The new results rule out that possibility. The universe must have erupted from a single explosive event that by itself accounts for at least 99.97% of the radiant energy in the universe.[109]

PART 2—*What Do We Really Know about the World?* The Science

Ross notes that a few scientists still prefer the "steady state" explanation for the existence of the universe. This is the theory that the universe is considered as having no beginning, and somehow a small amount of matter is constantly being created. But most have come to accept the big bang theory. A few other theories have more recently been devised, but all have serious problems and so far have obtained little support.

2. Only a Transcendent God Could Bring This Universe into Existence

Ross makes it very clear that he believes that, of all religions in the world, only the God of the Bible could have created the universe. He says he studied the world's religions and their holy books, and found that only the Bible proved scientifically and historically accurate.

He writes that "the Bible alone describes God as a personal Creator Who can act entirely independent of the cosmos and its four space-time dimensions. The God of the Bible is not subject to length, width, height, and time. He is the One who brought them all into existence."[110]

The Cosmic Background Explorer project leader declared, "What we found is evidence for the birth of the universe," and, "It's like looking at God." It is said that Stephen Hawking exclaimed, "It is the discovery of the century, if not of all time."

> This excitement was stirred by astrophysicists' recognition of undeniable proof for the big bang model of the universe. The big bang together with the equations of general relativity tell us there must be a simultaneous beginning for all the matter, energy, and even the space-time dimensions of the universe. This beginning occurred only a few billion years ago and places the cause of the universe outside, that is, independent of, matter, energy, space and time. Theologically this means that the Cause of the universe is independent of and transcendent to the universe.[111]

Ross reports that, as a result of this discovery, denial of theism is now rare among astronomers.

Ross believes that at least nine dimensions of space and time exist, and God controls them all and transcends them all. He explains that because God controls all these dimensions, He must be able to fully operate in them all. He may also operate in spiritual dimensions completely distinct from space and time. Ross believes that there must be a dimension unrelated to time, and that this is where God exists. God is "transcendent" because He exists in a dimension that "transcends" our four-dimensional universe, an eternal dimension not subject to time or to any space limitations.

Ross uses, as illustration, the New Testament account of Jesus after He rose from the dead and appeared to His disciples (John 20:19; Luke 24:36–43). Jesus entered a locked room because He put His body in a much higher dimension than ours. Ross thinks that at least a sixth space-dimension was needed to do this.

3. A "Just-Right Universe" Is Necessary to Sustain Advanced Life on Earth

> As of October 1993, twenty-five different characteristics of the universe were recognized as precisely fixed. If they were different by only slight amounts, the differences would spell the end of the existence of any conceivable life. To this list of twenty-five can be added thirty-eight characteristics of our galaxy and solar system that likewise must fall within narrowly defined ranges for life of any kind to exist... Three of the characteristics of the universe must be fine-tuned to a precision of one part in 10^{37} or better. That's supernatural![112]

Ross studied the cosmos—the great expanse of this universe as revealed by modern telescopes and mathematical computations of the data gained by carefully studying the heavens. He marveled at how many things fit together to make it possible for us humans to be on this earth.

He discovered that "four major building blocks must be designed 'just right' for life." The four building blocks are electrons, nucleons, atoms, and molecules. The universe must be exactly constructed to create all four of these. He says that, unless everything is skillfully fashioned, atoms will not be able to assemble into sufficiently complex molecules, and the human mind cannot comprehend the precise balancing of all these factors.[113]

He also found that a special celestial environment was needed. Just the right *galaxy*, the right *star* (the sun), and the right *planet* were essential for life. He continues, "As biochemists now concede, for life molecules to operate so that organisms can live requires an environment where liquid water is stable. This means that a planet cannot be too close to its star or too far away. In the case of planet Earth, a change in the distance from the sun as small as 2% would rid the planet of all life."

> For life to exist on a planet, the planet must be close enough to its star to maintain a temperature suitable for life chemistry. For advanced life to exist, the planet needs the gravitational pull of a single, large, and relatively nearby moon to stabilize the tilt of its axis (otherwise it would vary too much) and to assist in the removal of greenhouse gases.[114]

Ross also concluded that a "narrow window of time" was essential. Just the right age of the universe, galaxy, sun and planet were needed. The universe is gradually cooling and if it is either too hot or too cold biochemical processes cannot operate.

PART 2—*What Do We Really Know about the World?* The Science

Just the right age of the galaxy is also important. Ross believes that life is possible only in galaxies older than about ten billion years and younger than about twenty billion years.

Also just the right age of the sun is essential. The star nearby must burn at a near constant brightness and color for life on a nearby planet to be possible.

Our planet must be just the right age in order to support the advanced life that exists today. Ross insists that life cannot exist unless all these things—the universe, the galaxy, our star, the planet, and the moon—are "middle-aged," because only middle-aged systems are stable in astronomy.

> These factors would seem to indicate that the galaxy, the sun, the earth, and the moon, in addition to the universe, have undergone divine design. It seems apparent that personal intervention on the part of the Creator takes place not just at the origin of the universe, but also much more recently.[115]

This fortuitous combination of "just right" factors coming together to make our universe with advanced life on earth possible is often called the "Anthropic Principle."

4. Time

We have already noted Ross's view that discoveries in cosmology tell us "that a transcendent Creator *must* exist." He says one proof is that all known existence, including time, suddenly and simultaneously came into being from some source beyond itself.

Ross quotes Stephen Hawking as saying, "Time must have had a beginning." Stephen Hawking is one of the most famous names in modern cosmology. He came to this conclusion when it became evident from his research in astrophysics that space and time must have originated in the same cosmic bang that brought matter and energy into existence.

Ross argues that because of the fact that *time* had a beginning there is proof that God is transcendent, and operates beyond the dimensional limits of the universe. He adds that, by definition, time is that realm or dimension in which cause-and-effect phenomena occur. Therefore God exists outside of time and must have created it.

Ross reports his thinking on time into his ideas about the age of the universe, the age of earth, and that of the human race.

The Age of the Universe

Ross gives several reasons cosmologists believe in "continual cosmic expansion." One is the "law of redshifts." When a bright object is moving rapidly away from the observer the light waves from it gets longer and shifts toward the red end of the spectrum. This is the "red shift." Another is that distant galaxies appear from earth much closer together than nearby galaxies do. They believe their measurements are so exact that

What Can We Learn from the Cosmos?

they can now predict the universe has been expanding constantly for 13.73 billion years.

He says that the three easiest to understand methods for age-dating the universe involve the expansion of the universe, the burning of the stars, and the abundance of radioactive elements. These are presented in his book, *Creation and Time*.

> Because it takes time for light from distant sources to travel to the astronomer's telescope, the farther away astronomers look, the farther back in time they see. Therefore, the observations of galaxies closer and closer together as astronomers look farther and farther away confirms that the universe has been continuously expanding since the cosmic beginning.[116]

> The existence of planets and solar-type stars is possible only if the universe continuously expands—and does so at a just-right rate for a just-right duration. A universe that expands too slowly produces only neutron stars and black holes. A universe that expands too rapidly produces no stars at all and thus no planets.[117]

Ross is convinced that all the ways astronomers use to measure the age of the universe are in general agreement, and that it is somewhere around 15 billion years or so. He acknowledges that young earth creationists have at least five arguments for their belief that the universe has existed for a much shorter period of time. He believes he has answers to all five. One will be discussed in chapter 10.

The Age of Our Planet

Ross estimates the age of the earth at about 4.6 billion years.

> Every year earth's rotation period is slowed by a small fraction of a second. If the earth were much younger than its 4.6 billion years, it would be rotating too quickly for life. If it were much older, it would be rotating much too slowly.[118]

He estimates that primitive life could have survived on earth when it was about 0.8 billion years old.

Evidence for ancient life on earth.

> In early 1992 Christopher Chyba and Carl Sagan published a review paper on the origins of life. *Origins* is plural for a good reason. Research indicates that life began, was destroyed, and began again many times during the era before it finally took hold. Fully formed cells show up in the fossil record as far back as 3.5 billion years, and limestone, formed from the remains of organisms, dates back 3.8 billion years. The ratio of ^{12}Carbon to ^{13}Carbon found in ancient sediments also indicates a plentitude of life on Earth for the era between 3.5 and 3.8 billion years ago.[119]

PART 2—*What Do We Really Know about the World? The Science*

The Age of the Human Race

> Bipedal, tool-using, large-brained primates (called hominids by anthropologists) may have roamed the earth as long ago as one million years, but religious relics and altars date back only 8,000 to 24,000 years. Thus, the secular archaeological date for the first spirit creatures is in complete agreement with the biblical date.[120]

Then Ross adds that there is new evidence that indicates the various hominid species might have gone extinct before the appearance of modern humans.

In 2006 Ross added the following explanation:

> Though most anthropologists still insist that the bipedal primates were "human," the conflict lies more in semantics than in research data. Support for their views that modern humans descended from these primate species is rapidly eroding. Evidence now indicates that all bipedal primates went extinct, with the possible exception of Neanderthal, before the advent of human beings. As for Neanderthal, the possibility of a biological link with humanity has been conclusively ruled out.[121]

This is how Ross reconciles the apparent age of fossils of humanlike, erect-walking animals of seeming ancient origin. He is an old earth creationist, but here he is in agreement with young earth scientists in believing the human race is quite recent. He dates the creation of Adam and Eve between about ten to thirty-five thousand years ago, and considers the possible outside limits somewhere around six thousand and sixty thousand years ago.

How does Ross explain the difference between humans and hominids? He maintains that hominids did not have spirits like modern man does. God made man, and only man, in His image. This issue will be developed further in part 3.

To summarize, Ross is convinced that the universe is clearly well over ten billion years old, and the earth has existed for about 4.6 billion years. However, he thinks the human race has only inhabited earth for from six thousand years to sixty thousand years or so.

Not all astrophysicists share Ross's confidence in how well we understand the universe.

We started this chapter by quoting Ross's belief that all of the universe is observable today. But some experts express their doubts about how much we really understand.

There seems to be general unanimity regarding the picture that Ross presents regarding the age and expansion of the universe, even among those who reject the possibility that all of existence began with a big hot bang. An example follows.

Sean M. Carroll, senior research associate in physics at the California Institute of Technology, is reputed to have particular expertise in dark energy. He wrote the following in 2008:

> Some 14 billion years ago the cosmos was hotter and denser than the interior of a star, and since then there has been cooling off and thinning out as the fabric of space expands. This picture accounts for just about every observation we have made, but a number of unusual features, especially in the early universe, suggest that there is more to the story than we understand.[122]

Carroll remarked that "we do not know what constitutes space."

> The volume of space in the universe . . . appears to be growing without limit. In 1998 astronomers discovered that cosmic expansion is accelerating. The most straightforward explanation is the existence of dark energy, a form of energy that exists even in empty space and does not appear to dilute away as the universe expands. It is not the only explanation for cosmic acceleration, but attempts to come up with a better idea have so far fallen short.[123]

The existence of dark energy is theorized because cosmologists believe this is the only way to explain why the universe is expanding, and because the speed of expansion seems to be accelerating. There does seem to be empirical evidence for this theory.

As Timothy Clifton and Pedro Ferreira put it,

> Combined with our modern understanding of space, time and matter, the cosmological principle implies that space is expanding, that the universe is getting cooler and that it is populated by relics from its hot beginning—predictions that are all borne out by observation.[124]

Carroll is one of a minority of physicists who are looking for an alternative explanation to the big bang. He writes about the possibility of multiple universes besides our own—but this theory is very speculative and so far has few supporters.

5. How Testing Improves Theory Formation

Ross gives a survey of the development of modern cosmological theories. Newton's law of motion led to discovering planets like Uranus and Neptune. It also helped to explain the motions and positions of the visible stars, and explained more that scientists actually observe than any other cosmic model of the day. Then Le Verrier observed things that didn't fit Newton's theory. Inconsistencies were noticed in Mercury's orbit and in predictions about the velocity of light. Albert Einstein solved this puzzle with his special theory of relativity.

Albert Einstein's special theory of relativity, published in 1905, solved the puzzle of the observed consistency of light's velocity. In 1916, his general theory of relativity explained the mystery of Mercury's orbit. However, the solution for these two anomalies produced a radically different cosmic model—one that specified a continuously expanding universe. This new model showed that the universe had a beginning, that the cosmos was finite with respect to time.

The expanding universe model attracted scientific attention not so much because it explained the two small anomalies, but primarily because the model predicted phenomena that astronomers had not yet seen . . . For example, Einstein's first paper on general relativity predicted that gravity would bend space by specified amounts, which observers could either verify or refute.[125]

Ross reports that many, like Sir Arthur Eddington, found Einstein's conclusions "philosophically repugnant." This is because it challenged their belief that evolution had "an infinite time to get started." Eddington and others in the scientific community reacted and sought theories that would still allow evolution an infinite time to get started. They came up with three: the "hesitation model"; the "steady state model"; and the "oscillating model." Each was founded on different assumptions. None had any supporting physical evidence, and "when put to test after test, none of the observations matched the predictions."

Finally they came up with three that did fit Einstein's "unwanted (single) cosmic beginning in finite time."

1. Cold big bang models (proposing that the universe expands from an infinitesimally small but cold volume).

2. Hot big bang models (proposing that the universe expands from an infinitesimally small and nearly infinitely hot volume) dominated by ordinary matter—protons, neutrons, and electrons—that strongly interacts with radiation.

3. Hot big bang models dominated by exotic matter—particles such as neutrinos, axions, and neutralinos—that weakly interacts with radiation.[126]

At first these three theories did fit the observations.

As researchers continued to make observations and conduct tests, they eventually ruled out the first two sets of models. But new tests and observations reinforced the plausibility of the third set of models. Astrophysicists by now have established an array of amazingly detailed models that fit within the third category.

Today, the viable models are hot big bang models in which "dark energy" (the self-stretching property of the cosmic space fabric) dominates "exotic dark matter" (particles that do not strongly interact with photons) which in turn dominates "ordinary dark matter" (aggregates of protons, neutrons, and electrons that do not emit appreciable light.)[127]

What Can We Learn from the Cosmos?

Ross reports that Princeton cosmologists Masataka Fukugita and Philip James Edwin Peebles assembled the following inventory of the universe:

Dark energy (self-stretching property of cosmic space surface), 72 percent of total cosmic density.	Exotic dark matter (particles that weakly interact with light), 23 percent	Ordinary matter (protons and neutrons), 4.5 percent
Stars and stellar remnants, 0.27 percent	(The missing 0.2299 percent is part of the uncertainties for dark energy and exotic dark matter. Both are considered known to only 2 percent accuracy.)[128]	Planets, 0.0001 percent

It is amazing that they calculate that stars and planets account for less than 1 percent of the total composition of the universe. In this way they theorize a universe that is at least 95 percent invisible!

The Creation Model Can Be Tested

In 2006 Hugh Ross published *Creation as Science*. The subtitle is *A Testable Model Approach to End the Creation/Science Wars*. His chief purpose was to present testable hypotheses to guide scientific investigation that he believed would result in confirmation of his theory. He calls his theory the "RTB Creation Model." (RTB refers to Reasons to Believe.)

> In the creation and evolution conflicts, thoughtful people need to ask questions and apply tests. Does any Darwinian model offer an adequate explanation of life's beginning and history? Does any evidence support the claim of a "young" (as in 10,000-year-old) universe and/or earth? Do human beings possess unique spiritual characteristics or are they just advanced animals? Most important of all—is any position consistent with accumulating discoveries and increasing knowledge?[129]

He adds that, for a scientific model to be viable it must account for emerging discoveries, and have the ability to accurately predict future findings.

In *Creation as Science* Ross expresses confidence that the advancement of scientific testing and knowledge will increasingly provide support for the truth of the Genesis account of creation—and his RTB model.

Religious Freedom in Research

Ross quotes nine cosmologists on the implications of the big bang theory on religious issues. These include many famous names in the field: Sir Arthur Eddington, Sir James Jeans, Fred Hoyle and Hermann Bondi, John Gribbin, Robert Dicke, P. J. E. Peebles, Arne Penzias, and Stephen Hawking.

Gribbin wrote, "The biggest problem with the Big Bang theory of the origin of the Universe is philosophical—perhaps even theological—what was there before the big bang?"[130] Penzias said, "Astronomy leads to a unique event, a universe which was created out of nothing, one with the very delicate balance needed to provide exactly the conditions required to permit life, and one which has an underlying (one might say "supernatural") plan."[131] Hawking observed, "It would be very difficult to explain why the universe should have begun in just this way, except as the act of a God who intended to create beings like us."[132]

Ross agrees with most cosmologists that it is scientific to test religious arguments.

> By and large, cosmologists realize that their research provides a crucible for testing various philosophical and theological constructs . . . No religion or philosophy remains fully insulated from observational tests. In the words of China's most famous astrophysicist, Fang Li Zhi, "A question that has always been considered a topic of metaphysics or theology, the creation of the universe, has now become an idea of active research in physics."[133]

Ross reports that

> in practicing religious freedom, cosmologists grant to one another the right to use observations and measurements to demonstrate which theological propositions do or do not correspond with the observable universe. In principle, their desire for truth outweighs the desire to cling to an ideology.[134]

Summary and Conclusion

Ross is convinced that, as years go by and science advances, the realities of the universe will increasingly confirm that accuracy and inspiration of all that the Bible says about the natural world. He writes that "more than a century of progress in cosmology contradicts the fundamental belief of philosophical naturalism—the claim that all causes and effects are contained in nature."[135]

> In the face of such staggering fine-tuning, even nontheistic scientists have made bold concessions. One research team said, "Arranging the universe as we think it is arranged would have required a miracle . . . It seems an external agent intervened in cosmic history for reasons of its own."[136]

Ross believes that the laws of nature were established by God when he created the universe, and that the Bible is God's Word. Since both nature and the Bible are from God and reveal God, the discoveries regarding nature by science and the contents of the Bible are essentially in agreement.

Ross is careful to point out that there *are* disagreements—important disagreements—between the general theory of evolution and how the Bible is generally taught in many churches. He faults both sides. He faults evolutionists who believe in macroevolution because the scientific evidence is not really there. He also does this because he is convinced that solid discoveries by science demonstrate that the changes considered necessary at some points in the assumed evolutionary chain are contrary to nature and therefore scientifically impossible.

Of course Ross is far from alone in presenting this claim. Most of the scientists chosen as major sources for this book do this also, some of whom do not accept the claims of either Darwinians or Christians. As a general rule they reject macroevolution because of discoveries in their scientific specialties that they believe disprove it. Ross believes that when scientists reject the Genesis account of creation, they are really rejecting the narrow interpretations of those who claim the six days of creation are literal twenty-four-hour days. This is a claim Ross also rejects. Ross laments that both groups fail to appreciate the marvelous agreements between the Bible account and the scientific account because of their misunderstandings.

Ross apparently accepts all the current theories that govern the research in cosmology and astrophysics today. He believes Einstein's general theory of relativity has been demonstrated to be true by recent discoveries in the cosmos. This supports his belief that the big bang theory is a valid basis for understanding the universe and its creation.

1. If *all* assumptions of the big bang theory are correct, then based on observations about the universe gained from modern telescopes, the universe is apparently between ten and twenty billion years old. If these assumptions, or even some of them, are in error then this estimate is suspect. Ross seems to accept them all, even the assumptions by atheists like Sagan and Hawking. Even so, he believes clearly that the Bible account of creation is literal and accurate.

2. In stating that the earth is 4.6 billion years old, Ross believes that there have been several times when life became extinct and only fossils remain. The kinds of life that existed at a certain period were those adapted to environmental conditions at that time. (For example, conditions when the earth was rotating faster and when more of earth's surface was covered by water.) He sees no conflict between these ideas and Scripture.

3. Ross takes great pains to reconcile the Genesis account of the creation of Adam and Eve with the apparent existence on earth of humanlike creatures that

apparently existed on earth in antiquity. As this issue is less about scientific discoveries and more about how to reconcile the fossil record with the Bible account of the creation of the human race, it will be taken up in part 3.

All in all, I think Ross has done an excellent job of presenting recent discoveries in cosmology and astrophysics and showing how they demonstrate conclusively that our cosmos had a wise and powerful Creator. It is true that the big bang is not the only theory supported by astrophysicists and other scientists today. But Ross firmly believes that it has overwhelming support from all that we know about the cosmos. It is also the theory currently accepted by most cosmologists. In a later chapter we will consider challenges to the big bang theory by Harold Booher.

Ross believes that the realities of nature have disproved naturalism once and for all. He concludes that "either an impossibly lucky convergence of coincidences or something beyond the universe must account not only for such a scenario but also for life's beginning."[137]

7

Evolution Is Real but Limited

Have Darwin's Theories Passed the Test of Time?

One hundred fifty years have passed since *The Origin of Species* impacted Europe so dramatically. Have Darwin's ideas passed the test of time? Have developments in biology and in the other sciences confirmed or challenged his theories? I have done a lot of reading on this and there seems to be general consensus among those who have tried to report this objectively.

The simple answer is "both"—modern science seems to have validated some of his ideas and cast serious doubts about others.

The scientific establishment has clearly embraced his theories wholeheartedly. It is often reported that because of this it has been difficult if not impossible for scientists who question important parts of Darwin's theory to get their research findings published in mainstream scientific journals. But there are many biologists who, as a result of their research, question to what extent Darwinian evolution is even possible.

Biochemist Michael Behe reported the following in 1996. A well-known biologist, Lynn Margulis of the University of Massachusetts, often challenges molecular biologists "to name a single, unambiguous example of the formation of a new species by the accumulation of mutations. Her challenge goes unmet."[138]

Behe also wrote that "many students learn from their textbooks how to view the world through an evolutionary lens. However, they do not learn how Darwinian evolution might have produced any of the remarkably intricate biochemical systems that those texts describe."[139]

PART 2—*What Do We Really Know about the World?* The Science

The mass media and the scientific establishment have embraced Darwin's theories completely, but the scientific evidence is much more nuanced. In order to "follow the truth wherever it leads," we will need to look at the biological evidence more closely.

Two books on biology have had great influence in the English-speaking world. Let's meet the authors and learn what they have to say about Darwinism. The first is Michael Denton. The second is Michael Behe, whose research we will examine in the next chapter.

Michael Denton (1943–) is a medical doctor from Australia. He is now a senior research fellow in human genetics at the University of Otago in New Zealand. His specialty is molecular biology and he earned an MD and a PhD in developmental biology. He has an intimate knowledge of comparative anatomy and embryology. He seems well-qualified to write about how modern biological discoveries either support or refute Darwin's ideas.

Denton is not a creationist and does not completely reject evolution. He says he wants to find naturalistic explanations for the designs he sees in living organisms. This is because he does not believe in the logical alternative to naturalism—that is, in creationism. He does, however, recognize that the concept of creation can be explained scientifically, and that the concept has at least some scientific merit.

Denton is one of many scientists who recognize that creation is scientifically possible, but who themselves are not Christians. To become a Christian means much more than acknowledging that what the Bible says about the origin of the world and life on it is scientifically possible. It has clear implications for the believer's daily life.

Denton wrote *Evolution: A Theory in Crisis* in 1986. This book attracted much attention among scientists. He also wrote *Nature's Destiny* in 1998 and a chapter in *Uncommon Dissent* in 2004. He freely acknowledges the areas where Darwin's ideas have been demonstrated to be valid. But his research has led him to conclude that evidence is now available to show that the scope of evolution is severely limited. He also admits the failure of evolutionary explanations.

From his research, Denton concluded that Darwin's theory is correct regarding the emergence of new species, but that any evolution beyond that lacks scientific support.

But adequately supported or not, Darwin's general theory has affected civilization greatly. As Denton expressed it, "It is not hard to understand why the question of evolution should attract such attention. The idea has come to touch every aspect of modern thought; and no other theory in recent times had done more to mould the way we view ourselves and our relationship to the world around us."[140]

Denton says that when people accepted the idea of evolution over one hundred years ago it initiated an intellectual revolution. The results were more significant and far reaching than even the Copernican and Newtonian revolutions in the sixteenth and seventeenth centuries.

Denton, in *Evolution: A Theory in Crisis*, explains his reason for writing and reporting his research findings. He does so because he concluded that the problems of macroevolution "are too severe and too intractable [unmanageable] to offer any hope of resolution." He calls macroevolution "the orthodox Darwinian framework." It is what people commonly mean when they use the word evolution. This is the theory that all living things on earth came from one or only a very few ancestors. He believes that Darwinism, as it is believed and taught today "is no longer tenable." He says, "There can be no question that Darwin had nothing like sufficient evidence to establish his theory of evolution . . . The idea of evolution on a grand scale was entirely speculative . . . Yet despite the weakness of the evidence, Darwin's theory was elevated from what was in reality a highly speculative hypothesis into an unchallenged dogma in a space of little more than twenty years after the publication of *Origin*."[141]

Denton notes that there were social and political factors that provided a receptive climate for adopting Darwin's ideas.

Perhaps the simplest way to summarize Denton's conclusions is to say two things. First, that Darwin's special theory of microevolution appears to be "largely correct." Second, that discoveries in molecular biology have raised problems with his general theory of macroevolution that seem to have no solution.

Microevolution

Denton says that

> the fact that organisms can undergo a considerable degree of evolution under perfectly natural conditions has always been one of the most persuasive facts conducive to an overall evolutionary view of nature.[142]

Denton asserts that evolutionary biology can now provide a thoroughly worked out model showing how species formation occurs in nature. This is commonly called "microevolution." The term means small changes in biological organisms as a result of random mutations. It can also be done by the designing activity of a farmer or biologist.

Modern genetics of populations shows how a new species can develop. Let's consider a situation where geographical isolation prevents interbreeding between a "daughter population" and its "parent population." The isolated daughter population may have to adapt to new environmental conditions. In doing so it undergoes unique adaptive changes and develops into a subspecies. Eventually this subspecies develops to where it can no longer interbreed with its parent population. When that happens it is commonly considered a new species.

Denton says that the laws of nature that Darwin cited to explain this process also appear to be correct. These are that biological change over time occurs by (1) the natural random processes of mutation, (2) adaptation to conditions in the environment, and (3) the struggle for survival. Denton concludes that it is clear that Darwin's special theory was largely correct. However, years later Denton changed his view and concluded that the processes of mutation to adapt to the environment was at best only a secondary cause of biological change.

Macroevolution

Denton identifies Darwin's general theory as consisting of the idea that all life on earth originated and evolved by the gradual successive accumulation of fortuitous mutations. In other words, Darwin taught that simple life developed on earth spontaneously. This simple life then developed into many life-forms as a result of small changes one after another that helped these forms to survive better in their environments. Denton calls this a highly speculative hypothesis entirely without direct factual support. He asserts that the theory is far from being self-evident, like many zealous advocates of the theory want us to believe it is. This general theory is what is called "macroevolution."

Denton points out that as time passed, Darwin's *theory* of macroevolution became a *dogma*, and biologists began to accept the idea without reference to the facts. He says, "The fact that every journal, academic debate and popular discussion assumes the truth of Darwinian theory tends to reinforce its credibility enormously." He goes on to point out that Richard Dawkins has gone so far as to assert, "'The theory is about as much in doubt as the earth goes around the sun.' Now, of course such claims are simply nonsense."[143] Denton is surely right here! This is attested to within the ranks of evolutionists themselves. Many prominent evolutionists acknowledge that evolution as it is taught today is not supported by modern science, and they are looking for new answers.

Gaps in the Fossils Challenge Macroevolution

Denton says that a serious problem with the general theory is also seen in the fossils. He says of the fossil record that "the gaps have not been explained away." For macroevolution to be true all living things today must have descended from one or only a very few sources. Therefore there should be evidences in the fossil beds of ancient living things that fill in the "gaps" between the distinct life-forms we observe today.

Denton argues that one would naturally expect to find fossils of transitional creatures "that were a bit like what went before them and a bit like what came after. But no one has yet found any evidence of such transitional creatures." This is true even though he says "geologists have found rock layers of all divisions of the last 500 million years and no transitional forms were contained in them."[144] The fossils that were once assumed to be of such transitional creatures, like the "archaeopteryx missing link," were later found to not be true missing links at all.

Denton says that no "transitional creatures" or "natural links between the great divisions of nature" have ever been found. Because of this he firmly believes that it is extremely unlikely that random processes would ever bring together suddenly all the adaptations needed for the macroevolution that evolutionists argue occurred.

Besides arguing from the gaps in the fossil record, Denton uses the feather, the avian lung and the wing of the bat as examples. They are examples of "insurmountable conceptual problems in envisioning how the gaps could have been bridged in terms of gradual random processes."

> That the gaps cannot be dismissed as inventions of the human mind, merely figments of an anti-evolutionary imagination—an imagination prejudiced by typology, essentialism or creationism—is amply testified by the fact that their existence has always been just as firmly acknowledged by the advocates of evolution and continuity.[145]

Denton complains that the scientific literature speaks of uniform rates of evolution as if there was scientific evidence of it. He mentions the existence of the "ordered pattern of diversity" that is observed in the biological world. Then he says that no one has ever given convincing explanation of how "random evolutionary processes" could possibly have produced this diversity. Somehow this highly speculative theory is treated like reality by evolutionary biologists. But he says no one has ever explained how it could even have happened.

Molecular Biology Exposes Macroevolution as Impossible

Denton also asserts that there is evidence in molecular biology, his specialty, that throws doubt on even the *possibility* that macroevolution could occur. He explains in detail in his book some of the areas where he concludes macroevolution has not been demonstrated as having ever happened or even could have happened.

Another problem is that, although scientists have spent years proposing and testing theories of how life could have originated, all of these attempts have ended in failure. He says that evolutionary theory still tells us nothing about how new forms of life arise.

After giving a detailed explanation of how amino acids work, Denton concludes that there is simply no way of explaining the concept of evolution of amino acids by chance or natural selection. He was well aware of Stanley Miller's claim to have produced amino acids in his laboratory which Miller thought recreated conditions similar to the ancient earth, but concluded that Miller's claim is not supported by the scientific facts.

Denton also argues that the possibility of human organs evolving by chance defies both intuition and the mathematical laws of chance. He gives as an example of complexity and function the nerve cells in the human brain. He says there are a

PART 2—*What Do We Really Know about the World? The Science*

"thousand million million" connections among these nerve cells. Yet they function so well. He is skeptical that any kind of purely random process could ever have assembled such systems in the time hypothesized for this to happen.

Denton believes that the problem of random chance ever developing the cell structures and functions he found in humans is enough by itself to throw Darwin's theory in doubt.

> The Darwinian claim that all the adaptive design of nature has resulted from a random search . . . is one of the most daring claims in the history of science. But it is also one of the least substantiated. No evolutionary biologist has ever produced any quantitative proof that the designs of nature are in fact within the reach of chance.[146]

Denton acknowledges that many other biologists have come to similar conclusions. He has read extensively the literature in medicine and biology and concluded as follows:

> The intuitive feeling that pure chance could never have achieved the degree of complexity and ingenuity so ubiquitous in nature has been a continuing source of skepticism ever since the publication of the *Origins*, and throughout the past century there has always existed a significant minority of first-rate biologists who have never been able to bring themselves to accept the validity of Darwin's claims. In fact, the number of biologists who have expressed some degree of disillusionment is practically endless.[147]

In this way Denton presents example after example of discoveries in his field of molecular biology to demonstrate that the broad evolutionary change essential to Darwin's general theory—macroevolution—is not scientifically possible.

The conclusion that the gaps are there, and that biology today has discovered that they *cannot* be bridged, is to Denton a powerful argument that macroevolution did not in fact occur.

The Biblical Account of Creation Is Not Acceptable to Denton Either

Denton was aware of scientific research suggesting that life on earth was much older than six thousand years, but had been told that the Bible claimed that human life is only that old. He, like Darwin before him, rejected this possibility.

Denton also questioned what he thought was a Bible declaration that new species cannot appear. He wrote that it appears from observations (like Darwin's in the Galapagos Archipelago) that new species could arise naturally from "pre-existing species." However, the Bible nowhere uses the word "species."

Denton explained Darwin's disillusionment with the Bible account of creation in the following way:

> To Darwin and most of his contemporaries, particularly in the English speaking world, the only alternative [to evolution] seemed to be a very narrow type of special creationism which was not only unscientific but had also been discredited by the fact that the species barrier seemed to have been breached by perfectly natural processes.[148]

Denton also argues that to most geologists in Darwin's day it was becoming obvious that no known natural processes, such as water or wind erosion, could have shaped the earth's geological features in only six thousand years.

Denton listed the above seeming discrepancies between what Darwin believed early in his career as a Bible-believer and what he concluded on his famous voyage. Denton appears also to accept these objections to the Bible account as his own.

It is interesting that Denton, a molecular biologist who has written one of the most scholarly and destructive books against the general theory of evolution, is not a believer in the biblical account of creation either. There does not appear to be any reason to accuse him of any bias against Darwinism itself. His objections appear to be founded only on his scientific discoveries.

An Update on Denton

In a publication in 2004 Denton referred back to his 1986 book. He said that he still considered that *Evolution: A Theory in Crisis* "represents one of the most convincing critiques of the assumption that the organic world is the continuum that classical Darwinism demands."[149] His later research confirmed and strengthened his conclusion that the scope of evolution by random mutation and natural selection is severely limited.

Denton concluded in 2004, after years of seeking and experimenting and reading, that (1) the development of life and life-forms is directed by "natural law" rather than from outside these forms of life, and (2) not directed by the environmental influences proposed by Darwin. He does not fully reject the Darwinian idea that evolutionary change comes about when organisms adapt to their changing environments, but he is convinced that this is a secondary cause and that its effects are limited.

However, Denton does not address in this article what the source of "natural law" might be, or the source of all existence, if this universe and all in it actually did have a starting point. He thinks much of what guides biological change will eventually prove to be "inherent in the fabric of nature herself."

> Moreover, such a view lends itself . . . to the possibility of a teleological interpretation of life and indeed of the entire natural order. I believe it is the one route that can lead to a new synthesis between faith and reason.[150]

Denton seems to leave open the possibility that a god or gods brought this process about at the beginning; he just doesn't believe it was done by the God of the Bible.

When he says "teleological interpretation" above, he means the view that nature exists to accomplish some purpose, and this implies for many a personal Creator Who created it and watches over it so this purpose is carried out.

So Denton has come to believe that there is an adaptive order or evolutionary principle built into living organisms that guides their development. This is somewhat like the position that creationist Howard Van Till takes, as we shall see in a later chapter. But Denton does not accept the Genesis account of creation like Van Till does.

Conclusion

Denton shows that there is sufficient evidence to conclude that *limited* evolution has actually occurred. But he laments that so many scientists cling to the belief in Darwin's general theory in spite of the scientific evidences that contradict it. By Darwin's general theory he means macroevolution. This includes the belief that life arose from nonliving matter.

Denton's basis for concluding that the general theory of evolution—that all living forms on earth today came from one or more inorganic sources—is untrue is based on two large scientific conclusions.

The *first* is that the "gaps" have never been filled. Most fundamentally, the gap between the vegetable and mineral worlds, and the gap between the animal and vegetable worlds have not been bridged. But even the gaps between the large divisions within the animal world and within the vegetable world have not been bridged. No transitional creatures or natural links have been found in the fossil record or in existing life forms.

Not only have the gaps not been bridged, Denton claims that they have not even been narrowed. These divisions now are seen to be much greater than they were seen to be in Darwin's day. This is a result of advances in biological knowledge. He says this is so obvious that not even the most committed evolutionist could pretend that in some cases the gaps have decreased even a tiny bit.

Denton's *second* conclusion from the scientific evidence is that it is biologically impossible for life to develop from inorganic material.

> We now know, as a result of discoveries made over the past thirty years, that not only is there a distinct break between the animate and inanimate worlds but that it is one of the most dramatic in all nature, absolutely unbridged by any series of transitional forms and like so many of the other major gaps of nature, the transitional forms are not only empirically absent but are also conceptually impossible.[151]

Although scientists have spent years proposing and testing theories of how life could have originated, Denton concludes that all of these attempts have ended in failure. Denton says that not only is there absolutely no evidence that life developed from the inanimate world, there is no evidence that there ever was "a supposed primordial soup" where life might have begun. "In the final analysis we still know very little about how new forms of life arise. The 'mystery of mysteries'—the origin of new beings on earth—is still largely as enigmatic as when Darwin set sail on the *Beagle*."[152]

It seems that Denton, by accepting neither macroevolution nor the Bible account of creation, is left with a cosmos, and an earth teeming with life, and no explanation for the existence of anything!

> Ultimately, the Darwinian theory of evolution is no more nor less than the great cosmogenic myth of the twentieth century. Like the Genesis based cosmology which it replaced, and like the creation myths of ancient man, it satisfies the same deep psychological need for an all embracing explanation for the origin of the world which has motivated all the cosmogenic myth makers of the past, from the shamans of primitive peoples to the ideologues of the medieval church.[153]

By "cosmogenic myth" Denton means a theory of how the universe and all in it came into being. It is his dramatic way of emphasizing that Darwin's general theory has no basis in reality whatever.

8

Biology and the Complexity of the Cell

There is one book on the developments in modern biology that seems to stand out above all others. It was written by Michael Behe.

Michael Behe

Michael Behe (1952–) is a biochemist. He received his PhD in biochemistry from the University of Pennsylvania in 1978. He has served on the Molecular Biochemistry Review Panel of the Division of Molecular and Cellular Biosciences at the National Science Foundation. This shows how broadly his value as a research scientist has been recognized.

He wrote his famous book *Darwin's Black Box: The Biochemical Challenge to Evolution* in 1996, ten years after Denton wrote *Evolution: A Theory in Crisis*. The *National Review* called *Darwin's Black Box* one of the most important nonfiction books of the twentieth century. It has been said that with this book, Behe "makes an overwhelming case against Darwinism on the biochemical level" through an argument of "great originality, elegance, and intellectual power," and that no one has done this before.

Behe's conclusions about evolution are remarkably similar to Denton's, and include the results of scientific advances since Denton wrote. His conclusions are reported here to support, update, and strengthen what Denton wrote.

Behe says that while he was doing postdoctoral research in DNA at the National Institutes of Health near Washington, DC, he began wondering how the first cell came to exist. He later read a lot and felt he would find someone who knows how life could have started, but he didn't. Then he read Denton's book *Evolution: A Theory in Crisis* in the late 1980s. As a result of reading this book, he says, "I discovered my blissful assumption that *somebody* must know how Darwinian evolution produced life, even if I didn't, was quite wrong."[154] He then read the technical literature for answers, to discover only conjecture as to how life came to exist.

It is clear from what he writes that Behe believes in the scientific method, and seeks to keep separate his religious beliefs from his scientific conclusions.

Two large issues stand out in Behe's book. The first is his conclusion that Darwin's general theory *cannot* account for the existence of the basic building blocks of life as we know them today. His second point of emphasis is that these building blocks demonstrate that they were designed by an intelligent source. The proofs that Behe presents from biochemistry are generally considered by biologists to be even more powerful than those of Denton.

Behe's Assessment of Darwin's Theories

Microevolution

Like Denton, Behe addresses the issue of whether or not there is limited change that can be considered evolutionary in nature. He wrote it has recently become possible to gain evidence for microevolution on a molecular scale. He adds that recently some scientists have even begun to design new biochemicals using the principles of microevolution—mutation and selection.

He also notes that scientists from Princeton have actually observed the average beak size of finch populations changing over the course of a few years. He says this supports Darwin's conclusion that biological change over time by natural selection and mutation does occur. A newspaper article in July 2006 reported new changes in the beak shape of one group of Galapagos Finches. Because a more aggressive finch population migrated to their island and were eating the same food, these finches started eating other food and subsequently their beaks changed in shape. This demonstrates continuing change in the very geological location Darwin investigated over 150 years earlier.

There is no doubt whatsoever in Behe's mind that evolution on a limited scale occurs naturally. He also concurs with Darwin that the mechanisms he cited to explain how microevolution occurs in nature are real. These are random mutations and natural selection.

Macroevolution

Behe points out that in the biological meaning *evolution* is a process whereby life arose from nonliving matter and then developed entirely by natural means.

Behe, like Denton, has concluded that the scientific evidence demonstrates that large-scale evolution, the changes required to support Darwin's general theory, is in fact not within the realm of possibility.

PART 2—*What Do We Really Know about the World?* The Science

> For more than a century most scientists have thought that virtually all of life, or at least all of its most interesting features, resulted from natural selection working on random variation. Darwin's idea has been used to explain finch beaks and horse hoofs, moth coloration and insect slaves, and the distribution of life around the globe and through the ages... Darwin's idea might explain horse hoofs, but can it explain life's foundation?[155]

Behe graphically describes some of the marvelous functions of the cell. In the early 1950s scientists began to understand the shapes and properties of some of the molecules that make up living organisms. Gradually they learned how biological molecules work.

> The cumulative results show with piercing clarity that life is based on *machines*—machines made of molecules! Molecular machines haul cargo from one place in the cell to another along "highways" made of other molecules, while still others act as cables, ropes, and pulleys to hold the cells in shape. Machines turn cellular switches on and off, sometimes killing the cell or causing it to grow. Solar-powered machines capture the energy of photons and store it in chemicals. Electric machines allow currents to flow through nerves. Manufacturing machines build other molecular machines, as well as themselves. Cells swim using machines, copy themselves with machinery, ingest food with machinery. In short, highly sophisticated molecular machines control every cellular process. Thus the details of life are finely calibrated, and the machinery is enormously complex.[156]

Yet, with all that has been discovered recently about cells and how they function, *nothing* has been discovered to shed light on how cells came into existence. Behe says he surveyed the scientific literature on evolution, focusing his search on the question of how molecular machines—the basis of life—developed, and he found "an eerie and complete silence."

Behe explains that Darwinian theory is totally unable to account for the molecular basis of life. He demonstrates this by explaining about many complex bio-systems that could not have developed gradually, as Darwin's theory proposes. He continues,

> In the face of the enormous complexity that modern biochemistry has uncovered in the cell, the scientific community is paralyzed. No one at Harvard University, no one in the National Institutes of Health, no member of the National Academy of Sciences, no Nobel Prize winner—no one at all can give a detailed account of how the cilium, or vision, or blood clotting, or any complex biochemical process might have developed in a Darwinian fashion. But we are here. Plants and animals are here. The complex systems are here. All these things got here somehow: if not in a Darwinian fashion, then how?[157]

He reports that tens of thousands of scientists in the United States are interested in the molecular basis of life. Many are interested in evolution. In 1971, the *Journal of*

Molecular Evolution was started. Behe says he reviewed ten thousand of the articles published in this journal over a ten-year period. About 10 percent were concerned with the chemical synthesis of molecules thought to be necessary for the origin of life.

> To say that Darwinian evolution cannot explain everything in nature is not to say that evolution, random mutation, and natural selection do not occur; they have been observed (at least in cases of microevolution) many different times ... But the root question remains unanswered: What has caused complex systems to form? No one has ever explained in detailed, scientific fashion how mutation and natural selection could build the complex, intricate structures discussed in this book.[158]

Behe explains in detail several of the complex biological structures that he says could not possibly have evolved gradually. Among these are the blood clotting system, and the flagellum. He describes the flagellum as "an outboard motor that some bacteria use to swim." He says these could not have come about by the gradual changes envisioned by Darwinism.

Clearly Behe has done his homework! He has surveyed what has been published in English on the issue and presents overwhelming evidence. This evidence is that, as of the time he prepared *Darwin's Black Box* for publication, no research had been presented to show how Darwin's general theory—macroevolution—could have occurred. It is an interesting *theory*, but it has no support from scientific research.

From things Behe wrote, he clearly *wants* to believe in evolution, but has discovered overwhelming evidence that evolutionary theory cannot explain how it all began. (He seems to want to believe all animal life we observe today came from one common source. But from what he has written in *Darwin's Black Box* it is clear that the scientific proof is simply not there.)

Behe's Dramatic Conclusion

Darwin's Theory Cannot Account for the Existence of the Cell

Behe points out in *Darwin's Black Box* that Darwin knew that his theory of gradual evolution by natural selection carried a heavy burden, and also quotes Darwin's famous statement that his theory would "absolutely break down" if a complex biological organism was ever discovered that could not have possibly been formed by numerous, successive, slight modifications.

The main argument of Behe's book is that this *can* be demonstrated. He clearly believes that the *cell* is an *irreducibly complex* system—which cannot be explained by Darwin's theory and which essentially has to be a product of intelligent design.

In explaining what an irreducibly complex system is, Behe illustrates with a simple mousetrap. The function of a mousetrap is to immobilize a mouse. To accomplish this, it is composed of

1. a flat wooden platform to act as a base,
2. a metal hammer to pin down the mouse,
3. a spring with extended ends to press against the platform and the hammer when the trap is set,
4. a sensitive catch that releases when any slight pressure is applied, and
5. a metal bar that connects to the catch and holds the hammer back when the trap is set.

If *all* these five components are required for the trap to catch and hold a mouse, then this is an irreducibly complex system.

He says an irreducibly complex system is a single system, one that is composed of several well-matched, interacting parts that contribute to the basic function. To remove any one of the parts will cause the system to effectively cease functioning.

Behe describes and explains in his book many biological "irreducibly complex systems" which he demonstrates could not possibly have been formed by numerous, successive, slight modifications. He explains two "swimming devices" that enable cells to perform their essential functions: the *cilium* and the *bacterial flagellum*.

> The amount of scientific research that has been and is being done on the cilium—and the great increase over the past few decades in our understanding of how the cilium works—lead many people to assume that even if they themselves don't know how the cilium evolved, *somebody* must know. But a search of the professional literature proves them wrong. Nobody knows.[159]

> Because the bacterial flagellum is necessarily composed of at least three parts—a paddle, a rotor, and a motor—it is irreducibly complex. Gradual evolution of the flagellum, like the cilium, therefore faces mammoth hurdles.[160]

Behe thinks it is extremely unlikely that these could have been produced by "numerous, successive, slight modifications" such as is required by Darwin's theory.

Even simple phenomena like how blood coagulates in the human body cannot be explained by Darwinian theory. How the air cells of the lungs work cannot be explained either. Behe is convinced these systems are irreducibly complex and their development cannot be explained by any gradualistic explanations like Darwinians theorize.

The immune system is another irreducibly complex system. He says, "As scientists we yearn to understand how this magnificent system came to be, but the complexity of the system dooms all Darwinian explanations to frustration."[161]

Behe states that there are many other biochemical systems that are problems for Darwinism, but he cannot explain them all in his book. However, he does explain a great many examples of irreducible complexity.

What Is the "Edge" of Darwinian Evolution?

If Darwin's theory is severely limited, like Behe and Denton argue, where is the edge of evolution? How far do random mutations and natural selection go in developing new biological forms? What are the *limits* of Darwinism? What does the evidence tell us? Behe's more recent book was written to answer this question. It is *The Edge of Evolution: The Search for the Limits of Darwinism*, published in 2007.

His three main conclusions in this book seem to be (1) the biological machinery that he described in his former book are even far more irreducibly complex than he realized in 1996. (2) He is fully convinced that there is no possibility that the complex life forms in existence today could have "evolved" by any natural process. He concludes that they were clearly designed by an intelligent source. (3) "Common descent"—the concept that all living forms on earth today descended directly from one or a very few original organisms—is not possible by Darwinian means. However, Behe clings to the belief that it could somehow be possible by intelligent design.

Where, then, according to Behe, is the "edge of evolution"—the dividing line between what random mutations by natural selection can explain and what is obviously intelligently designed? Biologists divide living organisms into anywhere from seven to ten general categories. The simplest and most basic forms are first varieties, and then species, genus, families and orders. Above these are class, and then phylum and kingdom. At the top is domain. Behe has concluded that the *highest level* that Darwinian evolution can occur is "order." But he implies that further developments in biology may reveal that the "edge of evolution" is even lower, at the level between genus and families. If this were to happen, it would demonstrate that even cats and dogs could not have evolved from a common ancestor by Darwinian processes; that is, by random mutations and natural selection.

Now Behe is not saying that dogs and cats, or even dogs and monkeys do not come from a common ancestor. He is saying that they could not have done so by Darwinian processes. In other words, his studies in biology have led him to conclude that natural selection by random mutations could not have caused this—the cell is far too complex to allow for changes this large. But he also believes that biological processes are *designed* in such a way that dogs and monkeys, and even whales and humans, might have "evolved" from a common ancestor. However, he gave no proof from biology or science to demonstrate that this is actually possible.

The Limits of Change by Random Mutation and Natural Selection

Behe gives many examples in detail of how organisms do mutate (change) to survive in changing situations, and how biological complexity limits the possibilities of change. He concludes "that random mutation and natural selection can account for many relatively minor changes in life." As examples he mentions antibiotic resistance in rats or malaria, and also changes in the appearance of animals. He also includes the different sizes and shapes of dogs and the patterns of coloration of insect wings.

> Combining the reasoning from the past several sections, then, we can conclude that animal design probably extends into life at least as far as vertebrate classes, maybe deeper, and that random mutation likely explains differences at least up to the species level, perhaps somewhat beyond. Somewhere between the level of vertebrate species and class lies the organismal edge of Darwinian evolution.[162]

Behe thus places the "edge of evolution" by random mutation at somewhere right above the *species* level and no higher than the level of order. Behe's carefully researched conclusion here comes as no surprise. We noted earlier that animals can change biologically within a group of vertebrates, to the extent some can no longer interbreed with the others. This is considered by some what distinguishes separate species. In this narrow meaning of "species," change has been observed to where new species do develop. Behe is saying here that evolution cannot extend much farther than this—that organisms cannot change by natural selection and random mutation much beyond the development of new species.

We reported how Behe uses the examples of the cilium and the bacterial flagellum to illustrate irreducible complexity that defy an evolutionary explanation. He refers to them again in his new book as "irreducible complexity squared." He means that these organisms are far more complex than he knew when he described them in *Darwin's Black Box*.

Several evolutionary scientists have recently begun exploring a new theory. They recognize that the scientific evidence today does not support the gradual accumulation of small biological changes that can result in time in large changes. They are exploring the feasibility of "modules" or "self-contained biological features" that can simplify the process of large-scale evolution. Behe and many other scientists do not think this theory is at all realistic. Behe gives the following conclusion regarding this theory: "In sum, the new evolutionary writings have unintentionally done much damage to Darwin, but have not offered convincing alternatives or replaced him."[163]

Challenges to Behe

Behe's findings published in *Darwin's Black Box* have been challenged through the years by several scientists. Among those published are Russell Doolittle's in an essay in *Boston Review* in 1997 and Kenneth Miller's book *Finding Darwin's God* (1999). Behe carefully pointed out flaws in their interpretations of their experimental results.

Behe refers to Russell Doolittle "as a prominent biochemist, member of the National Academy of Sciences, and an expert on blood clotting." Doolittle argued that natural selection can explain how evolution can take place. The essence of his challenge is directed toward Behe's argument that the blood clotting system is irreducibly complex. A gene was taken out of one group of mice, causing one type of illness. Another gene was taken out of another group of mice causing another illness. These two groups were interbred, producing mice that appeared to be normal. Behe quotes Doolittle as saying, "Contrary to claims about irreducible complexity, the entire ensemble of proteins is not needed."[164] However, Behe points out that the rats produced by the interbreeding retain the sicknesses caused by the lack of the genes in question, even though they appear normal. He says that these rats had no functional clotting system at all, and the females died when they became pregnant. Therefore the actual results of the experiment in no way supports Doolittle's claim that Behe is wrong to call the blood clotting system irreducibly complex.

Kenneth Miller, in *Finding Darwin's God*, gives several examples of complex cells that *could have* developed gradually. He cites a study by Barry Hall where Hall produced a strain of *E. coli* in the laboratory. Miller does so to argue that "complex biological systems" can evolve. Behe responds by pointing out that Hall's cells could not survive "in the wild," and required "intelligent intervention" to survive. He states the study actually reveals the limits of Darwinism and the need for design.[165]

Behe acknowledged that Russell Doolittle and Kenneth Miller "advanced scientific arguments aimed at falsifying intelligent design." If the research findings had actually been as these two reported, "then they correctly believed that my claims about irreducible complexity would have suffered quite a blow."[166] But Behe showed that the actual scientific findings did not support their conclusions.

Summary

Michael Behe's famous book, *Darwin's Black Box: The Biochemical Challenge to Evolution*, became a best seller in America and beyond. It presented solid biochemical evidence that the complexity of the cell as we know it today could not have come into existence by Darwinian means. These means are random mutations and natural selection. Cells are irreducibly complex. It also explains how Behe searched the scientific literature to discover how life could have appeared on earth spontaneously, only to discover that "nobody knows."

His biochemical research led him to discover that life is based on machines made of molecules. Yet there is complete silence in the scientific literature how molecular machines—the basis of life—developed. He gives many examples of complex biological structures that he says could not possibly have evolved gradually by small adaptive changes.

Behe published *Edge of Evolution* eleven years after his famous *Darwin's Black Box*. But it carries the same message—the irreducibly complex biological organisms could not possibly have developed by the processes that Darwin proposed. The scientific evidence overwhelmingly demonstrates that they were *designed* to function and develop as they did.

From Behe's research he has concluded that it is possible for new species and possibly even whole families to evolve by means of random mutations. However, for *classes* of organisms to do so requires multiple coherent steps to develop. He is convinced that these coherent steps cannot occur without the process having been designed by an intelligent agent.

He concludes in both books that random mutation by natural selection cannot provide a pathway for evolution beyond a very limited degree. But he clings to the belief that common descent can account for large-scale evolution. His argument with Darwinism is only that random mutation and natural selection cannot possibly bring it about. Only intelligent design could bring it about.

The Biological Images of Evolution

Stanley Miller's Experiment

We met Lee Strobel in the introduction. He reported that the four "images of evolution" he studied as a schoolboy completely convinced him of the truth of Darwin's theory. Years later, as a journalist, Strobel talked with Jonathan Wells. Wells wrote *Icons of Evolution* in 2000 and is considered a world authority on the proofs of Darwinism. Wells received a doctorate in molecular and cell biology from the University of California, Berkeley, primarily studying vertebrate embryology and evolution.

Wells said that "everyone agrees that all organisms within a single species are related through descent with modification. This occurs in the ordinary course of biological reproduction." But then he said that Darwinism means much more than this. He told Strobel that

> Darwinism claims much more than that—it's the theory that *all* living creatures are modified descendants of a common ancestor that lived long ago. You and I for example, are descendants of ape-like ancestors—in fact, we share a common ancestor with fruit flies. Darwinism claims that every new species that has ever appeared can be explained by descent with modification. Neo-Darwinism claims these modifications are the result of natural selection acting on random mutations.[167]

It is this general theory that is usually equated with evolution, and this is where evolution is not supported by the biological realities.

Strobel then asked Wells about the "icons of evolution" that had impressed him as a boy, and are still used widely in school textbooks to present evidences of evolution. Wells says he has studied them carefully, and found that they are all "either false or misleading."

Biology and the Complexity of the Cell

It was Stanley Miller's experiment that had convinced Strobel that life was generated in a spontaneous way on earth—and this shows that a creator was not necessary. Miller tried to simulate the environment of the early earth in his experiment. Wells told Strobel that most geochemists ever since the 1960s have been convinced the early atmosphere was totally unlike what Miller used in his experiment. (Stephen Meyer says that Miller was able to create only two or three of the amino acids that form proteins, and there are actually twenty-three of these.)

Wells explained that later research concluded that the theory on which Miller based his research was wrong. If Miller has used the kind of atmosphere they now conclude was on earth at that time, Wells said, they would still get "organic molecules." But these molecules would be formaldehyde and cyanide—toxic chemicals that would kill life, not produce it!

Miller's experiment was in 1953. All attempts since then to demonstrate how life could have arisen spontaneously have reached a dead end. Wells said that the Miller experiment "has virtually no scientific significance." After talking with Wells, Strobel concluded the developments in science have clearly left Miller's experiment in the dust, "even if some textbooks haven't yet noticed."

Haeckel's Embryos

Strobel says that all young students of evolution are shown Ernst Haeckel's comparative drawings of embryos. These are Haeckel's famous drawings that depict the embryos of a fish, salamander, tortoise, chicken, hog, calf, rabbit and human side by side at three stages of development. They were used to support Darwin's theory by showing that the striking similarities between early embryos indicate that all organisms share a universal ancestor. Strobel also brought this icon up in his interview with Jonathan Wells.

The strongest single class of facts to support Darwin's theory that all biological organisms have a common ancestry is said to be striking similarities among early embryos. But Wells said that when he was a graduate student he compared actual photographs of embryos with Haeckel's famous drawings, and was stunned at the discrepancy. Wells told Strobel that these drawings were faked! He said that Haeckel was even accused of fraud by his colleagues. Evolutionist Stephen Jay Gould has said that he and other paleontologists have been long aware of this. But these drawings were still being used in school textbooks at least as late as Strobel's interview. Gould himself has complained about their use.

However, the faked similarities of Haechel are not the only indications of "universal ancestry." Strobel said he had learned in school that human embryos grow something very much like gills and was told this is a compelling evidence of evolution. He asked Wells if this isn't evidence that our ancestors lived in the ocean. Wells replied

PART 2—*What Do We Really Know about the World?* The Science

that vertebrate embryos develop lines in the neck like skin folds, but these are not gills or anything like gill-like structures.

Then they discussed homology in vertebrate limbs. Wells acknowledged the similarities, and observed they were known before Darwin. They are said to point toward a common archetype or design, and not toward descent with modification. The similarities could theoretically be either common descent or common design. But no mechanism has ever been discovered to show how it can be a result of common ancestry.

Then Strobel asked, "What about recent genetic studies that show humans and apes share ninety-eight or ninety-nine percent of their genes? . . . Isn't that evidence that we share a common ancestor?" Wells replied,

> No, it's just as compatible with common design as it is with common ancestry. A designer might very well decide to use common building materials to create different organisms, just as builders use the same materials . . . to build different bridges that end up looking very dissimilar from one another.[168]

Wells explained that humans and apes share 100 percent of what are called "body building genes," but they are anatomically different! He said this similarity of DNA is actually a problem for neo-Darwinism!

Conclusion

Denton accepts the Darwinian claim that new species can be formed by natural, gradual means. But he believes the fossil record shows the theoretical gaps in the fossil record demonstrate that this process cannot account for the existence of larger groupings in either the animal or vegetable worlds. There simply are no "transitional creatures" in the fossil record that might bridge these gaps. Occasionally discoveries of such creatures are reported in the media, but so far none have been proved of consequence. However, some geologists are facing this problem of gaps in the fossil records, and are exploring new theories that consider the possible effects of catastrophes to explain the massive extinctions and sudden appearance of new biological forms. This will be reported on later.

Denton also believes from his studies in molecular biology that it is not scientifically possible for the large divisions in either the animal or the vegetable world to have evolved from a single source.

Behe also is convinced that evolution on a limited scale occurs naturally, and that Darwin's explanations for microevolution are correct. But he says that, when the term *evolution* is commonly used, it means a process whereby life arose from nonliving matter and subsequently developed entirely by natural means.

Behe, like Denton, has concluded that the scientific evidence demonstrates that the changes required to support Darwin's general theory are actually not within the realm of possibility. The irreducibly complex systems in the human body could not possibly have developed by small adaptive changes. This demonstrates that evolution on the grand scale that Darwin envisioned in his general theory could not have occurred.

I have been told that Behe's arguments and discoveries on irreducibly complex biological systems have never been refuted even though scientists like Kenneth Miller claim that they have. My impressions from reading widely is that many prominent biologists acknowledge the problems Behe wrote about, and are looking desperately for a way to "save" Darwin's theory. They are looking for alternative evolutionary explanations for the weaknesses in Darwinism, in order to explain new discoveries in nature. But so far they have not been very successful.

By 2007, Behe knew even more about the irreducibly complex biological systems than when he wrote *Darwin's Black Box*. His message remains the same: that the irreducibly complex biological organisms could not possibly have developed by the processes that Darwin proposed.

He says that to understand Darwinism we should look at three distinct concepts. These are biological change by random mutation, natural selection, and common descent. Behe acknowledges that all three of these may be true, but he insists that Darwinism has been found to be untrue (falsified).

Simply put, he says that the *pathway* of evolution that Darwin envisioned does not match the biological realities, and that evolution by natural selection acting on random mutation is severely limited. It cannot proceed much beyond the development of new species.

Many other scientists, like Wayne Frair and Gary Patterson (whose writings are reported in part 3) also refer to complexity of the cell and its functions, and how this demonstrates that macroevolution is not scientifically possible.

Putting it all together, both Denton and Behe, from their knowledge and research in molecular biology, are firmly convinced that *limited evolution* happens by natural selection caused by random mutations. But they are also convinced that the grand scale of evolution that Darwin envisioned, macroevolution, cannot be explained by such means.

Strobel says that he was completely convinced of the truth of Darwin's theory because of four "icons of evolution" that so impressed him in his biology textbook. But when he discussed this with a scientist who is considered a world authority on the proofs of Darwinism, Jonathan Wells, he learned that none of the four icons have passed the tests of modern science. We discussed two of these above, and will mention the other two in the next chapter on geology.

9
Geology and an Old Earth

Background

Although geology seems to have little direct bearing on *origins*, it has much to do with the problem of determining the *age* of living organisms.

The central issue here is how to interpret the geological record, as it is revealed on geological formations and the fossil beds. It has two aspects. First is whether or not the Darwinian theory of gradual biological change over time is supported by the fossil record. Second is how to reconstruct the time periods in earth's history.

How to interpret the geological record divides evolutionists, but it also divides Christians. It is a battleground for Christians who accept much of the prevailing scientific opinion of the day and those who do not. Central to this battle is the matter of a universal flood. Was there one? And if so how long ago? This disagreement is attested to by the very many books written on this subject, several of which are referred to in this book in detail. These are books by Davis Young, Henry Morris, Gary Parker, and others. But here we will focus only on whether or not the fossils support Darwin and the apparent age of the earth.

The problem of the fossil record and how it is variously explained has become central to the task of reconciling science and Scripture. This is why the following arguments regarding geology and paleontology are offered to the reader. They are presented here in a simplified "digest" form. This is because I have spent many months studying what various authors have written about the discoveries in geology—and what they tell us about the ages of plant and animal life—and have concluded that there is not enough evidence to resolve all the issues presented. Surely there *are* answers, and some day we may know them with confidence, but that day has clearly not yet arrived.

Geology and an Old Earth

Up to now, we have discussed the discoveries of cosmology and biology as they relate to the conflict between Darwinian theory and the Bible account of creation. Now we turn to geology.

We noted that Michael Denton, when explaining issues in biology, also referred to issues related to geology. So did Hugh Ross, when explaining current thinking in cosmology. *Denton* did in explaining the large gaps in the fossil record. *Ross* did so to support his conclusions about the age of the earth and life on it.

The apparent age of geological features and fossil beds observed by Lyell and Darwin convinced Darwin that the earth was millions of years old. Darwin believed that the fossil record supported macroevolution in the sense that the fossil beds were laid down over a very long period of time and would show gradual change from simplicity to complexity and diversity.

The study of geology can be divided broadly into two areas. *Physical geology* studies the processes that operate on and beneath the surface of the earth. *Historical geology*, as it pertains to the topics covered in this book, can be defined as follows:

> Geology, as understood today, is a science that seeks to provide a history of the earth. It does so by examining minerals, rocks, and fossils.[169]

Some scientists liken historical geology to detective work—looking for clues that might help them figure out what happened long ago.

Here we will look at the views of two scientists, both of whom are Christians. Davis Young accepts modern geological theory, and seeks to reconcile the Bible account of creation and the flood to modern geology. John Klotz sought to reconcile discoveries in the field of anthropology with the Bible account of the creation of mankind. In the next chapter we will look at the discoveries of geology from a "young earth" perspective. Young earth scientists believe the fossils and geological formations are largely the result of a worldwide flood. Young and Klotz do not.

I. Davis Young

Introduction

Davis A. Young is the son of Edward Young, a well-known Old Testament scholar I had the privilege of meeting around 1962. He received from his father a deep interest in the teachings of the Bible. When he wrote his books he was professor of geology at Calvin College, a Christian school in Grand Rapids, Michigan.

Young studied geological engineering at Princeton University and mineralogy and petrology at Pennsylvania State University. He obtained his PhD in geology from Brown University. He wrote *Creation and the Flood* in 1977, *Christianity and the Age of the Earth* in 1982, and *The Biblical Flood* in 1995.

He expresses his overall position on science and the Bible as follows. "A basically healthy theology is no excuse for poor science . . . A basically healthy science is no excuse for poor theology."[170] I would agree with both of these statements, and think most of my readers would also.

Young appears to represent the mainstream of geological opinion today. Also, *Creation and the Flood* received a review grade of "A" by *Publishers Weekly* on February 27, 1995, so it appears to have received wide acclaim.

To better understand the issues, I will present the central problem as Young sees it, and explain his view of the science of geology.

Young's Position on Geology

Young accepts the testimony of modern science that the universe is over ten billion years old and the earth between 4.5 and 4.7 billion years old. He accepts the fossil record showing that life has been on this earth a very long time. He also accepts the Genesis account of creation. The theme of his books is on the discoveries of geology as they relate to the age of the earth and the Bible account of creation and the flood. His purpose in writing his books was to try to demonstrate how the seeming discrepancies between the Bible account of creation and geology can be reconciled.

Young says he is a methodological uniformitarian, as are most geologists today. He tells us methodological uniformitarians believe two things. First, "the laws of nature are invariant in time and space and the Earth processes of the past behave in accord with those laws just as they do now."[171]

Young acknowledges that the Bible-believing Christian cannot fully agree with this, because to do so would deny that God has performed miracles in which he suspended his laws. Then he explains that God is very economical with miracles. He believes that "miracles in Scripture are closely tied in with the history of redemption; they have little if any bearing on geological history."[172]

The second thing that methodological uniformitarians believe is that the invariance of natural laws does not mean that change has occurred at a constant rate throughout geologic history. Young insists that geologists must interpret the formation of rocks and other features in terms of processes that fit our present knowledge of the laws of nature. He says, "The rates of those processes must be consistent with the laws of nature, although not necessarily constant throughout time and not necessarily even slow."[173]

Methodological uniformitarians do not reject the possibility of catastrophes on a global scale. He says that such catastrophes must be understandable in relation to the laws of nature as we observe then today. Modern geologists acknowledge that the intensity of meteorite bombardment on the earth was greater in the distant past, that volcanic activity on the moon was more intense in the past, and that continental drift may not have occurred at all early in earth's history. They also acknowledge that rates by which glaciers form and vanish have varied enormously through time, and extinction of fauna and flora was rapid during one period.

Young gives a detailed analysis of how radiometric dating works, and how he believes it has become increasingly more reliable. Radiometric dating is also called radioactive dating. It is a technique used to date materials. This is usually done by comparing the observed abundance of a naturally occurring radioactive isotope and its decay products. It is done when decay rates are known.

Young does acknowledge that young earth scientists do not accept these methods as reliable, but Young himself clearly does. He figures that calculations based on the distribution of radioactive elements and their daughters in the earth, the moon, and in meteorites clearly suggest they were formed about 4.5 to 4.7 billion years ago. This is consistent with the estimates of astrophysicists. He also believes that radiometric dating is reliable for estimating the age of rock strata and fossils.

Summary of Young

Young's philosophy of science—as he applies it to the science of geology—seems similar to that of Ross, Denton, and Behe. "Methodological uniformitarianism," as Young describes it, says the laws of nature have acted the same in the past as they do now, but geologic change has not necessarily occurred at a constant rate. If there have been catastrophes that have affected the geologic record, as seen in strata and fossils, they have followed the laws we observe today. He says that most modern geologists accept these ideas.

Young does not reject the fact that God works miracles, but insists that geologic history must be understood apart from them. I don't think he would deny that the flood of Noah's day recorded in the Bible in Genesis chapters 6–8, was the result of a direct intervention by God.

Young sees the earth as over four billion years old, and believes the geologic record supports this. He says rock formations that exist today could not have been formed in the brief period of a year, as is argued by those who believe in a young earth and a universal flood. In this his thinking is no different than that of Lyell and Darwin.

II. John Klotz

John W. Klotz (1919–) received his PhD in biology from the University of Pittsburgh, and also studied genetics. He was a member of the Illinois Academy of Science, and professor of biology and chairman of the Division of Natural Sciences at Concordia Senior College (Fort Wayne, IN). His contribution to our study of geology comes from his research on fossils, and his philosophy of science from a Christian perspective. He wrote *Genes, Genesis and Evolution* in 1955.

PART 2—*What Do We Really Know about the World?* The Science

His Philosophy of Science

Klotz begins his book by emphasizing the tentative nature of scientific discovery.

> Only the Bible claims infallibility for itself. The scientist, for one, does not claim to have absolute truth. He says that what he presents is at best relative truth. Mavor, for instance, points out that no law or principle of science can ever be regarded as absolutely proved. All laws and principles, he says, are subject to modification with the accumulation of more data and the increase of knowledge.[174]

On the other hand, Klotz also says that Scripture does not claim to be a textbook of science, nor intend to give us a scientifically detailed account of creation. He says that Bible-believers do not need to feel apologetic or to feel on the defensive when they cannot answer all questions or solve all problems related to reconciling scientific discoveries with what the Bible says.

He emphasizes that Scripture is God's revelation of the truth about Himself, the universe, and its origin. Only God has the wisdom and omniscience to know the absolute truth. Science is only man's groping for the truth. He notes that "true science is the glimpse that God permits us to gain of the way in which the world operates. Science involves the fallible intellect and wisdom of man."[175]

Klotz reminds us that God is the First Cause, and established the laws of cause-and-effect. Therefore laws of cause-and-effect relationships are not the result of blind chance. God sustains the world. He believes God uses His divine power to keep this universe ticking, and without His sustaining hand would quickly collapse into chaos. Although God set up the laws of cause and effect and ordinarily works through them, He is not bound by them. He can also suspend their operation and intervene directly in the affairs of this universe.

Since God created the heavens and the earth, and established the scientific laws which govern the universe, He is outside the realm with which science deals.

> Thus we see that only Scripture claims to have absolute truth; science makes no such claim for itself. Because it must work with induction and deduction, both of which have their weaknesses, science can have only relative truth. If by faith we accept the statements of Scripture, it is only logical to prefer them to the theories of science. But we must be sure that we know what Scripture says. It is possible to read things into the Bible.[176]

Klotz warns, some do "read things into the Bible." He expresses this as follows:

> We must also be sure in discussing evolution that we are properly representing the statements of the Scriptures. There are those who insist that there have been no new species and can be no new species, because God's Word forbids the development of one species from another. As we shall see, this position is not tenable, for the Bible nowhere uses the term "species." There are also those who

Geology and an Old Earth

insist that the world can be only 6,000 years old, because Scripture says this. Again, as we shall see, the Bible nowhere tells us the exact age of the earth.[177]

It should be pointed out here that many advocates of young earth creationism today do not insist that new species cannot develop by natural law, or that the universe cannot be more than six thousand years old. But they do insist that the earth cannot be more than a few tens of thousands of years old, and that God directly created the larger biological divisions directly and separately. Klotz was speaking of the common young earth position of his day. He himself was a creationist—an old earth creationist.

The Nature of Scientific Laws

> When we carefully analyze [scientific laws and principles], we find that often they are not really explanations of observed phenomena but rather represent records of observations. For instance, the law of gravity does not really tell us why objects fall to the center of the earth or what these forces really are which cause bodies to attract one another. Newton himself after discussing the various phenomena associated with gravity said: "The cause of gravity is what I do not pretend to know."[178]

Klotz adds that the law of gravity as a scientific law helps us predict what will happen in the future. It tells us *what* will happen, but not *why* it happens. It does not tell us how this law came to exist. "Scientific laws merely record the ways in which God operates."

In this way Klotz tells us that the conflict between Genesis and science is a result of the tentative nature of scientific knowledge and the limitations of human knowledge. He says that scientists are by and large men of intellectual honesty. There is a minority who delight at scoffing at those who believe the Genesis account of creation and ridicule Christianity. He explains that most scientists and evolutionists are earnest seekers after the truth and are trying to discover how the world actually came into being.

Evolution and Dating the Fossils

Klotz acknowledges that limited evolution has occurred within the animal kingdom, and this may possibly have produced new species. But he suggests that these changes have occurred "within closed systems" and this may be within the limits consistent with a careful reading of Scripture.

> Certainly evolution is by no means proved, and it is not the only possible explanation for the organic diversity that we find. It is not unreasonable, then,

to assume that the changes which have occurred have been finite and limited and that they have occurred within closed systems, the "kinds" of creation.[179]

On the issue of evolution in general (macroevolution), Klotz questions whether it could have occurred, no matter how long the period. He notes that "some living mammalian species are believed to have existed for a million years without having evolved beyond the species level, and some living invertebrate species are believed to have existed in the same situation for 30 million years."[180]

Klotz, when writing in 1959, insisted that the methods used to date what appear to be fossils of humans are highly unreliable. He said that the Bible does not give a clear idea of how long mankind has been on the earth. His conclusion, from carefully considering the fossil record, is that "we have not had evolution in man, but we have had degeneration and deterioration."[181]

Klotz is different from Young in evidently seeking to keep his mind open to both the possibility the earth is young and the possibility it is old like Young believes. He makes clear that even if the earth is very old this in no way proves evolution happened. He concludes that it would be possible to have a very old earth and still have no evolution at all.

Klotz presented in his book a lot of specific detail about human and pre-human fossil remains. As the issue of human origins is dealt with in chapter 11, this discussion will be presented then.

Summary of Klotz

Klotz acknowledged that limited evolution has occurred within the animal kingdom, and this may possibly have produced new species. But he suggests that these changes have occurred "within closed systems" and this may be within the limits consistent with a careful reading of Scripture. He also said that the fossil record shows within the human race a *devolution*—"degeneration and deterioration"—rather than the evolution that Darwinians insist on.

The above conclusions of Klotz are quite similar to those of Davis Young in many ways. It seems Klotz believes that neither the fossil record nor Scripture reveals an estimate of the age of the human race that can be used with any confidence.

What about the Cambrian Explosion?

Darwin knew about the rich fossil deposits that were considered to have been laid down during the Cambrian period. Paleontologists today believe that *almost all* of the world's forty phyla—the highest classification of the animal kingdom—appeared suddenly during no more than a five million year timeframe. A phylum includes all animals with the same basic body pattern. One phylum includes all vertebrates and some invertebrates.

Darwin wrote the following about this phenomenon:

> As all the living forms of life are the lineal descendants of those which lived long before the Cambrian epoch, we may feel certain that the ordinary succession by generation has never once been broken, and that no cataclysm has desolated the whole world. Hence we may look with some confidence to a secure future of great length. And as natural selection works solely by and for the good of each being, all corporeal and mental endowments will tend to progress towards perfection.[182]

We will remember that Darwin also wrote the following about his theory:

> The several difficulties here discussed, namely—that, though we find in our geological formations many links between the species which now exist and which formerly existed, we do not find infinitely numerous fine transitional forms closely joining them all together;—the sudden manner in which several groups of species first appear in our European formations;—the almost entire absence, as at present known, of formations rich in fossils beneath the Cambrian strata,—are all undoubtedly of the most serious nature.[183]

But Darwin was confident that future discoveries would support his belief that scientists would eventually discover an unbroken sequence of fossils from primitive organisms to the complex organisms that exist today. He was so confident that he wrote the following:

> Why then is not every geological formation and every stratum full of such intermediate links? Geology assuredly does not reveal any such finely-graduated organic chain; and this, perhaps, is the most obvious and serious objection which can be urged against the theory. The explanation lies, as I believe, in the extreme imperfection of the geological record.[184]

Have discoveries since his day supported this confidence? The simple answer is no!

This sudden appearance of complex fossil life is called the Cambrian Explosion. Hugh Ross named this phenomenon "biology's big bang." He estimates that the Cambrian period was no more than two or three million years in length and possibly much shorter. He believes that the fossils show God's activity as he created millions of species of life, introduced new species and replaced and upgraded the species as they were going extinct by natural processes.

Geologists date the Cambrian period at about 540 million years ago, and Ross agrees. But creationist Gary Parker believes it was much more recent than that. Ross uses the cosmological evidence of the Cambrian Explosion to demonstrate the weakness of the traditional evolutionary interpretation of it—but he does not question the antiquity of the event.

Then What about the Icons of Evolution?

We will remember that it was four "icons of evolution" that Strobel says completely convinced him of the truth of Darwin's theory. One was Darwin's Tree of Life. Another was the archaeopteryx. Strobel reported that years later he asked Jonathan Wells, an expert on the "icons of evolution," about the scientific validity of these icons.

When Wells explained current scientific discoveries that relate to the "four icons of evolution," Strobel responded as follows: "I was left with an origin-of-life experiment whose results have been rendered meaningless; a Tree of Life that had been uprooted by the Biological Big Bang of the Cambrian explosion; doctored embryo drawings that don't reflect reality; and a fossil record that stubbornly refuses to yield the transitional forms crucial to evolutionary theory. Doubts pile on doubts."[185]

Darwin's Tree of Life and the Cambrian Fossils

Wells said millions of fossils have been found predating the Cambrian explosion. Microfossils have been found in rock dating back to three billion years. So there is extremely little likelihood that fossils will ever be found that predate Cambrian that show the same complexity as those found dating to the Cambrian era. He also said that nobody denies that common ancestry exists at some level. Common ancestry is certainly true at the species level, Wells said. But the evidence doesn't support it above the species level.

Strobel concluded that the Cambrian period was a time "in which a dazzling array of new life forms suddenly appeared fully formed in the fossil record, without any of the ancestors required by Darwinism."[186] All of this totally contradicts Darwinism, which predicted the slow, gradual development of organisms over time.

It is quite clear from all this that modern science has not "explained away" the problem with the Cambrian explosion as Darwin anticipated. Leading evolutionary paleontologists like Stephen Jay Gould have acknowledged the problem and sought for other possible explanations of the fossil record. But they have so far not formulated any theory that has found wide acceptance.

The Archaeopteryx Missing Link

Strobel in his book has an interesting explanation of this icon. We will refer to it only briefly here. Strobel asked Wells if it wasn't half a bird and half a reptile. Wells responded, "No, not even close."

> It's a bird with modern feathers, and birds are very different from reptiles in many important ways—their breeding system, their bone structure, their lungs, their distribution of weight and muscles. It's a bird, that's clear—not part bird and part reptile.[187]

Wells says further that paleontologists are pretty much in agreement that archaeopteryx is not an ancestor of modern birds. There are too many structural differences.

Newspaper articles still report on the archaeopteryx. A *Wall Street Journal* article dated October 23, 2009, said it was "easily the world's most famous fossil remains." This icon was called "the poster child of evolution." Today scientists frankly admit they know little about the biology of the archaeopteryx. How true this seems to be!

Conclusion

There has been relatively little advance in geological science since Darwin's day—compared to the rapid advances in biology and in cosmology. Even the rich fossil beds from the Cambrian period were known to Darwin. But there has been a clearer recognition that the fossil beds do not show the gradual evolution that Darwin anticipated. The Tree of Life that so influenced Strobel in his earlier years has been found to be based on Darwin's imagination alone.

Davis Young is a Christian trained in geology and one who seemingly accepts mainstream geological opinion, while also maintaining a strong faith in the inspiration of the Genesis account of creation. He believes in an old earth, and that life has been on earth for a long time. He is convinced that the natural processes that change geological conditions today are the same processes that formed the geological columns and fossil beds. So he does not accept the argument that a worldwide flood changed the face of the earth so drastically that natural laws worked differently before the flood than they do now.

Young does believe that God can work miracles, but is convinced that He does so rarely and does so primarily to accomplish His redemptive purposes on earth. Global catastrophes have occurred in the distant past, but they followed the laws of nature that operate today. In his old earth creationist views he is similar to Hugh Ross. In part 3 we will look at how Young reconciles the Genesis account of creation with modern geology.

John Klotz is a Christian trained in biology and genetics who studied the fossil record. He believes that the Bible is without error, and is "absolute truth." He contrasts this with scientific knowledge, which is limited and tentative. He warns against reading into the Bible things it does not say, or reading into the facts of nature our own assumptions and biases.

Klotz believes that God created all things, and is still active today keeping the universe functioning smoothly. God works through the laws of nature to accomplish His purposes, but sometimes suspends their operation to act directly on nature. He does not believe that evolution has been proved, but he acknowledges that evolution may be occurring on a limited basis. He questions the accuracy of dating methods and believes the age of mankind cannot be discerned either from the Bible or from the fossil record.

PART 2—*What Do We Really Know about the World?* The Science

It is clear that historical geology studies a record of past events embedded in geological formations and fossil remains that must be *interpreted*. Time frames are assumed by various methods that can only be tested by modern methods in a very limited fashion. But the geological strata and fossil beds that have been discovered around the world do tell us a few things that I believe we can rely on. First, there is much evidence of both gradual change and sudden change due to catastrophes. The sudden appearance of what several scientists call "fully formed" complex biological organisms that is called the Cambrian explosion appears to be one line of evidence that is very difficult to explain by Darwinian means. This is clearly acknowledged by well-known paleontologists like Stephen Jay Gould—who was called one of "our current most popular scientific spokesmen for evolution."

10
Geology and a Young Earth

Darwin was raised in a culture where many believed that the earth was only about six thousand years old. This is still believed by many today. Among this group are biblical scholars and also scientists. They look at certain verses in the Bible and interpret them in a very literal, some say narrow, sense. They conclude that since God created the universe and all in it, our planet cannot be more than around six to ten thousand years old. In this chapter we will consider some of the publications that have strongly influenced current thinking on this creation science or young earth view.

Background and History of Creation Science

Arguments about the relationship between the findings of natural science and the teachings of the Bible started with the birth of science itself. Long before the days of Galileo Galilei (1564–1642) and Charles Darwin (1809–1882) the issue was debated among scholars. It was a concern to Copernicus (1473–1543) and to Augustine centuries earlier. The philosophy of science of Aristotle (384–322 BC) had a great impact on the issue.

> Around 1830, professional geologists (that is, those with specialist expertise) tended to "harmonize" Genesis and geology by using geology to explain the sense in which the natural history of Genesis was true. They were opposed by nonprofessional "Scriptural geologists," who used Genesis to determine geological truths.[188]

The publication of Darwin's *Origin of Species* in 1859 brought the apparent conflict between geology and the Bible more fully to the attention of the public. George McCready Price (1870–1963) was the most influential geologist in the United States to promote "creationism." His greatest contribution to the debate was his publishing *The New Geology* in 1923.

By the early 1960s one historian predicted that the creationist movement was as good as dead. Then John Whitcombe and Henry Morris published *The Genesis Flood* in 1961. The historian Ronald Numbers called *The Genesis Flood* "the most impressive contribution to strict creationism since the publication of Price's *New Geology* in 1923." Numbers explained that

> in many respects, their book appeared to be simply a "reissue of G. M. Price's views, brought up to date," . . . Whitcombe and Morris went on to argue for a recent Creation of the entire universe, a Fall that triggered the second law of thermo-dynamics, and a worldwide Flood that in one year laid down most of the geological strata.[189]

The March 2003, periodical *Citizen*, by Focus on the Family (p. 19), gives six main tenants of what came to be called "creationism." I quote from this article:

- The universe, energy, and life were created from nothing.
- "Created kinds" of plants and organisms can only vary within fixed limits.
- Mutations and natural selection cannot bring about the development of all living things from a single organism.
- Humans and apes have different ancestors.
- The earth is young—in the range of 10,000 years or so.
- Earth's geology can be explained by catastrophic events, primarily by a worldwide flood.

"Creation science" is also called by the narrower term "young earth geology," and is part of the broader field of "creationism." As a response to Darwin's *Origin of Species*, a movement developed in America called "flood geology." It was so named because of the insistence that the flood of Noah's day produced almost all the vast sequence of fossil-bearing strata and many of the geologic features observed around the earth today. After about 1970 the field of study was broadened to include the study of cosmology and other sciences, and is now called "creation science."

Proponents of flood geology argued that the flood reported in the book of Genesis produced all of the fossil-bearing geological strata that are found around the world. They also argued that it produced many of the large geological features that are observed throughout the world today, such as the Grand Canyon in the United States. This is essentially the position of creation scientists today such as Gary Parker and the late Henry Morris.

This chapter will give a summary of my research into the writing of Henry Morris, as he was by far the most influential author in the movement to reintroduce and modernize flood geology. It will also present some findings by Gary Parker that appear to be more recent than Morris's research. It will include some of Parker's conclusions from his study of fossils.

I. Henry Morris: A Voice for Creation Science

Henry Morris (1918–2006) and John Whitcomb (1924–) grabbed the attention of Christians in the United States when they published *The Genesis Flood* in 1961. Morris was a professor of hydraulic engineering at the Virginia Polytechnic Institute in Blacksburg, Virginia. He is recognized as an expert in the field of hydraulic engineering. He is also a well-known voice in support of the young earth creationist position. John Whitcomb was a professor of Old Testament studies at Grace Theological Seminary in Winona Lake, Indiana.

To understand the convictions of what is commonly called "creation science" we will take a brief look at *The Genesis Flood* and then consider *Biblical Cosmology and Modern Science* (1970). We will also look at *Science, Scripture, and the Young Earth* (1989), coauthored by Morris and his son. Later we will consider more recent contributions to this view by Gary Parker.

In *The Genesis Flood* Morris and Whitcomb give their purposes for writing as twofold. It is first to ascertain exactly what the Scriptures say concerning the flood and related topics. They say they do this from a belief in the complete divine inspiration of Scripture, a belief that Scripture says exactly what it means, and a belief that a correct understanding of it yields authoritative truth in all matters with which it deals.

This declaration of faith puts them squarely in the camp of evangelical Christianity with its emphasis on verbal inspiration. It includes the acceptance of the Bible as one's sole authority for "faith and practice." Since Scripture *is* truth, and science merely *seeks* truth, in any apparent disagreement between the two, Scripture takes precedence in their thinking.

Yet this statement does not seem to set Whitcomb and Morris very far apart from Ross or Klotz or Bernard Ramm. It isn't even very different from Davis Young, with whom they vehemently disagree. The disagreements between Young on the one hand and Whitcomb and Morris on the other, regarding Scripture and its pronouncements on the fields of science, are not ones of doctrine but of interpretation.

> The second purpose is to examine the anthropological, geological, hydrological and other scientific implications of the Biblical record of the Flood, seeking if possible to orient the data of these sciences within the Biblical framework. If this means substantial modification of the principles of uniformity and evolution which currently control the interpretation of these data, then so be it.[190]

Again, though some might consider their wording somewhat confrontational, this position on science does not seem much difference from Ross or Klotz or even

Young. Young's emphasis is clearly different in areas where there may be conflict between what the Bible *appears* to be saying and what science *appears* to be revealing. Young would be as quick to revisit Scripture and seek a new interpretation as he would to revisit the geologic record. Not so with Whitcomb and Morris; they are out to defend what they consider to be the "traditional interpretation of the church."

In reading several books that support young earth creationism, I have become convinced that they see all of science from this position. And when reading books that support Darwinism, I have become convinced that many of these scientists see all of nature from a macroevolutionary orientation. This means that many in both groups show little or no openness to alternative explanations that might fit the facts better.

Matters of interpreting Scripture will be left to part 3. Here we are primarily concerned with understanding how young earth scientists interpret the scientific data in the field of Historical Geology.

1. The Genesis Flood

The full title of the first book is *The Genesis Flood: The Biblical Record and Its Scientific Implications*. I was surprised to find out, when reading this book, that it was written at least partially to counter the arguments for an old earth expressed by Bernard Ramm. Whitcombe and Morris refer to Ramm's ideas in at least forty pages of their book. A whole chapter is devoted to answering Ramm's arguments for an "*anthropologically universal flood.*" By this Ramm means that the flood was regional, but covered the entire area where mankind was dwelling.

As Ramm wrote as an authority on biblical interpretation his contributions will be presented in part 3 of this book. Here I will refer briefly to some important comments on the study of geology presented in *The Genesis Flood*.

Throughout his several books Morris contrasts a theory of "universal catastrophism" with the prevailing theory of uniformitarianism. In *The Genesis Flood* they state their conviction that only when the Genesis flood is given its proper place in the thinking of Christian men of science will a true historical geology be formulated. In other words, the flood of Noah's day explains geological history.

They present their main arguments from the geological record in chapter 5 of *The Genesis Flood*.

> Fossil deposits are still harder to account for on the basis of uniformity. We have shown that some kind of catastrophic condition is nearly always necessary for the burial and preservation of fossils. Present-day processes are forming very few potential fossil deposits, and most of these are under conditions of rapid, sudden burial, which are abnormal. Nothing comparable to the tremendous fossiliferous beds of fish, mammals, reptiles, etc. that are found in many places around the world is being formed today.

> And yet it is the fossils which are the basis of historical geology and the geologic time scale! It is the fossils which are considered to be the one sure proof of organic evolution, regardless of how they came to be buried. Nevertheless uniformity—modern processes—cannot legitimately account for the fossil deposits.[191]

They argue that most of the rocks containing fossils were deposited under conditions that don't exist today. Not only that; they insist there are many contradictions in the way interpretations are made when the strata is dated by the fossils found in them.

This is the heart of flood geology—fossils are not being deposited much at all in modern times, and the vast amounts of fossils in geologic layers around the world could not have been deposited and preserved by the natural processes we observe today. The very "proof" evolutionists use for insisting on a long drawn-out "organic evolution"—the existence of the fossil beds—could not have developed in the manner evolutionary theory describes. Only a catastrophic event, not a long period of gradual build-up, can account for these beds. Thus goes the main argument for flood geology.

They strengthen their arguments above by noting how common it is to have sedimentary beds that don't conform to assumed age layers. They further argue that geologists *date* fossils by the strata in which they are found and then date the strata by the fossils found in them on the basis of evolutionary theory. They call this circular reasoning, which they believe leads to erroneous conclusions. They say geologists also acknowledge that some fossils are found in layers where they shouldn't theoretically be found.

The authors' central argument is divided into three parts:

1. The geological evidence argues strongly against the theory of a gradual evolution of plant and animal life. Instead, it supports a theory of catastrophism.

2. This shows it is important to investigate the facts to find a better explanation for them.

3. This gives scientific support to their attempt to reconcile the scientific evidence with the account of Creation in Genesis.

It should be noted that the authors do not reject modern science in general, but only certain aspects of it. They call most of the branches of geology "true sciences in every sense of the word," and include mineralogy, geophysics, geochemistry, marine geology, and many others as true sciences. They add that they have no quarrel with the data of even historical geologists, but only with how those data are interpreted.

PART 2—*What Do We Really Know about the World? The Science*

2. Biblical Cosmology and Modern Science

This book by Henry Morris deals much more with scientific issues than does *The Genesis Flood*. In reading it carefully, the issues of *evolution, uniformitarianism* and *catastrophism* stand out as important and recurring themes.

Evolution

Scientists point to the fossil record to support Darwin's theory. But Morris faults geologists in general as not giving sufficient attention to what he calls "sedimentation mechanics." He admits this field has not advanced to where interpretations of sedimentary deposits inspire much confidence. He quotes the prominent British geologist Gerald A. Kerkut as saying that it was the discovery of various fossils and their correct position in relative strata and age that provided the main factual basis for the modern view of evolution. But Morris questions the evidence for evolution referred to by Kerkut. Older rocks are said to contain only primitive forms of life, and fossils become increasingly complex in more modern strata. This is considered circumstantial evidence of evolution. But Morris says which rocks are older is determined by the assumed relative antiquity of the fossils they contain.

In a later book Morris quoted several geologists who believe in evolution who acknowledge that the arguments used to date fossils and the strata they are in are "circular" arguments. One was well-known Canadian paleontologist J. A. Jeletzky, who argued that very few fossils are considered as having practical value for dating purposes and these are primarily fossils whose relative ages were determined by other means than radioactive dating. So it seems that Morris did have sufficient support from evolutionary geologists for his claim that the way the fossils are dated is based on circular reasoning, and this clearly weakens the main geological argument for Darwinian evolution.

Morris believes that the real foundation for the theory of evolution is the marine fossils found in geologic columns on land. These are considered by geologists to have been deposited over hundreds of millions of years and uplifted in relatively recent times to form the mountains.

Morris says that there have been many violent geological catastrophes all over the world throughout geologic time, many of which were far stronger than any in the modern world. He argues that geologists are aware of this, and know that the uniformitarian framework is utterly incapable of accounting for them. He believes this suggests a strong religious bias *against* the concept of one great deluge like the Bible reports and *for* an evolutionary interpretation of geological history.

Now it should be pointed out that Morris's argument here has nothing to do with the age of the earth. It has everything to do with the age of *life on earth*. If the deposits were made fairly recently then they cannot be used as proof of evolution, according

Geology and a Young Earth

to Morris' logic. This is because the deposits could have been made by a recent catastrophic flood. Geologists talk of evidence of "evolution" obtained by the depth of various strata and their fossils. They say that this demonstrates a gradual evolution of life from primitive to complex. Morris argued this evidence would be in doubt if there has been a recent catastrophic flood.

Uniformitarian versus Catastrophic Sedimentation

Morris points out that our earth appears unique among all planets in its abundance of water. There is considerable geologic evidence that all planet earth was once covered by water, supporting what Peter wrote in the Bible in 2 Peter 3:5. About 71 percent of earth's surface is covered by water today. Morris says that most of the rocks on the surface of the earth were originally laid down by moving water.

He insists that there are only two possible types of explanations for the scientific evidence that all the surface of the earth was beneath the sea at least once in the distant past. One is called "catastrophism" and the other "uniformitarianism."

The explanation called "catastrophism" means to Morris a tremendous cataclysm of water, pouring down from the skies and up from the subterranean deeps. This produced a year-long flood that caused a great amount of erosion and sediment deposits that could explain most of the sedimentary deposits on the earth. He is clearly referring to the flood reported in Genesis 7 and 8.

The explanation called "uniformitarianism" is that of very slow processes. These processes include weathering, erosion, river flow, the forming of deltas, and land masses being forced upward or downward. These processes combine over many millions of years would produce the geological formations.

Both theories acknowledge the same results—the geologic formations both sides agree exist today. But it is primarily a difference in how much *time* was involved in accomplishing these results.

> In either case, the bulk of the work was accomplished prior to recorded human history, and therefore the process is not subject to scientific examination. It is completely impossible to prove, scientifically, whether catastrophism or uniformitarianism provides the true explanation.[192]

Yet Morris argues that the modern processes of sedimentation in general appear to be quite incapable of accounting for the sedimentary rocks discovered in the geologic column.

> The Biblical creationist, of course, has no objection whatever to the concept of the uniformity of *natural law*, as prevalent in the present cosmos. It is the assumption that present *processes* (which operate within the framework of uniform natural law) must always operate at the same rates as at present with which he takes issue.[193]

Morris explains that, although the natural laws involved in geological formation do not change, process rates vary considerably. They do so because the rates depend on multiple factors, and when even one of these factors changes the rates may change considerably. Sediment erosion, transportation and deposition may take place slowly or rapidly, because they can be affected by many causes. Fossil beds can develop rapidly, even catastrophically, on the basis of uniform natural law, as well as slowly over millions of years.

The Necessity of Catastrophism

Morris pointed to seven "remarkable phenomena" that are seen in the earth's sedimentary rocks and suggest catastrophic deposition. These include *fossil graveyards* where millions of fossils are grouped together, *ephemeral markings* of rain prints and bird tracks, and *preservation of soft parts* like soft tissues of organisms. He said these soft organisms seldom leave fossil remains of any kind.

Evidence of a Single Depositional Epoch

> It can be said that in general catastrophism provides a very adequate framework of interpretation for most, and probably all, the features of the known geologic column. Uniformitarianism, on the other hand, while satisfactory as a framework for some parts of the data, seems utterly inadequate to account for most of them.[194]

Morris added that historical geologists still prefer uniformitarianism and great ages as a general framework, but are willing to recognize any number of intense and widespread floods and other local catastrophes that occur within that framework. Notice here that Morris recognises that the effects of catastrophes—local and widespread—are acknowledged by uniformitarian geologists as real. This is a point that Young also emphasizes. The issue seems to be how much effect they have had on the fossil beds around the world, and whether they were caused by one huge flood or many smaller ones.

> It is concluded . . . that the concept of one great hydraulic cataclysm, accompanied by great volcanic and tectonic activities, on a worldwide scope, provides a much more realistic model to explain the sedimentary strata and the fossil record, than does the philosophy of evolutionary uniformitarianism, with its utterly unscientific multiplication of hypothesis and manipulation of the data.[195]

In this way Morris presents example after example to demonstrate his belief, which is based on his understanding of the Bible. This belief is that a universal flood caused all major geological formations in existence today, and explains the geological evidence. He also argues that the uniformitarian belief that these formations could only be produced by millions of years of change does not adequately explain the facts.

3. Science, Scripture, and the Young Earth

Introduction

Henry Morris and his son John, a geological engineer, say that they wrote this book specifically to respond to what they call the "anti-creationist reaction of many Christian evangelical intellectuals" to *The Genesis Flood*, and to later books by Morris and others defending "creation science." They stated that this book is a response to Davis Young's *Christianity and the Age of the Earth*. They say that today literally thousands of scientists who started out as old earth evolutionists have now become young earth creationists, and that many of them have scientific credentials equal to those of Davis Young. (Morris will be used in the following discussion to refer to both Henry Morris and his son.)

Creationists like Morris say they recognize that there are still unresolved *geological* problems in their theory. But they claim that this is not justification for Young to attack it because his theory also has serious unresolved geological problems. Morris quotes from professional geological journals to show that uniformitarian geology is not accepted by all secular geologists. But he also adds that creationists have always acknowledged that there are various unresolved scientific questions in the concept of the young earth and global flood. Many of these at least have plausible—though not yet confirmed—solutions. This seems a commendable acknowledgement that no theory has as yet all the answers.

Is Catastrophism an Adequate Explanation?

Morris, as expected, divides earth history into only three geological periods: the pre-flood period with very limited or no fossil deposits, the flood period that produced most of the deposits, and the post-flood period with recent deposits.

Morris acknowledges that Young and many other uniformitarians acknowledge catastrophes do play an important part in geology. But Morris and other young earth creationists go much further in that they insist on the supreme important of one universal catastrophe in explaining geological history. Geologists like Young believe that uniformitarianism *includes* local catastrophes in its framework, and that these do not *necessarily* argue for a worldwide flood.

Morris reports that Young and other geologists assume that, when very thin layers of sedimentary rock are piled on top of each other, each layer shows the passing of a year. Geologists found a place in Wyoming, the Green River Formation, where several *millions* of very thin layers are found. They assume this is proof this formation is millions of years old. But Morris argues that such layers, called "laminated sediments," can form very quickly. He says that "this conclusion was still further confirmed in the after-effects of the 1980 volcanic explosion at Mount St. Helens . . . Up to 600 feet of *laminated* sediments were deposited in a single day, looking exactly like stratified beds normally interpreted to represent long ages."[196]

Radiometric Dating

Morris reports that the basic difference between Young and other uniformitarians and young earth creationists is based on three "fundamental assumptions" that the former accept in radiometric dating. He says, "None of these assumptions can be proved correct, or even tested."

Morris quotes Frederick Jueneman, whom he says is not a creationist, as follows:

> The age of our globe is presently thought to be some 4.5 billion years, based on radio-decay rates of uranium and thorium. Such "confirmation" may be short-lived, as nature is not to be discovered quite so easily. There has been in recent years the horrible realization that radio-decay rates are not as constant as previously thought, nor are they immune to environmental influences. And this could mean that the atomic clocks are reset during some global disaster.[197]

Morris continues,

> The fact is, there is no scientific method whatever, by which it can be *proved* that the earth is old. All such attempts are forever blocked by the fact that the necessary assumptions (known initial conditions, constant process rate, an isolated system) are not only unprovable and untestable, but also unreasonable and impossible, in the light of God's Biblical revelation of supernatural creation and a world-destroying flood.[198]

Here Morris emphasizes a point that is very important. Geologists *assume* that the conditions that existed when the geological strata and fossils were laid down were similar to conditions today. But how can they really know this? If the conditions were very different, as Morris argues, then current dating methods are indeed unreliable.

Now, if Henry Morris and John Morris are correct in the above declaration that there is no *scientific* method to prove the earth is old, then the argument is over! But if they are only half right—that the geological assumptions are "unprovable and untestable"—then the discussion is far from over. There still remain the arguments from cosmology, presented by Ross, that the earth must be over four billion years old. We will return to this matter later, in part 3.

In chapter 10 of *Science, Scripture and the Young Earth*, Morris presents evidence that he believes proves the earth is no more than ten thousand years old. This, of course, is contested by most geologists. It should also be pointed out here that even some young earth scientists today say that the earth may be many times older than this.

In the appendix of *Science, Scripture, and the Young Earth*, Morris outlines a few current projects the Institute for Creation Research is pursuing to seek further validation of their theories about the age of the earth.

4. Conclusion of Morris

After publishing *The Genesis Flood* with John Whitcomb in 1961, Henry Morris became the chief spokesman for "flood geology" or what came to be known as "creation science." He clung to what creation scientists consider "the traditional view of the church" on the origins of the universe and all it contains. Morris's view is that the universe is no more than ten thousand years old, that God in six literal twenty-four-hour days created the earth, plant and animal life, and created mankind on the sixth day. Plants, animals, and humans were created in their present "mature" condition, and there has been no advancement in their condition as described by the theory of evolution.

Morris acknowledges that fossil beds may have the appearance of being much older than ten thousand years, and geological formations may appear to have been built up over a much longer period of time. But he argues that these things can be explained better by understanding that they were caused by a universal flood, the flood in Noah's day that is reported in the book of Genesis.

Morris was an expert in hydraulic engineering, and so had a detailed understanding of the effects of how water affects and forms geological conditions.

He agrees with geologists that the natural laws involved do not change, but says that process rates that affect sedimentation and fossil deposition may take place slowly or rapidly. He argues that *catastrophism* explains the fossil beds and geological formations on earth far better than does uniformitarianism. To Morris catastrophism, as an explanation of existing geologic formations, means "one great hydraulic cataclysm, accompanied by great volcanic and tectonic activities, on a worldwide scope."

In *The Genesis Flood*, Morris directly responded to the arguments for an a*nthropologically* universal flood in Bernard Ramm's *The Christian View of Science and Scripture*. In *Science, Scripture, And the Young Earth*, Henry and John Morris say they were writing to respond to Davis Young's *Christianity and the Age of the Earth*.

Morris clearly saw himself as a "defender of the faith" from Christians who did not accept the literal interpretation of creation in Genesis as a process covering just six twenty-four-hour days. He also attacked evolutionary thinking, but his strongest remarks were toward Christians like Young and Ramm. He attacked both the arguments made for a longer creation process and arguments for the flood in Genesis as being possibly less than a worldwide phenomenon.

Yet Morris also acknowledged that the scientific issues that separate him and "old earth creationists" and methodological uniformitarians like Young are issues of how to interpret the geological record. Neither position can be proved scientifically—at least for the present. He would accept the "continental drift theory" if and when it is "proved"—though it is clear he did not consider this likely. He put no trust in "radiometric dating," and expected further research to demonstrate conclusively that the long periods of geologic time suggested by this method are not reliable. Now we will turn to the ideas of Gary Parker.

II. Gary E. Parker: Another Voice for Creation Science

Gary Parker (1941–) earned his doctorate in biology from Ball State University, with a related major in geology (paleontology). He is a member of the American Society of Zoologists. He has served as chairman of science departments at two Christian colleges, and at the Institute for Creation Research.

He says he began his teaching career at a secular university, teaching evolution "enthusiastically." He thought he was helping his students to shake off "pre-scientific superstitions, such as Christianity." Then he attended a Bible class, became a Christian, and spent three years re-examining the evidence for evolution. He says he came to the conclusion from his scientific observations, that "the biblical framework" fits what we actually observe in nature better than evolution does.

Parker's contribution to our study of geology is twofold. First, he has personally studied the geologic columns in the Grand Canyon to try to determine how they were formed. He has also studied the effects of the eruptions of Mount St. Helens in the 1980s. Second, his research is more recent than that of Henry Morris, and includes the science of paleontology.

Parker has written several books, among which is *Creation Facts of Life*, written in 1994. The information below is taken from this book. We will look now at Parker's view on what the fossils tell us, and then consider how he thinks the geologic column was formed.

1. The Witness of the Fossils

Parker says that fossils help answer two types of questions. First, the questions about *what kind* of plants and animals once lived on the earth. Then, about *how fast* the fossils and the layers of rocks around them were formed.

a. What Kind of Fossils

Parker reports that the fossil record shows clear separations between the large groupings. He says that most scientists have given up looking for fossils that link the major invertebrate groups, because even in the deepest fossil-rich deposits the groups are separate. Invertebrates include all animals without backbones or spinal columns.

He also notes that

> decline and even extinction, *not* evolution, is the rule when we compare fossil sea life with the sort of marine invertebrates we find living today. In fact, all major groups, except perhaps the groups including clams and snails, are represented by *greater variety* and *more complex* forms as *fossils* than today.[199]

This shows that the fossil record reveals greater complexity of sea life existed earlier than exist today. Parker then comments that this evidence is absolutely crushing to evolution.

The oldest and richest fossil deposits are considered to have been laid down during the Cambrian period, and this is now estimated as beginning about 540 million years ago. Parker reports that the fossil-rich portion of the geologic column has at the bottom what appears to be the sudden appearance of a multitude of complex and varied life forms. As we learned earlier, this is what geologists call the Cambrian Explosion. Parker says this sudden appearance of a huge number of fossil types and the scarcity of fossils deemed "pre-Cambrian" is one of the great mysteries of geology, especially to Darwinists.

Darwin Was Right

As we noted in chapter 4, Darwin insisted that there should be fossilized remains that show very gradual change over long periods of time because simple and primitive organisms are assumed to develop slowly into more complex and modern ones. Yet he could give no satisfactory answer to why these fossils had never been discovered. Darwin was particularly concerned that fossils have not been discovered from before the Cambrian period. We will remember that Darwin acknowledged if the fossil record did not eventually demonstrate the slow and gradual development of life from the simple to the complex that this would put his whole theory in doubt.

Parker says that Darwin was right; the lack of evidence in the fossil record of "intermediate links"—before or even after the Cambrian deposits—is a very serious objection to Darwin's theory. Parker, along with many other biologists and geologists today, points out that well over a century has passed since Darwin's day—but the intermediate links have *still* not been found or are ever expected to be found!

But Parker carries his argument still further. First, he reports that geologists have now discovered great stretches of Precambrian sedimentary rocks that contain fossils, and says that Precambrian fossils strongly support the creation concept.

Parker gives several arguments to support the creationist position that the fossil record is far better explained by creation than by evolution. Among these are the fact that no intermediate links have ever been discovered, and that the earliest fossils found for the Cambrian period exhibit complexity.

> My wife, Mary, and I have found soft-bodied jellyfish and members of the earthworm group (annelids) in the famous Ediacara beds of South Australia. What lessons do we learn from the "oldest" animal fossils? Once a jellyfish, always a jellyfish, once an "earthworm" (annelid), always an "earthworm." Most people think of segmented worms as fish bait, but to a biologist, they are marvelously complex. The "lowly" earthworm, for example, has five "hearts," a two-hemisphere brain, and a multi-organed digestive system.[200]

Parker says firstly that there are no "in-between forms" among the fossils, or any that are harder to classify than plants and animals that are living today. Then he says, secondly, that the Cambrian period produced fossils of animals that show every bit the same complexity and variety as do animals today.

Parker quotes David Raup as a recognized authority on fossils. He says Raup was curator of the famous Field Museum of Natural History in Chicago, which is reputed to possess 20 percent of all known fossil species. Raup referred to Darwin's argument that the poorness of the then-known fossil record was not proof his theory was wrong because eventually the "gaps" would be filled. Raup pointed out that this did not come true, because these gaps have never been filled, or even narrowed.

Then Parker says Raup summarizes recent discoveries.

> Well, we are now about 120 years after Darwin, and knowledge of the fossil record has been greatly expanded . . . Ironically, we have even fewer examples of evolutionary transition than we had in Darwin's time. By this I mean that some of the classic cases of Darwinian change in the fossil record, such as the evolution of the horse in North America, have had to be discarded or modified as a result of more detailed information.[201]

Raup, an evolutionist, recognized that biological change does occur, but this appears to simply be variation within kinds and not large change brought about by natural selection.

Parker summarizes this section by saying that "genetic studies suggest that mutation-selection *could not* lead to evolutionary change; the fossil evidence seems to confirm that it *did not*."[202]

Parker debunks the attempts to find links between apes and mankind. He gives examples of fossils that were originally thought to be of "primitive man," but these bones and teeth were found later to be from other animals. The famous "Nebraska man's" fossil tooth turned out to be that of an extinct pig.

Views of Modern Evolutionists

Parker refers to a *Newsweek* magazine article reporting on a gathering of leading evolutionists from around the world in 1980 in Chicago. He quotes from this magazine as follows:

> Evidence from fossils now points overwhelmingly away from the classical Darwinism which most Americans learned in high school . . .
>
> The missing link between man and the apes . . . is merely the most glamorous of a whole hierarchy of phantom creatures. In the fossil record, missing links are the rule . . . The more scientists have searched for the transitional forms between species, the more they have been frustrated.[203]

Parker tells us the lack of fossils to support Darwin's theory led well-known scientists Stephen Gould and Niles Eldredge to write that gradual evolution "was never 'seen' in the rocks." Gould and Eldredge claim that the idea of gradual evolution did not come from evidence or logic, but expressed the cultural and political biases of nineteenth-century liberalism. Parker summarizes this by declaring,

> Believe it or not, when it comes to fossils, *evolutionists and creationists now agree on what the facts are.* The overwhelming pattern that emerges from fossils we have found is summarized in the word *stasis*. *Stasis* and *static* come from the same root word, a word that means "stay the same." Gould and Eldredge are simply saying that most kinds of fossilized life forms appear in fossil sequence abruptly and distinctly as discrete kinds, then show relatively minor variation within kind, and finally disappear.[204]

Parker quotes from several evolutionists who attended this 1980 meeting in Chicago. One reported that he was convinced from what the paleontologists say that small changes do not accumulate. If this is true, there can be no such thing as macroevolution! Though no one denies that small changes do occur.

Another evolutionist at the 1980 meeting expressed his summary of the conference by saying that species stasis is the "single most important feature of macroevolution." Parker points out that if the concept of "evolution" has any meaning at all it means "change. It seems evident that the conclusions made at this conference had a lasting effect on the scientific community. A book published by Stephen Meyer and others in 2003 refers to the conclusions drawn in this 1980 meeting as demonstrating that the mechanisms of microevolution cannot explain macroevolution.

Parker summarizes with four observations and two interpretations of the fossil record.

The fossil record shows each kind of plant and animal life existed "complete, fully formed, and functional" from the beginning. It does not show the progression from simpler to more complex forms Darwin envisioned.

These fossils can be classified into the same groups that exist today.

Sometimes there is difficulty classifying a fossil into a group. In these cases it is no more difficult to do this for the most ancient fossils than it is to classify existing forms of life.

Each kind of fossil shows broad "ecologic and geographic" variation. This means that the fossils show variation in their locations and in the relationship with their environment just as living organisms do today.

His conclusions regarding fossils are first, these facts should lead an open-minded person to consider, as a possible cause, that these groups of living things were *created*—created in such a way that variation and change would only occur within "kinds." This is exactly what the book of Genesis teaches. Not only that, but second it is clear that the fossil record reveals greater variety, and occasionally even greater size of animals than exist today. He says this suggests a "corruption" like that reported in Genesis 3, and a universal flood like reported in Genesis 6–8.

Parker presses his argument for creation a step further.

> The real test of a scientific theory is its ability to predict in absence of observation. When it comes to fossils, *creation has passed the scientific test with flying colors*. The original Darwinian theory of evolution and the neo-Darwinist . . . views, have been disproven twice, both by genetics and by the fossil record.[205]

As support for the above statement, Parker refers to Denton's book *Evolution: A Theory in Crisis*. He quotes Denton as we did earlier. "We now know, as a result of discoveries made over the past thirty years, that not only is there a distinct break between the animate and inanimate worlds but that it is one of the most dramatic in all nature, absolutely unbridged by any series of transitional forms and like so many of the other major gaps of nature, the transitional forms are not only empirically absent but are also conceptually impossible."[206]

Every once in a while popular magazines, like *Time* or *Scientific American*, report of a dramatic new fossil discovery that is said to demonstrate scientific evidence of macroevolution. Some time ago *Time* reported the finding of a fossil of a large "fishapod" on an island in Canada. Estimated to live some 375 million years ago, it is said to have had a fish's scales, fangs and gills, a suggestion that it had lungs, and what looked like a primitive hand with "five fingerlike bones." The article comments that fishapod is exactly the sort of transition animal Darwinian theory predicts, and shows new physical traits gradually emerging to help it thrive in a novel environment.

The *Time* article goes on to say the fishapod thrived in the marshy floodplains of large rivers that opened between 410 million and 356 million years ago. It reports that this discovery did not surprise most biologists or even the scientists who believe in intelligent design because they concede that transitional creatures have been showing up in the fossil record for some time.

This article is titled "Our Cousin the Fishapod" and sub-titled "An Ancient Fish with Primitive Fingers That Fills an Evolutionary Gap and Shows Darwin's Theory in Action."

A companion article in *Time*, by Michael Novacek, a curator of paleontology at New York's American Museum of Natural History, says, "Many who reject evolution in favor of divine creation claim that the fossil record doesn't contain the so-called transitional species anticipated by Darwin's theory. This ancient, walking fish is yet more evidence that such an argument is simply wrong; all sorts of missing links preserved in exquisite detail have been and will be discovered."[207]

But is fishapod really a transitional creature that illustrates large-scale evolution? If paleontologists could point to fossil beds that show a clear transition from simple organisms to very complex ones with only a few gaps here and there, then their claim of modern discoveries of "missing links" that complete the chain would have more validity. In spite of their assumptions that fossils like the fishapod did or even could actually walk, the evidence presented does not really show this. Scientists like Denton and Gould point out that the gaps in the fossil record are great and the fossils show only very limited biological evolution. Will these huge gaps ever be filled, anywhere in the world?

Is fishapod a freak, or a link that survived when all its immediate descendants left no fossils? There are other cases of "links" in faraway places like Latvia, but little or no evidence of the "chain" to which they are said to belong. They may simply have been unusual specimen of their biological classes that were poorly adapted to survive and produced no offspring.

The point Parker is making is that the fossil record does not show a linkage of these so-called transitional creatures to any later fossils to demonstrate that they evolved into a new biological form.

b. How Fast Do Fossils Form and Get Buried?

Parker explains that "surprisingly enough, just about everybody—creationist, evolutionist, and everyone in between—agrees that individual fossil specimens themselves begin to form very, very rapidly!"

> Most fossils are formed when a plant or animal is quickly and deeply buried, out of reach of scavengers and currents, usually in mud, lime or sand sediment rich in cementing minerals that harden and preserve at least parts of the dead creatures. Evolutionists and creationists agree: the ideal conditions for forming most fossils and fossil-bearing rock layers are *flood conditions*. The debate is just whether it was many "little floods" over a long time, or mostly the one big Flood of Noah's time.[208]

"Flood conditions!" Parker sees the fossil record as a testimony to a universal flood. He reports that "the evolutionist who discovered the Ediacara jellyfish said the fossils must have formed in *less than twenty-four hours*. He didn't mean one jellyfish in twenty-four hours, he meant millions of jellyfish and other forms had fossilized throughout the entire Ediacara formation, which stretches about three hundred miles or five hundred km from South Australia into the Northern Territory, in less than twenty-four hours! In short, floods form fossils fast!"[209]

Parker appeals to our everyday experiences with the way concrete is made to explain how fossilization occurs. Cement companies crush rock, and sell cementing minerals. The buyer adds water and a chemical reaction occurs, and rock forms again very quickly. He says that concrete is simply artificial rock. In the same way Parker thinks the jellyfish must have landed on wet sand, which turned into sandstone before the jellyfish had time to rot.

2. The Witness of the Geologic Columns

Parker found rocks in some places that were filled with fossils buried on top of one another in layers or "geologic columns" two miles thick. Evolutionists see the geologic column as an example of evolution displayed chronologically, from simple life forms near the bottom and gradually becoming more complex as one goes up the column. The lower the fossils are in the column the older they are assumed to be. Parker says this is only an *idea*, and we do not find the complete sequence anywhere.

He gives the walls of the Grand Canyon as an example. "There really are fossils out there; they really are in sedimentary rock layers, and those layers really are stacked on top of each other, over 1.5 miles (2km) deep across the Arizona-Utah border, for example."

> According to creationists, the geological systems represent different ecological zones, the buried remains of plants and animals that once lived together in the same environment. A walk through Grand Canyon, then, is not like a walk through evolutionary time; instead, it's like a walk from the bottom of the ocean, across the tidal zone, over the shore, across the lowlands, and into the upland regions. Several lines of evidence seem to favor *this ecological view*.[210]

Parker sees this as a striking illustration that the Grand Canyon was caused by a huge flood, even though most scientists assume it was formed by erosion over millions of years. He says the fossils in the lowest strata are of living things that are like those existing now at or near the bottom of the ocean, and those much higher are of animals that live on land.

One line of evidence is the matter of what are called "misplaced fossils." Some fossils are found too low in the geologic column to fit evolutionary theory. Some are found too high. He remarks that sometimes whole geologic systems are misplaced.

Geology and a Young Earth

Another is the problem of "paraconformity." This is where fossil columns show huge gaps in the assumed evolutionary fossil progression with no geological evidence of a disconnect at all. In a column Parker studied, a whole theoretical "evolutionary period" of twenty-five million years was absent. But in the missing gap between the two hypothetical periods, "Only millimeters separated them, and there was no change in color, no change in texture, not even a bedding plane. There was *no physical evidence at all* for those hypothetical 25 million years of evolutionary time."[211]

This means that there was no evidence between fossils representing these hypothetical periods of assumed evolutionary development to show that they had not been originally laid in the order they appear in now. It looks as if the living things that became these fossils jumped the evolutionary ladder by twenty-five million years!

Parker thinks that to assume these layers are from different ecological areas and were swept there by a huge flood is a better scientific explanation for so-called paraconformity and misplaced fossils than the evolutionary assumption is.

Mount St. Helens's Water

Parker describes how dramatically the mud flow changed things when Mount St. Helens erupted. He said that Mount St. Helens was a small volcano that didn't produce a lava flow when it erupted.

> The volcano sent mud and debris hurtling down into Spirit Lake, sloshing a wave nearly 900 feet (300 m) up its initially tree-studded slopes. The wave sheared off trees with enough lumber to make all the houses of a large city! The trees were sheared off their roots and stripped of their leaves, branches, and bark. The "forest" of denuded logs floated out over the huge lake. As they water-logged, many sank vertically down into and through several layers of mud on the lake bottom. Many features of the lake-bottom deposits are reminiscent of coal deposits.[212]

> What supplies the power for volcanic eruptions anyway? Water. Yes, *water*— *Superheated* water found in the underground liquid rock called magma. If some crack develops to release pressure, the superheated water flashes into steam, *generating colossal power*—a power to blow islands apart, power that dwarfs mankind's nuclear arsenal. About 2/3 of what comes out of the average volcano is water vapor, what geologists call "juvenile water." How much water could be released by volcanic processes? Most evolutionists believe all the earth's oceans were filled by outgassing of volcanic water![213]

According to Parker, a geologist who was looking for a way to start a worldwide flood would have difficulty finding a better mechanism than by breaking up the "fountains of the great deep." By this he means water coming up out of the earth as a result

PART 2—*What Do We Really Know about the World?* The Science

of volcanic activity. (This refers to Gen 7:10–12, where he says subterranean waters came forth and along with the heavy rains caused the flood of Noah's day.)

The Grand Canyon Water

Parker believes that this is what caused the Grand Canyon—a universal deluge during the flood of Noah's time.

> Many people have the completely mistaken notion that the Biblical Flood covered the whole earth almost instantly, stirred everything up, and then suddenly dumped it all. Not at all! According to the Biblical record, Noah was in the Ark for over a year. It took about five months before "all the high mountains under the whole heavens" were covered, and it took several more months for the water to subside as "the mountains rose up and the valleys sank down" at the end of the flood.[214]

Parker is correct here. Creationists may disagree over how extensive an area the flood covered, but Genesis 7 seems to be saying the flood waters kept rising for 150 days before beginning to recede. And it was a year and ten days before Noah looked down at the earth and saw it was dry.

Although geologists in general and young earth scientists both recognize that the fossil-bearing rocks were laid down under catastrophic conditions, only the latter attribute them to a one-year period during a universal flood.

Parker said he led university students through the Grand Canyon fifteen times on week-long backpacking trips. They studied the rock formations carefully. Parker said he once assumed the rock formation of the Grand Canyon showed a record of a *lot of time*, but now he looks at it as evidence of a *lot of water* instead. In other words, Parker sees it as being caused by the flood in Genesis.

His arguments are too detailed to present here. But he gave examples of similar cases that are now recognized by evolutionists. These are cases where it appears that great catastrophes formed geologic features once thought caused by millions of years of gradual erosion—like an ice dam suddenly breaking, causing the cascading water to form a deep canyon between what are now the states of Washington and Oregon.

Using Mount St. Helens to explain the Grand Canyon, Parker said,

> Right before our eyes, a small volcano (which never ever produced a lava flow) had stacked up horizontal bands of sediment, and cut channels through it, *forming a 1/40th size "scale model" of Grand Canyon in just five days!* All sorts of features once thought to take millions of years of time, were formed, instead, by a lot of water in just five days![215]

Henry Morris also referred to the explosion of Mount St. Helens. He noted that in 1980, "up to 600 feet of *laminated* sediments were deposited in a single day, looking exactly like the stratified beds normally interpreted to represent long ages."[216]

Parker adds,

> There are surely many other questions to be researched, but the weight of evidence we have available now (and that's as far as science can go) seems to suggest strongly that the horizontal rock layers at Grand Canyon were formed rapidly, *not* by a lot of time, but by a *lot of water instead*.[217]

Conclusion on Geology

Thus Parker sees the fossils as a witness *for* creation, and *against* Darwin's general theory—against macroevolution. He is convinced that there is simply no evidence in the fossil record to support the theory that life on earth developed from the simple to the complex, or from few life forms to many. The fossil record does not support the evolutionary theory that life on earth is millions of years old either.

The witness of the geologic columns is considered by Parker to be consistent with the witness of the fossils. They do not support the concept of a long-time progression from simple life-forms to more complex ones. This is because the fossils in the columns actually suggest ecological variation. By ecological variation I mean that Parker sees the lowest fossil layers as representing life on the bottom of the ocean, and then a progression of fossils going upward from those on lowlands to highlands. He believes this is consistent with the theory that these fossils are actually flood deposits.

The witness of the explosions of Mount St. Helens in the 1980s also suggests that the Grand Canyon could have been formed very quickly by a gigantic flood. Parker concludes that it must have been formed by a universal flood—the one reported in Genesis as having happened in Noah's day.

3. The Witness of Biology

Darwin Was Right!—about Natural Selection

Parker states that he can claim that scientific evidence supports what the Bible says about origins only if he can deal fairly and honestly with natural selection.

He admits that some of Darwin's discoveries and those of later biologists do seem to provide "strong evidence of natural selection." He states that there is evidence that Darwin was right about natural selection, but that the concept can fit the creation model as well as the evolution model. He says that Edward Blyth, a creationist, discovered natural selection before Darwin did.

Parker concludes that natural selection only produces limited change, and that the weight of evidence points to "variation within the created kinds." He puts it this way because the Bible suggests that God created living things so they would reproduce "after their kind."

But Darwin Was Wrong about the Beneficial Effects of Mutations

Parker asks, "Can mutations produce real evolutionary changes? Don't make any mistakes here. Mutations are real; they're something we observe; they do make changes in traits. But the question remains: do they produce evolutionary change? . . . No!"[218]

Parker lists three things here to prove his point. First, mutations are so rare that the *series* of changes needed for evolutionary change is mathematically impossible. This is a problem acknowledged by many evolutionists. Second, he says that almost all the mutations we know are identified with a disease. A mutation breaks down the existing genetic order whereas evolution is supposed to build it up. Third, mutations point back to creation. He asserts that mutations presuppose creation because they are only changes in genes that already exist. (I have read research reports where viruses mutate quickly to their own benefit. But this is far from the kind of evolution Darwinians claim occurred. So this is not covered by Parker's analysis here)

Parker feels that many people think natural selection and evolution are the same. But he says that Harvard's Stephen Gould clearly recognizes the difference between them. Parker considers Gould a "leading anti-creationist." Parker reports that Gould says the theory of a series of minor changes by mutations to produce "profound structural transitions" needed for macroevolution is "effectively dead, despite its persistence as textbook orthodoxy." (This may be why Gould, a paleontologist, was seeking a new theory that can more realistically account for the fossil record.)

Evidently many evolutionists cling to the belief that macroevolution by mutation and natural selection is a viable theory. But Parker refers to the results of the gathering of leading evolutionists in Chicago in October 1980. He reports that "the large majority of evolutionists at the Chicago conference agreed that the neo-Darwinian mechanism of mutation-selection could no longer be regarded as a scientifically tenable explanation for the origin and diversity of living things."[219]

Parker quotes from several leading evolutionists who have totally rejected mutation as an explanation of evolutionary change. This includes Nobel Prize winner Albert Szent-Gyorgi, who said that the origin of complex traits by random mutation has the probability of *zero*.

Parker argues that Darwin's natural selection does *not* explain the *origin* of species or traits, but only their *preservation*. He describes Lewontin as an evolutionist and outspoken anti-creationist, but one who honestly recognizes the same limitations of natural selection that creation scientists do. He quotes Lewontin as saying that, rather than improving organisms, natural selection operates essentially to enable them to maintain their present state of adaptation.

Then Parker asks if there could be enough variation in two created human beings to produce all the variation among human beings today? He answers his own question by saying there is no problem with this biologically. He gives the example of skin coloring. All human beings (except albinos) have exactly the same skin-coloring agent. All of us have the same basic skin color, just different amounts of it. He estimates it wouldn't take a million years to get all the variations of skin color we see among people today, only *just one generation*!

> Evolutionists *assume* that all life started from one or a few chemically evolved life forms with an extremely small gene pool. For evolutionists, enlargement of the gene pool by selection of random mutations is a slow, tedious process that burdens each type with a "genetic load" of harmful mutations and evolutionary leftovers. Creationists *assume* each created kind began with a large gene pool, designed to multiply and fill the earth with all its tremendous ecologic and geographic variety. (See Genesis, chapter 1.)[220]

Parker argues that all we have are our assumptions about how life started and developed. None of us were there when it happened. He says, "The evolutionist assumption *doesn't* work, and it's *not* consistent with what we presently know of genetics and reproduction. As a scientist, I prefer ideas that *do* work and *do* help to explain what we observe, and that's creation!"[221]

Conclusion on Biology

Parker's major points here appear to be the following:

There does appear to be a strong basis for believing in limited biological change by natural selection. But this actually is not evidence that evolution is correct; it fits the concept of creation also.

"Genetic variability" was built into creation from the beginning. For example, it is possible for all the genes necessary to produce all the variations among humans today to have been present in Adam and Eve when they were created. There is no biological proof whatsoever that new genes can be produced.

Although leading biologists recognize "the marvelous fit of organisms to their environment" and perfection of structure, they can explain by evolution only their preservation from generation to generation and cannot explain their origin.

Mutations happen, but evolutionary biologists now recognize that they cannot produce large-scale evolution (macroevolution).

This all reinforces Parker's belief that there is biological proof for creation and there isn't biological proof for Darwin's general theory—macroevolution.

III. An Old Earth or a Young Earth?

What exactly is the primary issue between young earth scientists like Morris and Parker and old earth scientists like Young in the field of geology?

Modern geologists acknowledge that the intensity of meteorite bombardment on the earth was greater in the distant past, that volcanic activity on the moon was more intense in the past, and that continental drift may not have occurred at all early in earth's history. Also, that the speed by which glaciers formed have varied enormously through time, and extinction of fauna and flora was rapid during one period.

It is precisely at the point of *how long* it took to produce the geologic formations that exist today where Young and other modern geologists part company with young earth geologists.

Leading creation scientists such as Morris and Parker accept the two tenants of methodological uniformitarianism—that the laws of nature have remained the same but that the rate of change has not. Then where is the conflict? In two primary areas: First, in interpreting the geologic record, as we have already shown. And second, in believing in God's direct intervention in geological processes by working miracles. This is something that Young takes a different position on than young earth scientists do, who sometimes resort to God's intervention in order to support their theories. Young does not reject the fact that God works miracles, but insists that geologic history must be understood apart from them. I don't think he would deny that the flood of Noah's day was the result of a direct intervention by God.

Young made clear he believes that the flood in Noah's day actually occurred. But he wanted in his books to show Christians that they could believe in a historical flood without accepting a flood geology theory that he believes is inconsistent with the scientific data related to creation. He argued that pre-flood geography was essentially the same as post-flood geography and this demonstrates no "wholesale reorganization of terrestrial surface features" occurred like flood geologists insist.

But the unsolved question seems to be "what proof of a worldwide flood would satisfy Young and other geologists?" They admit there is evidence of local flooding in much of the world. Morris and Parker see the Grand Canyon as an example of the effects of a worldwide flood; Young and other geologists do not. It seems that both sides are interpreting the geological evidence according to their presuppositions. As we saw in chapter 1, this is "normal" for humans.

In this chapter we reported on some of Morris's criticisms of the writings of some old earth creationists like Ross and Young. Here, to balance this, we will report some criticisms of Morris and other young earth scientists.

The controversy about the age of the earth is related to both cosmology and geology. Here we will report on this very briefly.

First, Cosmology

Scientists like Morris and Parker believe that the universe was created recently. How do they reconcile this belief with the cosmologists' theory that it is billions of years old? They do so by a central argument that they call "the appearance of age." They argue that God could have created the light waves already in transit. God could have created the universe in such a way that it was expanding rapidly from the beginning and was already large. Most cosmologists believe they are peering back billions of light-years to where the universe began as a very tiny, very hot entity. But many young earth creationists think it only seems this way to them because it was actually very large and expanding rapidly at the moment of its creation. In this way they reject the currently popular big bang theory.

Hugh Ross responds to these arguments as follows:

> The overlooked fact here is that star light and galaxy light give direct indications of their travel distances. The spectral lines (light waves of various frequencies) of stars and galaxies are broadened in direct proportion to the distance they travel.[222]

Ross is saying here that the frequencies of light waves reveal that they have actually been traveling for billions of years. One response to this is that astronomers may not understand the qualities of light as well as they think. They need to assume the existence of dark matter and even dark energy in order for their mathematical equations to work. This suggests their concepts and measurements of distances and time may not be as accurate as they assume. It is not unscientific to assume there may yet be new major developments in cosmological theories. There may be factors that influence the speed of light and our measurements of distances in space that are not understood yet. Perhaps the final word on this issue is yet to be written.

The young earth response to such criticisms typically is that God could have created the universe with light already in motion in such a way that Ross's objection here is not valid.

Geology

Perhaps Davis Young's overall criticism of *The Genesis Flood* can be summed up with his conclusion on the sedimentary rock record.

> The sedimentary rock record is replete with rocks that formed in glacial, lake, desert, reef, evaporite, and other environments that are utterly impossible to reconcile with the Flood hypothesis. These deposits all formed in environments that take considerable time to develop. They all point to the fact that the earth is far more than just a few thousand years old as creationists would have us believe.[223]

My son Dan majored in geology and earth science at his university. He read a lot about the unusual rock formations like Young referred to above. He concluded that most of the conclusions were just "educated guesses," and that geologists were not in unanimous agreement about them.

The issue of the age of the earth is important for understanding historical science. But it is not directly related to the purpose of this book. If developments in science actually do progress to where this can be answered, then I think we should accept their findings—as we "follow the truth wherever it leads." But, as we shall see in part 3, there are ways to reconcile the clear teachings of the Bible with a belief in either an old earth or a relatively recent earth. The Bible is crystal clear in its central truths regarding God, man, sin and salvation. But, as it is often said, the Bible was never intended to be a textbook on science.

IV. General Summary

A large number of books were written in the last half of the twentieth century in support of "flood geology," or "creation science" as it is now called. Foremost in this movement have been the books by Henry Morris.

Young and Morris differ primarily in how old the earth is and how long life has been on it. Both believe in creationism, and both accept the authority and accuracy of the Bible. Young accepts the current prevailing theories in geology. Morris does not. Their *approach* to historic geology and their attitude toward geologic dating methods are very different.

Morris writes to defend what he considers to be the traditional Christian view of the origin of the universe and against the teachings of evolution and uniformitarianism in the field of geology. He does so as a hydraulic engineer and a believer in a literal interpretation of the Bible. He believes that the traditional view of the Bible story of creation is that the universe was created in six days of twenty-four hours each, and that the universe is only a few thousand years old. He also believes that the flood reported in Genesis 7 and 8 was a worldwide deluge that deposited the fossil beds that exist around the world, and also caused the geologic columns.

Morris argues that geologists who believe the fossil beds were laid down over millions of years cannot prove this, and cannot prove that the geologic columns were built up over long periods of time. He also acknowledges that he is not able to prove scientifically that they are wrong, and that a worldwide flood deposited them in the last six thousand years or so. But he does give examples that he is convinced demonstrate that his flood theory appears a better interpretation of the geological evidence than do the theories of evolutionists and of "old earth creationists" such as Bernard Ramm and Davis Young.

Morris states that the differences between him and the others is not in the belief in natural laws, but in what processes—acting according to known natural laws—produced the fossil beds and geologic columns.

Morris recognizes that he cannot solve all the geological problems that have bearing on this theory. But he does challenge Young's "geological proofs that a recent creation is impossible." Among these are Morris's challenge to the validity of radiometric dating.

Gary Parker writes in support of Morris's position. He argues from his personal observations of the famous Ediacara fossil beds in Australia, the Grand Canyon, and from the effects of the explosions of Mount St. Helens. He also quotes from recent writings of well-known evolutionists to support his arguments.

Parker asserts that the fossil beds are a strong argument for creation and against evolution. He supports this conclusion by observing that pre-Cambrian fossils include those of earthworms and jellyfish that are fully as complex as those living today. Yet evolutionists say these beds are four hundred to over five hundred million years old. Parker also claims that fossils from the Cambrian period have more variety and complexity than do marine invertebrates today. Besides this, he says that no clear examples of "transitional forms" are found in the fossil beds. Parker supports Morris's arguments that a sudden catastrophic event explains the fossil beds and geologic columns better than the theory of gradual change over time does.

Parker also explains certain biological facts to support his arguments. He argues that the diversity within types are a result of being created possessing a large gene pool. He says that the evolutionist's argument that the gene pool increased with evolution is contrary to what is now known about biological processes. Mutations are real, but some evolutionary biologists are now coming to realize that they do not explain macroevolution. To Parker, the facts of both biology and geology support creation and a young earth, just like they did to Morris. Parker, however, seems open to the view that life on earth is older than just six thousand years.

Morris and Parker give many arguments to support their assertion that a worldwide flood explains the fossil beds and geologic columns better than the current uniformitarian theories of geologists today do. These arguments do seem persuasive. Yet we can agree with the young earth creationists that their conclusions are based on their interpretations of how these phenomena occurred, and that neither their conclusions nor those of old earth creationists like Young can be proved by science today.

An Unfair Criticism

Young earth creationists are often accused of "not believing in science." But, as we have seen in this chapter, Morris and Parker say that they believe in the verified findings of science as it relates to the way nature is observed *today*. They differ from the majority of scientists by insisting that we cannot assume conditions in the far distant past were the same as now. I recently heard an astrophysicist who believes in the recent creation of the universe say that he accepts all the discoveries in cosmology regarding the universe as it is observed now. He accepts the results of "operational science," including distances between stars within a parameter of two thousand light years. He also accepts Einstein's laws of relativity, the laws of thermodynamics, the laws of gravity, etc. But he insists that attempts to reconstruct what happened in the distant past are not objective science. He says that he accepts science that is *testable*, but not that based on the past, as they are now untestable events.

Are they right? Only God knows for sure. Only God was alive then to witness what occurred at creation. But I think we should give scientists like Morris and Parker credit for building an equally impressive "edifice of logic" as Darwin did—while starting their research with opposite assumptions.

It is interesting to me that Darwin and Parker had the same experience of reversing their worldview from that formed early in their lifetimes. Darwin began his career as a believer in the Bible teaching on nature, only to change dramatically as he pursued his career as a naturalist. Parker says he began his career as an ardent evolutionist, only to change dramatically after studying the Bible.

We shall see later that philosophers of science like Stephen Meyer argue that we should accept the explanation that best fits the scientific data—until a better explanation comes along. So let us continue our quest for the best explanation. In the next chapter we will look at how evolutionists and creationists understand how the human race developed.

11

The Origin of the Human Race

> God created man in His own image; in the image of God he created him; male and female He created them.
>
> —Genesis 1:27

> There is no evidence that man was aboriginally endowed with the ennobling belief in the existence of an Omnipotent God.[224]
>
> —Charles Darwin

> The Biblical idea of the origin of man and the evolutionary origin of man are mutually exclusive. At this particular point there must be conflict between Christianity and modern science as long as science maintains its current attitude with regard to the origin of man.[225]
>
> —Davis Young

Introduction

In the quotation above, Davis Young points out the center point in the conflict between advocates of creationism based on Bible teaching and advocates of neo-Darwinism (macroevolution). It is focused on how the human race came into existence—the origin of mankind.

The implications of this are broad and profound. Are you and I simply unplanned "accidents" in this world—only the product of random mutation and natural selection? Or were we made by a Creator Who has personality, and wants a personal relationship with each one of us? Are we created by a holy, righteous God Who rewards righteousness and punishes unrighteousness? Are we the objects of the unselfish love of a holy God Who wants us to live in harmony with Him? Has God provided a salvation for us that promises eternal life beyond the grave? Or are we on this earth by pure chance, not accountable to anything beyond ourselves, with our existence ending at death?

Is this natural world all there is to existence? Is the Darwinian view of nature accurate? Or is there a spiritual world that I am also a part of—whether I know it or not? I really do not know any question more important for us today to answer than this. This is at the core of our worldview—how we understand and evaluate life's experiences and choose "what's best for me."

Strobel gives himself as an example of one who embraced a worldview that excluded God. As he expressed it,

> Rather than facing this "unyielding despair" that's implicit in a world without God, I reveled in my newly achieved freedom from God's moral strictures. For me, living without God meant living one hundred percent for myself. Freed from someday being held accountable for my actions, I felt unleashed to pursue personal happiness and pleasure at all costs.[226]

Readers may respond to Strobel's last sentence in the above quotation in various ways. Some may feel like he did and also feel free from ever being held accountable for their actions. They may decide to do things they would be ashamed of if other people knew about them—feeling that no punishment will ever come to them. So much wrongdoing in our societies is surely done by those who hope they will never have to pay for their bad behavior.

Strobel doesn't clearly say how his new freedom affected his social and moral attitude. But he says it increased his desire to succeed as a journalist. He said he would let nothing, and no one stand between him and his ambitions. I wouldn't want to work closely with someone who held that attitude—would you?

But can naturalism really explain human nature? This is another major question that Darwinism and naturalism in general must answer to be viable as an all-embracing scientific theory.

Can Darwinism and Naturalism Explain Humanity?

Darwin wrote *The Descent of Man* to explain his theory of the origin and nature of the human race. It is based on the philosophy of naturalism; that nature is all there is and there is no spiritual world or anything beyond the grave. To Darwin, humans, just like all organisms, are a product of random mutations, natural selection, and the struggle for existence.

Darwin ended *The Descent of Man* with these words:

> We must, however, acknowledge, as it seems to me, that man with all his noble qualities, with sympathy which feels for the most debased, with benevolence which extends not only to other men but to the humblest living creature, with his god-like intellect which has penetrated into the movements and constitution of the solar system—with all these exalted powers—Man still bears in his bodily frame the indelible stamp of his lowly origin.[227]

Ernst Mayr, the distinguished evolutionary biologist, wrote the following: "To be sure, man is, zoologically speaking, an animal. Yet, he is a unique animal, differing from all others in so many fundamental ways that a separate science for man is well-justified. When recognizing this, one must not forget in how many, often unsuspected, ways man reveals his ancestry. At the same time, man's uniqueness justifies, up to a point, a man-directed value system and man-directed ethics."[228]

Darwin wrote in *The Descent of Man* that evolution shows that man is descended from a hairy quadruped, who probably lived in trees—but Darwin gave no persuasive scientific evidence for this conclusion.

The Common Neo-Darwinian Explanation

The *Discover* magazine of Disney, in their September 2003 edition, included an article titled "Great Mysteries of Human Evolution."[229]

This article treats the following subjects: Who was the first hominid? Why do we walk upright? Why are our brains so big? When did we first use tools? How did we get our modern minds? Why did we outlive our relatives? What genes make us human? And, have we stopped evolving? The term "hominid" is used by neo-Darwinians to mean ancient primates with some humanlike characteristics and include direct ancestors of the human race. It is also used by scientists who are convinced from the evidence of the fossils that hominids had no biological connection with modern humans.

The article begins with a popularized overview of how old humans and their ancestors are. Our ancestors, according to this article, stood upright over four million years ago, and began using tools at least 2.5 million years ago. They began developing a brain capable of language and abstract thought 2 million years ago. Tools came before language.

It reports that paleoanthropologists have found only a few teeth and skull fragments of the most ancient fossils to theorize from. They reported that they found a species in the sand dunes of the Sahara and estimate it as from six million to seven million years old and think they were the first hominids. Other experts, however, believe this species was an ancestor of apes and not of humans.

The oldest human fossil of our species is considered to be that of two adults and one juvenile that were dug up in Ethiopia in 2003. They are considered to be about 154 thousand years old, and were named *Homo sapiens idaltu*. This discovery supports a theory, said to be based on genetic evidence, that traces the ancestry of all humans today to a small group of Africans that lived some 150 thousand years ago.

They believe the earliest hominids were short, had tiny brains, and could neither speak nor make tools. But experts still disagree over how these early hominids lived and what caused the evolutionary changes they assume occurred. Speaking of Lucy, the famous hominid found in Ethiopia in 1974, the article observed that some paleoanthropologists studied Lucy's skeleton and concluded she walked much as we do, and yet others argued she moved awkwardly on the ground and spent much of her time in the trees.

Did Lucy and other hominids walk on earth as we do, or spend their lives in trees? All scientists can do is infer this from the bone structures they observe in the fossil remains.

Paleoanthropologists generally accept the existence of hominids. This article says that the best clues we have to our upright origins may come from living apes, although no one knows for sure how much chimpanzees have evolved from the last common ancestor they shared with us. This statement is among many that suggest just how much speculation there is about "our common ancestors" and how little agreement as to how to interpret the fossil findings. It seems there is general agreement on the *theory* that humans and apes have a common ancestor, but considerable disagreement on how to interpret the *scientific findings* in a way consistent with the theory.

The article reports that the human brain is "grotesquely huge" and a typical mammal the same size has a brain about one-seventh as large. The reason for this is vigorously debated among scientists. They add an interesting comment:

> Hominid kids, then as now, needed years to develop large brains, during which time they depended on adults for high-energy foods. It's possible that the basic shape of the human family as a group of parents, siblings, and grandparents formed to feed the brains of their children.[230]

In other words, the fact that human children need much care and high-energy foods over many years may be the chief reason for the development of the extended family that include parents, aunts and uncles and grandparents of children.

The ability to manipulate tools is affected by the development of certain parts of the brain and the size of our hands compared to other primates. But this article acknowledges that scientists still have almost no clues as to how that evolutionary change took place. Chimpanzees and other apes have been observed making rudimentary tools. This article says,

> Scientists don't know yet how that modern mind came into existence . . . Instead, they have to infer what those ancient minds were like by looking at the things they made . . . Archaeologists have documented an explosion of expressions of the modern mind after roughly 50,000 years ago in the form of jewelry, elaborate graves, bone-tipped spears, and other new kinds of tools.

> The bones of the people who made these things look like our own. They were members of *Homo sapiens*, complete with long, slender arms and legs, a flat face, a jutting chin, and a high forehead that fronted a big brain.[231]

The article reports that all humans alive today are descendants of *Homo sapiens* that lived in Africa 150 thousand years ago. It says two other hominid species existed at the same time. One was the Neanderthals, who lived in Europe. The other was *Homo erectus*, which had existed in Asia for 1.5 million years. It says both of these species disappeared soon after *Homo sapiens* spread from Africa into Europe and Asia. (The prefix *Homo* refers to beings considered to be in the genus "homo" or human.) But if the existence of cultural artifacts is considered proof of the existence of human beings, the consensus seems to put this at about fifty thousand years ago.

In one crucial part of the genetic code humans and the common chimpanzee are said to have gene sequences that are 99.4 percent identical. "The early evidence already suggests that perhaps several thousand human genes have undergone intense natural selection since our ancestors split with the chimp lineage."[232]

Another important question this article considers is "Have we stopped evolving?" The article says that

> it has been an amazing run: Over 7 million years our lineage has evolved from diminutive apes to the planet's dominant species. We've evolved brains that are capable of things never achieved on our planet, and perhaps in the universe. Why shouldn't we continue evolving more powerful brains? It's easy to think that we'll just keep marching ahead, that in another million years we'll have gigantic brains like out of some episode of *Star Trek*. But scientists can't say where we're headed. It's even possible that we've reached an evolutionary dead end.[233]

Scientists think that human brains have not gotten larger in the last 160 thousand years, and that big brains have their drawbacks. One is that the head of the baby at birth is so large delivery is difficult, and there is a limit on how wide a woman's pelvis can become and not hinder her upright walking.

This article is representative of many that report fossil findings and explain their meaning based on the framework of macroevolution. They begin with the assumption that the human race has evolved over millions of years from primitive forms of life. One of the ways they seek to differentiate non-humans from humans is based on how similar their DNA is. For example, they think they have found DNA from *Homo neanderthalensis*, or Neanderthals. These beings are thought to have lived in the area from Spain to Israel about 230 thousand to 228 thousand years ago. DNA suggests they were not related to modern man.

They summarize by saying the oldest fossil considered to be human was recently found in Ethiopia, and is considered to be around 150 thousand years old. However they consider "modern man" to be only about fifty thousand years old. This is based on archaeological evidence of advanced civilization. But researchers in the field are

PART 2—*What Do We Really Know about the World?* The Science

not in agreement as to the actual age of many of the fossils, nor in their assessment of how old advanced civilization is. They assume that civilization developed slowly over tens of thousands of years. As the *Discover* magazine article put it, "What we don't know about our evolution vastly outweighs what we do know. Age-old questions defy a full accounting, and new discoveries introduce new questions."[234]

There have been many journal and newspaper articles published on this matter since the *Discover* article was published in 2003. But few seem as forthright as this article in attempting to present a frank assessment of the strengths and weaknesses of the fossil evidence for human evolution.

One of these articles is a May 2009 feature article in *Scientific American*. It is titled "What Makes Us Human?" The author, Katherine Pollard, a biostatistician, began the article reporting that she lined up the human DNA sequence with that of the common chimpanzee. The report says that "a humbling truth emerged: our DNA blueprints are nearly 99 percent identical to theirs. That is, of the three billion letters that make up the human genome, only 15 million of them—less than 1 percent—have changed in the six million years or so since the humans and chimp lineages diverged."[235]

It is clear from this article that the assumption is made at the beginning that humans and chimpanzees had a common ancestor, and no alternative explanation for the results is considered in the article. There has been much research in the twenty-first century on what percent of human DNA is shared with chimpanzees—and the focus of attention has narrowed to the specific functions of the DNA that differ.

In *The Case for a Creator* Strobel relates a discussion with Jonathan Wells where he asked about this similarity in DNA. He asked, "What about recent genetic studies that show humans and apes share ninety-eight or ninety-nine percent of their genes? . . . Isn't that evidence that we share a common ancestor?" Wells responded by saying the humans and apes share 100 percent of "the so-called body building genes." Yet they are anatomically different! Wells said the similarity of DNA is actually a problem for neo-Darwinism. Does this similarity prove common ancestry? Wells answered,

> No, it's just as compatible with common design as it is with common ancestry. A designer might very well decide to use common building materials to create different organisms, just as builders use the same materials . . . to build different bridges that end up looking very dissimilar from one another.[236]

In my neighborhood are many houses which are different from each other in obvious ways but have somewhat similar floor plans. Their garages are in the same place and look generally similar. Their front doors and paths to the doors look generally similar. It appears that the kitchens and bedrooms are in similar places. Some houses look from the outside to have exactly opposite layouts—like mirror images. But some are larger than others. They have different types of siding, and are of different colors. It seems obvious that the same general design and floor plan was used over and over again, but with changes to suit different tastes and needs. I assume from this that they were designed and built following the instructions of the same general contractor.

This similarity in body-building genes shows that we are not simply determined by our genes. What makes us different from chimpanzees is different, and cannot be explained by evolution.

Recent Newspaper Articles on the Subject

Paleontology is certainly an active science. Discoveries of important fossils are still being made. In 2009 scientists made a new discovery in the Ethiopian desert. They found the bones of what they estimate to be at least thirty-five members of what they consider to be a new species of an "ancestor" to modern humans. They named the species *Ardipithecus ramidus*, and abbreviated the name to "Ardi." The find is considered important because the fossils showed "a less protruding mouth equipped with considerably smaller, blunter canine teeth than the chimpanzee's impressive fangs." The October 3–4, 2009, *Wall Street Journal* account of this discovery reported that the smaller, blunter canine teeth suggested that these animals were not as violent or aggressive as chimpanzees and were our "kinder, gentler ancestors." (Notice the assumption here that they were our ancestors, even though the report gave no basis for this assumption.) This find was considered so important by *Parade* magazine that they listed it as one of the three "breakthroughs of the year" in 2009.

The *Wall Street Journal* article went on to explain that gorillas and bonobos are considered our closest living relatives, and they have a close-knit family life and rarely kill each other. It said, "The once-popular killer ape theory is crumbling under its own lack of evidence, with 'Ardi' putting the last nail in its coffin."

The following were reported in the *Gazette-Times* of Corvallis, Oregon, on the dates indicated:

- "Genes for Human Brain Still Evolving, Scientists Say" (*Chicago Tribune* article of September 9, 2005).

This article reports "two brain-building genes, which underwent dramatic changes in the past that coincided with huge leaps in human intellectual development, are still undergoing rapid mutations, evolution's way of selecting for new beneficial traits." Two genes in the human brain were identified—one said to be related to music, art, religious expression, and tool-making that appeared around thirty-seven thousand years ago. The other is considered related to writing, which appeared in Mesopotamia about 5,800 years ago. The researchers report that "our studies indicate that the trend that is the defining characteristic of human evolution—the growth of brain size and complexity—is likely still going on."

- "Neanderthals May be Less Ancient than Thought" (Malcolm Ritter, Associated Press, September 14, 2006).

This article reports that small bands of Neanderthals occasionally took refuge in a large cave in Spain. "Now a study says charcoal from their fires indicates that Neanderthals were still alive at least 2,000 years later than scientists had firmly established before." These charcoal samples are considered from twenty-four thousand to twenty-eight thousand years old. Before it was thought they died out in Europe between thirty-five thousand and forty thousand years ago. (It is interesting how convinced the scientists are that they were Neanderthals, just from bones of butchered animals and charcoal fires.) This reminds us again that the assumptions of anthropology are still subject to change.

- A related article is dated December 15, 2005, and entitled "Scientists Find Humans in N. Europe Longer Ago" (Thomas Wagner, Associated Press).

It reports that ancient flint artifacts found in eastern England are said to show humans lived there seven hundred thousand years ago. It is thought that at that time England was connected to the European mainland and the winters were mild. Before this discovery it was assumed that humans existed north of the Alps no earlier than five hundred thousand years ago.

There does not seem to be any valid scientific reason to assume that these "humans" were related in any way to present-day mankind—especially since the "modern mind" is considered to be only fifty thousand years old.

- "Fossil Skeleton of 'Lucy' Toddler Found" (Associated Press, September 21, 2006).

Lucy seems to be "the most famous fossil find in human evolutionary history." Lucy appears to be a partial skeleton of a three-and-a-half-foot-tall adult of "an ape-like species." It was found in Ethiopia in 1974. It is estimated to be over three million years old. Recently they found what appears to be a three-year-old female of the same species in Ethiopia. The skeleton is fairly complete. They surmise that Lucy could walk on two feet, but are not sure if she could climb trees and move around in them easily. The new find appears to have had the ability to climb trees. It is considered 3.3 million years old.

The finding of Lucy and the skeleton of a child of what might have been the same species is interesting, to be sure. But, again, the debate is about whether or not beings that seemed to walk on two feet were really human, whether or not they used fire or made tools.

Scientists Are Still Seeking Explanations for Large Human Brains

A newspaper report in October 2012 reported the following: "Cooked food might be a big reason humans were able to grow such large brains compared with their body size, scientists say." They suggest this because they believe our brains are so large because they have so many neurons, and if we ate only raw food we would need to feed more than nine hours a day. The thinking is that cooking food releases nutrients locked in raw food, making digestion easier.

Solomon wrote a few thousand years ago that "of making many books there is no end, and much study is wearisome to the flesh" (Eccl 12:12). Certainly news of more findings will entertain us in the years ahead, and conclusions must always be made knowing that "more research is needed."

John Klotz

In chapter 9 we read about Klotz's philosophy of science and his comments on the nature of scientific laws. Now we will consider his research in the field of paleontology related to human fossils.

Klotz noted that scientists in his day were in general agreement that both the Cro-Magnon and Neanderthal races are *Homo sapien*. They also agree that some of the fossil remains of Neanderthal man and Cro-Magnon man are "fairly ancient," are clearly human, and show no signs of being more primitive than man is today. He says that "one of the problems with these supposedly old fossils is the fact that some of them appear to be fossils of giants. The question, then, arises as to whether man is the degenerate descendant of a race of giants."[237] His conclusion, from carefully considering the fossil record, is that "we have not had evolution in man, but we have had degeneration and deterioration."[238]

Klotz explains that the busts of Neanderthal man and Cro-Magnon man in the American Museum of Natural History are only artists' ideas of what these people actually looked like. Although these artistic renditions clearly show an advance from Neanderthal to a more human-looking Cro-Magnon, the fossils of skulls and bone fragments do not. Klotz says that we simply have no idea what they looked like. (Few readers may have visited this museum, but pictures of these busts have been widely shown in magazines and newspapers.) Klotz refers to anthropologist Earnest Albert Hooton as follows:

> Hooton says that you can model on a Neanderthal skull either the features of a chimpanzee or those of a philosopher. He concludes by saying that the alleged restorations of ancient types of man have very little, if any, scientific value and are likely only to mislead the public.[239]

Even though the artist's conceptions of ancient man are highly unreliable, they are freely used in TV productions on "natural history." But, then, when it comes to TV programs that claim to depict human history, the scenes of living things crawling out of some kind of "primordial soup" are even more deceptive. We reported earlier that Denton and Behe have concluded from their scientific research that life *could not have* begun that way. Many evolutionary scientists accept this conclusion, and are now considering alternative theories for the origin of life.

PART 2—*What Do We Really Know about the World? The Science*

Klotz studied the sites where many of the human fossils were found, and concluded that in a large number of cases it is very difficult to determine how old the fossils are. Yet he also concluded that "*Homo sapiens* himself is of great antiquity." By "great antiquity" Klotz means many thousand years, not millions of years. He recognized that limited evolution has occurred within the animal kingdom, and this may possibly have produced new species. But he says this may be within the limits consistent with a careful reading of Scripture. He also notes that the fossil record shows within the human race a *devolution*, a degeneration and deterioration, and not the evolution that Darwinians insist on.

Klotz, clearly believes the fossil record of Cro-Magnon and Neanderthal can be reconciled with the Genesis account of the creation of the human race. He also feels that it is probable that man has been on this earth for thousands, rather than millions, of years. Paleontologists today seem in general agreement that Neanderthals were not related to humans.

Davis Young

We also looked at the research and beliefs of Davis Young in chapter 9. He is a creationist who holds an old earth position based on his research in geology and belief in the Bible. He says that

> the Christian anthropologist who believes the Biblical doctrine of the special creation of man is ... confronted with a real dilemma. His problem is that he cannot identify from the fossil remains the time when true man first appeared on the earth.[240]

This is because he is unable to know with any confidence if certain so-called transitional forms were genuinely human or not. He says the skulls of Cro-Magnon man appear to be virtually indistinguishable from skulls of modern man. So they are considered to be *Homo sapiens*. But he adds that in reality it seems there is no way to verify whether these skulls were of members of the human race or even how long ago these creatures lived.

Jonathan Wells

In *The Case for a Creator*, Strobel relates a discussion with Jonathan Wells about the "Legend of Java man." Strobel says this is the best-known human fossil, and one of the only human fossils most people know. He was considered to be part ape and part human, a link between apes and modern man.

The bones called "Java man" were dug out of a riverbank in late 1891. The American Museum of Natural History made a bust of what scientists thought he must have looked like. They made him with a sloping forehead, heavy brow, jutting jaw and receding chin. But Wells explained that the truth is only a skullcap, a thigh bone, and three teeth were found. Later investigation showed that the thigh bone didn't belong with the skullcap. The skullcap was judged to be distinctly human. So they concluded that "Java man" apparently was a human with a brain capacity well within the range of those of humans living today.

Wells gave an interesting example of how difficult it is to reconstruct an animal from fossil bones. He related that the *National Geographic* magazine hired four artists to reconstruct a female figure from seven fossil bones found in Kenya. One reconstruction looked like a modern African American woman. Another looked like a werewolf. The third had a heavy, gorilla-like brow. The fourth had a missing forehead and jaws that looked something like a beaked dinosaur.

Wells commented that "this lack of fossil evidence also makes it virtually impossible to reconstruct supposed relationships between ancestors and descendants." Then he quotes Henry Gee, the chief science writer for *Nature* magazine, as writing, "The intervals of time that separate fossils are so huge that we cannot say anything definite about their possible connection through ancestry and descent."[241]

Evolutionary Psychology

I read a couple of books on evolutionary psychology a few years ago because the term "evolutionary psychology" interested me. In graduate school I majored in the social sciences and especially in personality theory (human nature). I wondered how the theory of evolution, with its commitment to naturalism, could even theorize about the origin of human personality and the aspects of human nature that go beyond the physical.

One of these books is *The Blank Slate: The Modern Denial of Human Nature*, by Stephen Pinker. Stephen Pinker is a well-known "evolutionary psychologist" who challenges the common belief that people simply respond to outward stimuli, based on their response habits. Pinker refers to current views on how people are different from one another. He says, "The three laws of behavioral genetics may be the most important discoveries in the history of psychology. Yet most psychologists have not come to grips with them."[242]

He explains these laws as follows:

The First Law says that all human behavioral traits *can be* inherited.

The Second Law is that the effect on personality of the genes we inherit is larger than is the effect of being raised in the same family. In other words, heredity has more influence than does early environment.

The Third Law says that the effects of the genes we inherit and our early family environment added together do not fully account for our behavioral traits. Pinker says that about half of what makes people unique cannot be accounted for by either heredity or family environment.

PART 2—*What Do We Really Know about the World?* The Science

This third law certainly suggests that humans have free will to some extent, and can choose how they will act and respond to situations. This is what the Bible teaches. But Pinker, as an evolutionist, cannot accept what the Bible teaches. He said that "cognitive and emotional faculties have always been recognized as nonrandom, complex, and useful, and that means they must be products either of divine design or of natural selection." Yet he also maintains that modern science has "made it impossible for a scientifically literate person to believe that the biblical story of creation actually took place."[243]

He also notes that most Americans continue to believe in an immortal soul, made of a non-physical substance, which can leave the body. He quotes Ashley Montagu as writing that "man is man because he has no instincts, because everything he is and has become he has learned, acquired, from his culture, from the man-made part of the environment, from other human beings."[244] Pinker says that Russell Wallace was the co-discoverer with Darwin of natural selection, but Wallace parted company from Darwin by claiming that the human mind could not be explained by evolution and must have been designed by a superior intelligence.

The second book that really caught my attention is *Created from Animals: The Moral Implications of Darwinism*. This book brings up the question whether any clear code or concept of morality can arise from Darwinian evolution. The author, James Rachels, is considered an "evolutionary moral philosopher."

In his book Rachels argues that philosophers largely ignore the problem and that evolutionists don't see the issue correctly. They think that Darwinism poses no threat to traditional values. He says, "I shall argue that Darwin's theory does undermine traditional values. In particular, it undermines the traditional idea that human life has a special, unique worth . . . Abandoning the idea that human life has special importance does not leave us morally adrift; it only suggests the need for a different and better anchor."[245]

Rachels quotes Darwin as writing this in 1838: "Man in his arrogance thinks himself a great work worthy the interposition of a deity. More humble and I think truer [is] to consider him created from animals."[246]

Rachels insists that "if Darwin is correct, there are no absolute differences between humans and the members of all other species—in fact, there are no absolute differences between the members of *any* species and all others."[247] He thinks that animals like monkeys have moral natures in the same sense that humans do.

Rachels argues that Darwinism undermined the idea that man is made in the image of God. This reminds us of the change that Lee Strobel said happened to him when he concluded that living without God meant living one hundred percent for oneself, with no fear of someday being held accountable for his actions.

What makes humans different from all other animals? Many natural scientists assume this is *language*. Chimps have been taught to use sign language somewhat successfully. I recently heard of a case where someone raised a chimp in a human family starting soon after birth. One scientist believed the chimp became a companion who shared common emotions and complex thoughts and communicated them through behavior and sign language. The interactions were filmed. Then another scientist who studied the films concluded that this never really happened. What did appear to be evident is that this chimp identified himself with humans and not with other chimpanzees.

Many animals evidently can communicate by sounds to some degree, but evidently not to the point where they can discuss complex thoughts or even culture.

Summary

> Lord, what is man, that You take knowledge of him? Or the son of man, that You are mindful of him? (Ps 144:3)

In this chapter we considered what neo-Darwinians now say about the origin of the human race. We also reported what the paleontologist and geologist we introduced in chapter 9 wrote about this. This is because the origin of the human race is the central issue in the conflict between Darwinism (macroevolution) and the Bible account of creation.

First we presented an article from the *Discover* magazine and other news reports that present the current neo-Darwinian perspective. Fossil finds considered as those of ancestors to modern man are thought to be from six to seven million years old, though there certainly is no agreement as to whether these were ancestors of humans or apes. Some scientists believe the earliest hominids were short, had tiny brains, and could neither speak nor make tools. The word "hominids" is used to include beings that some evolutionary scientists consider direct ancestors of humans and others do not. What some consider our ancestors are thought to have stood upright about four million years ago, and first used tools some 2.5 million years ago. They began developing a brain capable of language two million years ago. A common theory suggests that the direct ancestors of all humans living today came from a small group of Africans that lived some 150 thousand years ago. But it is reported that among paleontologists there is much disagreement in how to classify and date these fossils.

There seems to be general *theoretical* agreement among evolutionists that humans and apes have a common ancestor, but considerable disagreement on how to interpret the *scientific findings* in a way consistent with the theory. When our first ancestors appeared is hotly debated.

The *Discover* magazine article also speculates why human brains are so much bigger than those of any other animal of similar size, and says we don't know how we got our "modern minds." It reports that the genes that make us human are, in one crucial part of the genetic code, 99.4 percent identical in sequence to the common chimpanzee. It also says we don't yet know if we humans have stopped evolving or will continue to do so.

Newspaper and magazine articles appear frequently to report new fossil findings that are said in some cases to confirm current theories and in other cases to challenge them. Scientists are busy continuing their research, like they should, but theories of macroevolution are firmly clung to even when new findings are thought to be inconsistent with them.

John Klotz concluded that the methods used to date what appear to be fossils of humans are highly unreliable. He studied the sites where many of the human fossils were found, and concluded that in a large number of cases it is very difficult to determine how old the fossils are.

Klotz wrote that scientists in his day were in general agreement that both the Cro-Magnon and Neanderthal races are *Homo sapien*. Some of their fossil remains are "fairly ancient," are clearly human, and show no signs of being more primitive than man is today. Some are of giants. His conclusion, from a careful study of the fossils, is that "we have not had evolution in man, but we have had degeneration and deterioration."

He says that evolution is not in any sense proved, and other explanations of the findings are possible. He questions whether new species could have evolved in no matter how long a period. He insists that even if the earth is very old this in no way proves evolution happened.

He points out that the Bible does not give a clear idea of how long mankind has been on the earth. He clearly believed the fossil record of Cro-Magnon and Neanderthal can be reconciled with the Genesis account of the creation of the human race. He also felt that it is probable that man has been on this earth for thousands, rather than millions, of years.

Davis Young agrees with Klotz in this. He says the skulls of Cro-Magnon man appear to be virtually identical to skulls of modern man, but it seems there is no way to verify if they were of members of the human race or even how long ago they lived.

Thus Young and Klotz are in agreement that Cro-Magnon appears to be human, and that it is impossible to conclude scientifically how old the human race is. Both agree that it is not possible to determine this from Scripture either.

Evolutionary psychologist Stephen Pinker and evolutionary moral philosopher James Rachels deny the uniqueness of the human race. But they cannot explain many special characteristics about how mankind thinks, acts, and dreams.

A recent encyclopedia of science quotes two scientists as remarking that "with regard to human consciousness, scientific naturalism in its current form will need to undergo radical conceptual revision to account for important features of the world and human experience."[248]

Nicholas Christakis, professor of sociology and medicine at Harvard University, wrote in an article in the December 19, 2011, *Time* magazine, that the origin of human consciousness is "the hardest problem in all of science."

Next we will look at the growing field of intelligent design theory to see what it contributes to our quest to "follow the truth wherever it leads."

> So much of the modern struggle against evolution is due to the subconscious realization that what matters about man is that which distinguishes him from the animal world, not that which unites him to it. Even if some form evolutionary theory can eventually be proved to be correct, it will not be able to explain what we call "personhood," because personhood is the image of God in us.[249]

12

Is Intelligent Design Science?

Is intelligent design a *scientific theory*? Does it meet the requirements as well as, better than, or not as well as Darwinism?

Background

The idea that this universe must have been designed and is not a result of chance is very old. It has been an important issue in the philosophy of science (how to understand nature) since at least the days of Plato and Aristotle. It was originally an argument regarding the existence of the cosmos, but in the last three centuries it has been applied to biological theory.

Four professionals have been at the forefront of the recent debates about intelligent design in the United States. They are biologist Michael Behe, philosopher of science Stephen Meyer, mathematician William Dembski, and law professor Phillip Johnson.

Advocates of intelligent design point out the issues where Darwin's theory has failed to stand the test of new discoveries in science. I believe three of these issues deserve special attention. One is Darwin's assumption that life appeared on earth spontaneously—in "some warm little pond." Second is Darwin's famous statement that his theory would be disproved if it could be shown that "any complex organ existed which could not possibly have been formed by numerous, successive, slight modifications." Third is his admission that anyone who rejects his view that the fossil record as discovered in his day was very incomplete "will rightly reject the whole theory." Darwin was confident that new discoveries of fossils would soon show the gradual progression of incremental changes in organisms that he assumed existed when he developed his theory. It has clearly failed to show this. A fourth problem is often mentioned—how

What Is "Design"?

Behe explains that design is simply the purposeful arrangement of parts. This would include features that perform a function (as in a mousetrap) and features that add to the appearance of an object (as in the artistic designs on a vase).

William Dembski says, "In its most general form, the design argument infers from features of the physical world an intelligent cause responsible for those features."[250] For the purpose of this discussion, something can be considered designed when it clearly has a purpose and there appears to be no chance that it came into being in any purposeless way.

Phillip Johnson, speaking of the Intelligent Design Movement (IDM), says, "The defining purpose of IDM is to advance the argument that neo-Darwinism has failed to explain the origin of the highly complex information systems and structures of living organisms, from the first cells to new body plans."[251] The argument is that the origin and development of life cannot be explained by natural causes and processes alone. Therefore it is necessary to consider that some intelligent cause played an essential role in this process.

Part 1: Chance of Design? Arguments for Design from Biology

When we discover something that seems to have been designed, how can we know it has been? Behe says that you can tell a mechanical object was designed "because you see that the components of the system interact with great specificity to do something."

> In order to reach a conclusion of design for something that is not an artificial object (for example, an arrangement of vines and sticks in the woods to make a trap), or to reach a conclusion of design for a system composed of a number of artificial objects, *there must be an identifiable function of the system.*[252]

Darwin and the Design Argument

James Rachels tells us that

> when Darwin was at Cambridge, William Paley's *Evidences of the Existence and Attributes of the Deity* was required reading for all students. Paley's work, first published in 1802, was the classic presentation of the "design" argument. In it he declared, "The marks of *design* are too strong to be gotten over. Design

must have a designer. That designer must have been a person. That person is God." Only a short time before setting out on the *Beagle*, Darwin had studied this reasoning, and had decided that it was irrefutable.[253]

What Darwin discovered during his voyage on the *Beagle* led him to reject the doctrines of the church regarding creation. He seemingly rejected Paley's argument at the same time. This is understandable. People who embrace a religion in their childhood often reject all of it later when even one of its principle teachings is rejected.

But Paley's argument was simply that the intricate, fine-tuned universe, and the wonders of the human body demonstrate the existence of a "designer." The argument does not in its essential nature depend on believing the creation account in the Bible.

Paley's central argument began with an illustration. If a person found a watch, even in an uninhabited desert, he would immediately conclude that it had a maker and was not the result of some process of nature. The harmony of its many parts would convince him that it had been made by a skilled watchmaker. Paley gave the following example: Since the human eye is much more intricate and perfect in its design than a watch, it is clear that it also had a maker.

Darwin responded as follows:

> It is scarcely possible to avoid comparing the eye with a telescope. We know that this instrument has been perfected by the long-continuing efforts of the highest human intellects, and we naturally infer that the eye has been formed by a somewhat analogous process. But may not this inference be presumptuous? Have we any right to assume that the Creator works by the intellectual powers like those of a man?[254]

Darwin used this reasoning to reject Paley's argument, but it really proves nothing. There are many philosophers and scientists who acknowledge that the universe and living things were created by an intelligent and powerful entity, but who do not accept the creation account in the Bible. Michael Denton is one of these. Michael Behe also separates the scientific issue of intelligent design from any attempt to identify the designer of life.

Paley had a broad knowledge of scientific matters, but he was also an Anglican clergyman. Behe thinks his enthusiasm to promote his belief in God led him to overreach in his illustrations and invite counter-arguments. But Behe asks the reader to appreciate Paley's argument.

> But, exactly where, we may ask, was Paley refuted? Who has answered his argument? How was the watch produced without an intelligent designer? It is surprising but true that the main argument of the discredited Paley has actually never been refuted. Neither Darwin nor Dawkins, neither science nor philosophy, has explained how an irreducibly complex system such as a watch might be produced without a designer. Instead Paley's argument has been

sidetracked by attacks on its injudicious examples and off-the-point theological discussions.[255]

A. Scientific Support for Design in Nature

1. Denton's Research Supports Design

Since Paley made his famous argument that likened the eye and other parts of the body to machines like a watch, philosophers such as David Hume and Darwinian biologists like Richard Dawkins have challenged this comparison. But both Denton and Behe argue that biology has now demonstrated that this comparison is very accurate. Denton wrote in his 1986 book,

> It has only been over the past twenty years with the molecular revolution and with advances in cybernetic and computer technology that Hume's criticism has been finally invalidated and the analogy between organisms and machines has at last become convincing... In the atomic fabric of life we have found a reflection of our own technology. We have seen a world as artificial as our own and as familiar as if we had held up a mirror to our own machines.[256]

Denton states that Darwin, even while attempting to convince the world of the validity of evolution by natural selection, was confiding to friends that he doubted this could bring about complicated adaptations of what he called "organs of extreme perfection."

Denton gives many examples of "organs of extreme perfection." He mentions the elegance of the design of the mammalian kidney, and says that it defies both intuition and the mathematical laws of chance to think that such organs could "evolve" by chance.

Denton believes that the problem of random chance ever developing the cell structures and functions he found in humans is enough by itself to put Darwin's theory in doubt. He wrote of the very intricate connections among the nerve cells of the retina, which enable the eye to carry out many types of data processing of visual information. He says that modern technology and NASA have learned to copy only a few of these functions.

This is also true of developing "articulated legs" for space exploration. NASA has found it very challenging to develop the very sophisticated computer-control programs that are essential for developing machines that smoothly crawl, walk or run. A newspaper article in December 2009 reported that researchers at Oregon State University called the common cockroach a "biological and engineering marvel." These researchers were trying to build the world's first legged robot capable of running effortlessly over rough terrain.

Denton refers to problems "involved in developing chemical coding devices," and says those that exist in nature are vastly more efficient than the microfilms used today. Scientists have not yet learned to do this, but cells that store hereditary information in living things do this now. "The capacity of DNA to store information vastly exceeds that of any other known system." He adds,

> In the near future, one of the major technological challenges facing our culture will be the problem of a new source of energy. A solution to the problem of extracting solar energy was solved three and a half thousand million years ago when life began on earth. The solution is the chloroplast, which is a micro-miniature solar energy plant which converts the light of the sun into sugar—the hydrocarbon fuel which ultimately energizes every cell on earth. (This includes the fossil fuels we use.)[257]

Denton gives several other examples of cells doing things humans are not yet able to duplicate. He believes that the scientific evidence from molecular biology is that "random processes" could not possibly have constructed a functional protein or gene. He is convinced that they had to have been designed by an intelligence far greater than ours. He says the following about human intelligence:

> Human intelligence is yet another achievement of life which has not been equaled in our technology, despite the tremendous effort and some significant advances which have been made in the past two decades towards the goal of artificial intelligence—a goal which may still be further away than is often assumed.[258]

Denton then refers to Paley's argument that a machine such as a watch was something we would never consider as due to natural processes. Denton wrote the following: "Paley was not only right in asserting the existence of an analogy between life and machines, but was also remarkably prophetic in guessing that the technological ingenuity realized in living systems is vastly in excess of anything yet accomplished by man . . . The conclusion may have religious implications, but it does not depend on religious presuppositions."[259]

2. Behe's Research also Supports Paley's Argument

Behe is a biochemist and reports the following: "In recent years some scientists have even begun to design new biochemicals using the . . . principles of microevolution—mutation and selection. The idea is simple: chemically make a large number of DNA or RNA, then pull out of the mix the few pieces that have a property the designer wants."[260]

Behe explains that the fact that biochemical systems can be designed by intelligent agents for their own purposes is agreed on by all scientists. Because biochemical systems can be designed, and people who didn't see or otherwise know of the designing can detect it, Behe concludes that the question of recognizing whether a given system was designed or not comes down to showing that evidence supports design.

If a biological structure can be explained in terms of biological reproduction, mutation, and natural selection, then he says we cannot conclude it was designed. Behe, in referring to his famous book, *Darwin's Black Box*, says,

> Throughout this book, however, I have shown why many biochemical systems cannot be built up by natural selection working on mutations; no gradual, direct route exists to these irreducibly complex systems, and the laws of chemistry work strongly against the undirected development of the biological systems that make molecules. If natural laws peculiar to life cannot explain a biological system, then the criteria for concluding design becomes the same as for inanimate systems.[261]

Behe says that we can be as confident that the biochemical systems he used as examples in his book were designed by an intelligent agent as we can that a mousetrap was designed. He refers back to cases he explained earlier where irreducibly complex systems were clearly designed and not results of natural random processes. He gives many examples. The function of the cilium is to be a motorized paddle, and this function is impossible if even one of the component parts is missing. The function of the blood-clotting system is like a strong but transparent barrier, and all the components work together. If even one component is missing the task cannot be accomplished. He argues that the cilium and the blood-clotting system were designed for their specific functions is as clearly as that the mousetrap was designed for its specific function.

The same is true of the many other examples from microbiology that Behe gives in his book. He reasons that because the functions of these biological systems depend critically on the intricate interactions of their parts we must conclude that they, like a mousetrap, were designed.

A Challenge to Behe's Argument for Intelligent Design

Behe's book, *Darwin's Black Box*, became a scientific best seller, and attracted far more attention than did Denton's *Evolution, a Theory in Crisis*. Perhaps it is because it aroused so much attention it invited criticism from various directions. Behe explains a common theoretical argument.

> In discussions about intelligent design, no objection is more frequently repeated than the argument from imperfection. It can be briefly summarized: if there exists an intelligent agent who designed life on earth, then it would have been more capable of making life that contained no apparent flaws; furthermore it would have done so.[262]

Behe refers to arguments by biologist Kenneth Miller of Brown University. He refers to Miller's book *Finding Darwin's God*, which presents a few instances where Miller refers to Behe by name. Miller tries to refute Behe's arguments in *Darwin's Black Box* that the cell is so irreducibly complex that it could not have developed incrementally from simple life like Darwin proposed. Behe refers to Miller's argument that if there is an intelligent designer, organisms should perfectly perform their tasks. Miller says this would not be true if they are a result of modifications through evolution. He explains that the eye is not as effectively designed as it could have been, and there is even a blind spot in the retina. Miller takes this as evidence against intelligent design.

Behe responds that "Miller elegantly expresses a basic confusion!" He answers, "The most basic problem is that the argument demands perfection at all. Clearly, designers who have the ability to make better designs do not necessarily do so."[263]

Behe argues that the "intelligent designer" was under no moral or ethical obligation to make as effective and "perfect" a creation as he could. But I believe the Bible gives a much more persuasive reason for the imperfections biologists observe today. Theologians and philosophers like Francis Schaeffer study more than just the story of creation in Genesis 1 and 2 to explain human nature and society as we observe and experience it today. Schaeffer points out that it is necessary to look at Genesis 1–11.

The Bible identifies the "intelligent designer" as God. Genesis 1 and 2 outline the events of creation, centered on the human race and earth as man's habitat. Genesis 3 reports that the first human couple, before giving birth to any children, sinned by disobeying their Creator. The Bible clearly teaches, in both Old and New Testaments, that this first act of sin changed man's nature. They did not completely lose "the image of God" of their former perfect, sinless state, but something foreign to that nature entered them and was transmitted to all their descendants. This change-due-to-sin brought about a permanent change in the nature of humans, an imperfection that they did not have when first created. This imperfection affected both their moral and physical natures. There came a "curse" not only on human nature, but also on the animal and plant kingdoms.

Thus, according to the Bible in Genesis 3, what biologists and other scientists today observe in organisms today may not be a flaw in the original design. It may be a deterioration, a degeneration, from the original design.

Behe says you cannot know the nature of the designer simply by finding something he has designed. He naturally does not refer to the teachings of the Bible at all. So, of course, he does not refer to the biblical concept of original sin. But he does give other arguments in support of his conviction that imperfections in the design do not weaken the argument from design.

Behe concludes that "when it is considered by itself—away from logically unrelated ideas—the theory of intelligent design is seen to be quite robust, easily answering the argument from imperfection."[264]

Kenneth Miller reports a debate that he had with Behe.

> In a 1995 debate, I presented him with molecular evidence indication that humans and the great apes shared a recent, common ancestor, wondering how he would refute the obvious. Without skipping a beat, he [Behe] pronounced the evidence to be convincing, and stated categorically that he had absolutely no problem with the common ancestry of humans and the great apes.[265]

Miller says that the creationists who were witnessing this debate were dismayed, because Behe's words stand in clear opposition to the position of proponents of intelligent design like Phillip Johnson and creationists like Henry Morris. But this is consistent with what Behe has written about being open to the possibility that all biological organisms came from only a few original sources. His argument from biology is that only the activity of an intelligent designer could have produced the prolific number of biological designs that we observe around the world today—it could not have come about by random mutations and natural selection alone. Behe's position is that there was intelligent intervention at certain points in evolutionary development when new biological designs appeared.

3. More Recent Developments Strengthen the Design Argument

In *The Edge of Evolution*, written in 2007, Behe expands on his arguments for intelligent design by reporting on advances in biological research since he wrote *Darwin's Black Box*. He points out that "relatively recent results from various branches of physical science—physics, astronomy, chemistry, geology, molecular biology—actually points insistently toward purposeful design in the universe . . . They paint a vivid picture of a universe in which design extends from the very foundations of nature deeply into life."[266] He says this was "unexpected and surprising" to Darwinians.

Origin of Life and Design

> If the design hypothesis is a leading contender as an explanation for the fine-tuning of the laws of the universe, then by the same reasoning it also must be regarded as a leading explanation for the finer physical, chemical and astronomical details and events that make life possible.[267]

Behe points out that astronomers have tried for decades to figure out how the moon originated by some process of natural law and have failed to do so. He likens this failure to attempts by researchers to figure out how life came into existence by the laws of nature.

Behe explains that natural law acting on the universe by itself cannot explain the fine-tuning of earth. He argues that in the same way no plausible law-like model for the origin of life is currently being considered. Behe reports that physicist Paul Davis studied the issue for a couple of years and concluded that a fully satisfactory theory of the origin of life requires some radically new ideas.

Behe quotes Nobel laureate Francis Crick as follows: "An honest man, armed with all the knowledge available to us now, could only state that in some sense, the origin of life appears at the moment to be almost a miracle, so many are the conditions which would have had to have been satisfied to get it going."[268] Then Behe adds,

> Let me be clear. I am not saying the origin of life was simply an extremely improbable accident. I am saying the origin of life was deliberately, purposely arranged, just as the fundamental laws and constants and many other anthropic features of nature were deliberately, purposely arranged. But in what I'll call the "extended-fine-tuning," view, the origin of life is merely an additional planned feature culminating in intelligent life.[269]

In other words, the final object in the mind of the designer was "intelligent life," by which Behe clearly means humanity. He notes that "as with the origin of life, it may be possible for scientists to select proximate physical conditions in the laboratory, and deliberately cause batches of certain mutations to occur at the right times, and that would be a scientifically interesting project. But without the intimate involvement of a directing intelligence, they would not come about in nature."[270] So Behe is convinced that the designer continued to be active during the time the designed organisms were developing and "evolving."

Behe says that a designer who is capable of fine-tuning "the laws and constants of nature is immensely powerful." He could have designed everything from the beginning to produce intelligent life with no further involvement on his part, or he could continue to fine-tune as the process of developing this life unfolded. He says that nature clearly reveals the purposeful design of life. But he makes clear that the purposeful design of life is *not* compatible with Darwin's proposed mechanism of evolution. By this he means *random* variation and natural selection—the theory that attempts to explain the development of life without the need for "guidance or planning by anyone or anything at any time."

Then Behe faces the common criticism that the design theory is not scientific. He argues as follows:

> Then is design a philosophical conclusion, a scientific one, or maybe both? Here's where serious opinions diverge. On the one hand, some people think that, if a conclusion implicates an intelligent agent (even merely a human mind), or if it threatens to point beyond nature, it's better classified as a philosophy. On the other hand, I regard design as a completely scientific conclusion . . . as a rough-and-ready definition, I count as "scientific" any conclusion that relies heavily and exclusively on detailed scientific evidence, plus standard logic.[271]

He continues his argument,

Darwinism implicitly entails the strong, broad, basic claim that, given enough chances, random mutation and natural selection can build the sorts of complex machinery we see in the cell. Intelligent design implicitly entails an equally strong, broad, basic prediction, that random mutation *cannot* do so. Design denies not only that *some* specific piece of machinery (say, the bacterial flagellum) would be produced by random mutation, but that *any* complex, coherent molecular machinery would. Although random processes can account for small changes, there are real limits. Beyond these limits, design is required.[272]

An Example of How Design Is Different from the Effects of Random Mutations

Behe gives a graphic example to show the contrast between the effects of random mutations and the purposeful organization of the cell.

> Complex interactive machinery—whether in our everyday life or in a cell—can't be put together gradually. But some simple structures can. One example from our large world is a primitive dam. Because gunk accumulates, the drain in my family's kitchen sink slows and stops every so often. It doesn't much matter what makes up the garbage—bits of food, paper, big pieces and small . . . Even large rivers can get clogged by the gradual accumulation of debris. Depending on your circumstances, that might be a favorable development. Sometimes a clogged river or stream might accidentally do some animals some good if, say, it forms a reservoir. Slowing and eventually damming the flow of water doesn't require sophisticated structures—just a lot of debris. Genetic debris can accumulate in the cell, too. If it accidentally does some good, then it can be favored by natural selection.[273]

In other words, random mutations can pile up in an organism and thereby produce noticeable change in it and in its descendants. But this never produces evidence of design. Random mutation cannot produce what Behe calls "discrete structures."

Behe calls the following "a very important point."

> Randomly duplicating a single gene, or even the entire genome, does not yield new complex machinery; it only gives a copy of what was already present. Although duplicating genes can be used to trace common ancestry, neither individual gene duplications nor whole genome duplications by themselves explain novel, complex forms of life.[274]

Recent research shows that mutations in simple organisms like viruses seem to sometimes enhance the effectiveness of the organism, but Behe says they do not do so in *complex organisms*.

Summary

Denton and Behe, speaking from their personal knowledge of genetics, have concluded that certain biological structures could not have come into being in their present state of complexity and function in any gradual manner as required by Darwin's general theory of evolution. They conclude that these structures must have been designed by an intelligent designer. But neither of these biologists will, on the basis of their scientific discoveries, identify this designer.

Their arguments are persuasive, and at this point my friends who are biologists say they have not heard of any counter-argument that Behe hasn't effectively answered. The counter-attacks by Darwinians have been many, but at least to date they do not appear to shake or weaken the arguments for design. The mass media has generally taken the side of the evolutionists. So have many recent court cases on this issue. However, the scientific evidence does not seem to support the evolutionists.

B. The Need for a New Paradigm

1. The Failure of the Old Paradigm

Denton noted that

> the influence of evolutionary theory on fields far removed from biology is one of the most spectacular examples in history of how a highly speculative idea for which there is no really hard scientific evidence can come to fashion the thinking of a whole society and dominate the outlook of an age.[275]

Denton states that "in the final analysis we still know very little about how new forms of life arise. The 'mystery of mysteries'—the origin of new beings on earth—is still largely as enigmatic as when Darwin set sail on the *Beagle*." He laments that

> increasingly, its highly theoretical and metaphysical nature was forgotten, and gradually Darwinian concepts came to permeate every aspect of biological thought so that today all biological phenomena are interpreted in Darwinian terms and every professional biologist is subject throughout his working life to continuing affirmation of the truth of Darwinian theory.
>
> The fact that every journal, academic debate and popular discussion assumes the truth of Darwinian theory tends to reinforce its credibility enormously.[276]

Denton says that if we reject Darwinism then in effect, there is no scientific theory of evolution. He explains that the Darwinian model is still the only model of evolution ever proposed which invokes well-understood physical and natural processes as the cause of evolutionary change. In this way Denton argues that the reason scientists still hold on to a belief in evolution is because no alternative explanation for the world as it is today is available to take its place when creation is rejected.

He remarks that

> the entire scientific ethos and philosophy of modern western man is based to a large extent upon the central claim of Darwinian theory that humanity was not born by the creative intentions of a deity but by a completely mindless trial and error selection of random molecular patterns.[277]

In my own study of evolutionary biology and geology I have discovered several very specific places where the theory cannot explain the realities of nature. As Denton points out above, it cannot explain how life could have developed by natural causes. It cannot explain how complex organisms could have "evolved" from simple organisms, as Behe has clearly shown. The fossil record, as dramatically illustrated by the Cambrian Explosion, does not support the concept of slow, incremental evolution.

2. The Need for a New Paradigm

Denton reported that scientists cling to the belief in Darwin's general theory in spite of the scientific evidences that contradict it. Behe agrees. Behe points out that many—perhaps most—scientists believe in God and in the supernatural. But they have been taught that anything beyond natural causes should not be considered in scientific research.

Behe presents an interesting illustration:

> Imagine a room in which a body lies crushed, flat as a pancake. A dozen detectives crawl around, examining the floor with magnifying glasses for any clue to the identity of the perpetrator. In the middle of the room, next to the body, stands a large, gray elephant. The detectives carefully avoid bumping into the pachyderm's legs as they crawl, and never even glance at it. Over time the detectives get frustrated with their lack of progress but resolutely press on, looking even more closely at the floor. You see, textbooks say detectives must "get their man," so they never consider elephants.[278]

Behe likens this to fellow biologists.

> There is an elephant in the roomful of scientists who are trying to explain the development of life. The elephant is labeled "intelligent design." The straightforward conclusion, based on all the evidence, is that many biological systems were designed. But many scientists feel obligated to restrict their search to "unintelligent causes"—laws of chance and necessity. Like the detectives in the illustration, they are not prepared to look at what all the evidence reveals or conclude what it indicates.[279]

Both Denton and Behe say that evolutionary biologists are facing greater and greater problems. This is because the evolutionists are seeking a natural cause for life and not finding it. It is becoming increasingly clear that the scientific evidence doesn't match up with evolutionary paradigms. Behe says that intelligent design is "the elephant in the room" that alone can explain the evidence.

The "Unwritten Rule" Limits Scientific Thinking

Behe explains an "unwritten rule" that most scientists follow. "Science must invoke only natural causes. And explain by reference only to natural law." He quotes Richard Dickerson, a prominent biochemist and a member of the elite National Academy of Sciences. Dickerson wrote, "Operational science takes no position about the existence or nonexistence of the supernatural; it only requires that this factor is not to be invoked in scientific explanations."[280]

Behe explains that Dickerson himself believes in God, as do most scientists. But Dickerson believes science should not evoke the supernatural, including a "higher intelligence," in seeking causal explanations. Science is involved in making hypotheses, careful testing, and replicating results. Behe replies, "But how can an intelligent designer be tested? Can a designer be put in a test tube? No, of course not. But neither can extinct common ancestors be put in a test tube. The problem is that whenever science tries to explain a unique historical event, careful testing and replicability are by definition impossible."[281] In other words, evolution and creationism are alike in the sense that neither can be tested by the methods used to verify theories that can explain what is happening today.

Science *can* observe the effects of a past event, such as a comet's lingering effects on earth today. Behe argues that, in a similar way, science can see the effects that a designer has had on life.

Although Behe states that many scientists believe in God, he is not directly supporting a religious argument for the existence of life. He believes in the scientific method, and argues from the facts of science. He insists that biochemical machines must have been *designed*. He says that you cannot know who the designer is by studying the designs in nature. He declares as a scientist that life must have had an "intelligent designer." But he considers his belief that this "designer" is God a personal matter. This is because science itself cannot describe or identify the designer, though it can conclude one acted in the past.

Behe explains the "most powerful reason for science's reluctance to embrace a theory of intelligent design." He thinks it is because even many well-respected scientists "don't *want* there to be anything beyond nature." He says, "They don't want a supernatural being to affect nature, no matter how brief or constructive the interaction may have been." He says that sometimes this leads to rather odd behavior. He gives as examples Einstein and Eddington and Hoyle. These men resisted a scientific theory that flowed naturally from the cosmic data that support the big bang theory. This is because they didn't want to accept the philosophical or theological conclusions that unavoidably arise from acknowledging it.[282]

Behe reports that tens of thousands of graduate students, post-doctorate associates and professors have worked long hours to uncover the secrets of the cell. He says that "the result of these cumulative efforts to investigate the cell—to investigate life at the molecular level—is a loud, piercing cry of '*design.*' The result is so unambiguous and so significant that it must be ranked as one of the greatest achievements in the history of science."[283]

He points out that many students learn an evolutionary worldview from their textbooks. But they do not learn how the theory of evolution they are taught could have produced the intricate biological systems that those texts describe. He surveyed the professional literature, and says this.

> Molecular evolution is not based on scientific authority. There is no publication in the scientific literature—in prestigious journals, specialty journals, or books—that describes how molecular evolution of any real, complex, biochemical system either did occur or even might have occurred. There are assertions that such evolution occurred, but absolutely none are supported by pertinent experiments or calculations. Since no one knows molecular evolution by direct experience, and since there is no authority on which to base claims of knowledge, it can truly be said that . . . the assertion of Darwinian molecular evolution is merely bluster.[284]

Then Behe concludes his argument as follows:

> If a theory claims to be able to explain some phenomenon but does not generate even an attempt at an explanation, then it should be banished . . . molecular evolution has never addressed the question of how complex structures came to be. In effect, the theory of Darwinian molecular evolution has not published, and so it should perish.[285]

Behe says some people have the philosophical commitment to the principle that nothing beyond nature exists. He says this commitment should not preclude the consideration of a theory that flows naturally from observable scientific data.

Clearly the old paradigm has failed to explain the designs of nature by naturalistic means. A new paradigm is needed to replace it.

3. A Different Approach Is Needed

Evolutionists often ask Behe, Johnson, or Dempski to produce a "different theory." They mean a different historical theory that answers the same questions that Darwinism does. Phillip Johnson makes the point that this demand reflects a fundamental misunderstanding. He says that "evolutionary science has been trying to build a materialist model of biological creation." They have tried to formulate a theory about how life began, and how it changes or evolves. Then Johnson says something significant:

> There is nothing wrong with the (Darwinian) model *per se*; it is probably the best that can be built on materialist assumptions. It is the assumptions that are wrong. The task for now is to provide not a better materialist model but a better set of assumptions that will lead science to discard the wrong questions and embrace the right questions . . . *If the evolutionary scientists were better informed or more scientific in their thinking, they would be asking about the origin of information.*[286]

Johnson says that scientists need to acknowledge the true character of the information that governs the life processes. In order to do so, Johnson argues, scientists will have to face the fact that they do not know how a combination of chance and natural law could do the creating. This means they must acknowledge that their present theory of the history of life is very incomplete and may be completely erroneous. He says their claiming to have knowledge is "not a triumph for science unless it is true knowledge." He adds that admitting that we don't have answers is better than dogmatically retaining the wrong answers.

Biologists are already seeking to understand and modify design in their field. Johnson tells us how biochemists say they can "reverse engineer" specific biochemical structures to determine what functions they perform. He calls this "the language of intelligent design."

4. Intelligent Design: The New Paradigm

It Is an Inescapable Scientific Fact that Life Was Designed by a Higher Intelligence!

Behe says, "The philosophical commitment of some people to the principle that nothing beyond nature exists should not be allowed to interfere with a theory that flows naturally from observable scientific data."[287] Behe concludes his arguments for the need of a new paradigm by saying there is no solid defense against the strange conclusion that an intelligent agent designed life. But there have been many strange conclusions in the past few hundred years. Scientists have concluded that the earth is spinning even when we don't feel it spin; that space is curved; and that the universe had a beginning.

Behe argues that accepting the scientifically-verified truth that the cell had an intelligent designer should help scientists move beyond the now-disproved theory of Darwinian macroevolution and encourage them to develop more realistic scientific paradigms.

Molecular Biology Proves the Existence of a Designer, but Does Not Identify Him

Behe strongly supports Paley's argument that the human body could not have come into existence unless it had been designed by an intelligent designer. But he does not use this as an argument for belief in the God of the Bible like Paley did. Behe makes an interesting inference. It is that you cannot know the character of the watchmaker simply from finding his handiwork in some solitary place. In the same way you cannot know the character of the designer of life only by studying his handiwork. (I beg to differ here, but not on the basis of biology or other natural science. I disagree on the basis of social science, and on the basis of something in human nature that reflects the nature of our "Designer." We can know something accurate about our Creator simply because man was originally created in His likeness in His intellectual, social and moral nature. There is something about us humans that reflects the nature of God.)

Behe also says that the argument for intelligent design does not depend on *when* life was intelligently designed. Some scientists confuse the theory that life was intelligently designed with the theory that the earth is young because some religious groups strongly advocate both of these ideas. Behe maintains that the young earth view is not a part of intelligent-design theory, and says he has no reason to doubt that the universe is billions of years old like physicists say it is.

Behe does not say that he has concluded as a scientist that the Genesis account of creation is true. Denton goes farther; he says he personally does not accept the Genesis account. But how could biology or any other science make such a claim in the name of science? There are *two belief systems* that claim to explain how life began. These are creation by an eternal God, and evolution by "blind chance." Creation seems clearly a more logical explanation than "blind chance." But to identify who is the "creator" is not a proper *scientific* pursuit.

5. Summary

Darwin was aware of the argument that what he called *Organs of extreme Perfection and Complication* are evidence of creation. Darwin said this was one objection to his theory, but he felt he had an answer to this. It seems clear now that the scientific facts do not support him in this.

Paley argued that the intricate functions of living organisms demonstrate that they were designed by an intelligent source, a "designer" who has personality. Although Darwin rejected Paley's argument, Denton and Behe say it has never been refuted. They also say that it is the only plausible conclusion a biologist can make. Denton says "the puzzle of perfection" of living organisms proves there was a designer. Behe says the irreducible complexity of organisms proves there was a designer. Behe also says that the arguments against design are not sound. The chief argument some make is that not all organisms on earth appear to be "perfect" in function and in design. Behe argues that a designer had no obligation to make all organisms perfect.

I believe the Bible has a better answer to this argument. Genesis 1 and 2 report that God created everything "good" in its original state, but in Genesis 3 adds that man sinned, and this brought "imperfection" to God's original creation. This means that what biologists study today is not organisms as first designed, but a degenerated state from that first design, a degeneration that resulted from human sin.

Other challenges have been made to Behe's argument that irreducible complexity requires design. Few of these challenges have presented empirical findings to support arguments against design. Behe has pointed out that those that have presented it have really not demonstrated scientifically verifiable data to support their claims.

Denton and Behe also say that scientists and scientific journals have elevated Darwin's general theory to a "dogma." By this they mean that discoveries since Darwin's day have not produced the proofs for the theory needed to substantiate it. These discoveries have also demonstrated that the cell is too complex to have ever evolved from simpler life-forms by the processes Darwin envisioned. Denton branded Darwin's general theory as "the great cosmogenic myth of the twentieth century." Behe surveyed the professional literature, and discovered many articles arguing that molecular evolution occurred. But he says there is not one article "that describes how molecular evolution of any real, complex, biochemical system either did occur or even might have occurred."

Both Behe and Denton insist that the scientific evidence overwhelmingly demonstrates that complex biological organisms were *designed* to function and develop as they did. Behe, in both books, focuses his attention on a possible *pathway* to explain how Darwinian evolution could possibly occur. He concludes in both books that random mutation by natural selection cannot provide a pathway for evolution beyond a very limited degree.

Behe expresses his conclusion as follows:

> Darwin and design hold opposite, firm expectations of what we should find when we examine a truly astronomical—a hundred billion billion—number of organisms. Up until recently, the magnitude of the problem precluded a definitive test. But now the results are in. Darwinism's most basic prediction is falsified.[288]

A new paradigm is needed. One that rejects the "unwritten rule" that science must invoke only natural causes, and explain by reference only to natural law. Natural law alone has failed to explain the complexities of life on earth.

A more persuasive paradigm is needed, one that can explain what the paradigm of macroevolution has failed to explain. Behe and others insist that it is an inescapable scientific fact that life was designed by a higher intelligence. Behe argues that molecular biology proves the existence of a designer, but does not identify him. He says you cannot know the character of the designer of life only by studying his handiwork.

A designer exists! He is intelligent and has what we call personality. His personality is seen by his intelligence, his choices in creating life as he did, and in his ability to act in our cosmos. It is also seen by the fact that humanity was created possessing intelligence. It takes intelligence to create intelligence. Science may not be able to "discover his nature," as Behe contends—but the Bible reveals much about God as Creator and is called God's "self-revelation."

Part 2: Into the Twenty-First Century

There has been much debate recently whether or not the theory of intelligent design is really science. This is because it allows for the consideration of causes that are not naturalistic. On the one side of this argument is the insistence that science can consider only naturalistic causes. This side insists that Darwin's theory is still the best argument of naturalism to explain the biological realities that we observe today. On the other side of the argument is the insistence that Darwin's theory and intelligent design theory be considered *equally* as explanations how living organisms developed to where they are now. Then the two can be compared to see which better explains the realities of nature as we understand them today.

As we begin the twenty-first century the scientific arguments that support the theory that the cosmos and life on earth have been designed by a higher intelligence seem to be getting stronger and more convincing to many scientists in all fields of research.

Both the 150th anniversary of Darwin's *Origin of Species* and the 200th anniversary of Darwin's birth were celebrated in 2009. To commemorate this memorable year two important books on historical science were published. One is *The Greatest Show on Earth: The Evidence For Evolution*, by Richard Dawkins. The other is *Signature in the Cell: DNA and the Evidence for Intelligent Design,* by Stephen Meyer. These two books have contrasting styles and conclusions. Both agree that the study of historical science is primarily a matter of "detective work." This means that trying to reconstruct how the living world developed to what it is today is primarily a matter of looking for the clues that remain. It is an attempt to figure out what happened and "who did it." But these two books are polar opposites in the way they seek to explain how life developed to its present complexity.

Dawkins retired in 2008 from Oxford University, where he was a professor of biology. He is an active lecturer and debater in the United States. He is the author of several books, the most famous being *The God Delusion*. An Englishman, Dawkins is a fellow of both the Royal Society and the Royal Society of Literature. He has received many honors and awards for literature, including the International Cosmos Prize of Japan. No awards in original scientific research are mentioned though. This supports my impression that he is an eloquent and persuasive writer, but in his treatment of science he is more entertaining and persuasive than objective. This entertaining style is seen throughout the book.

A. The Greatest Show on Earth

This book is called "a stunning counter-attack on advocates of 'Intelligent Design.'" It is said to explain "the evidence for evolution while exposing the absurdities of the creationist argument." But Dawkins does not address the more scientific arguments for intelligent design in this book, even though he has publically debated Stephen Meyer and other proponents of design theory. He does not mention Meyer or Dembski or Johnson, or even Behe in this book. Years earlier he wrote challenging Behe's irreducibly complex argument in biology. But Behe successfully used biological science to defended his argument against Dawkins' attack.

In *The Greatest Show on Earth*, Dawkins gives a "popularized" version of the science and his debates. He does not deal directly with the hard science that cannot be explained by Darwinian interpretations. For example, he argues that Darwin's theory of macroevolution is proved by the fossils—but he is very selective in the specific fossils he uses to try to demonstrate this.

Dawkins is obviously trying to find a way to explain how complex organisms could have evolved without the necessity of recognizing that the cell has been *designed*. He is clearly trying to counter the arguments of Behe and Meyer and others that the cell shows it has been intelligently designed.

Dawkins refers to public debates with people who are clearly not scientists. He refers also to challenges by individuals whose arguments he correctly calls "silly." But in this book he does not address the challenges of *the scientists* who dispute macroevolution and the scientific data they present. He goes so far as to claim the following: "Evolution is a fact, and this book will demonstrate it. No reputable scientist disputes it, and no unbiased reader will close the book doubting it."[289]

Here Dawkins resorts to a semantic device that is very common in popular presentations of evolution. "Evolution is a fact" is not disputed by natural scientists in the sense that *some limited degree of evolution* by Darwinian means has been demonstrated as real—what is commonly called "microevolution." But it is surely disputed by many "reputable scientists" in the sense of "macroevolution." Dawkins dismisses his critics by referring to comments of individuals with no formal background in science who occasionally make unsophisticated comments to him during or after his lectures.

Dawkins makes fun of creationists, without acknowledging when they have a valid scientific point. His book may be "great literature," but it is not objective science. He does not mention that some evolutionists, like Stephen Jay Gould, acknowledge that scientific research has revealed areas where Darwinism is not supported by the facts.

Is Intelligent Design Science?

A Contrast of Two Evolutionists: Openness to the Realities of Nature

How open are we to the realities that challenge our beliefs? This was initially discussed in chapter 1. In this regard the examples of Gould and Dawkins are very instructive.

Richard Dawkins, in all his books, argues that Darwin's theory of macroevolution has been "proved" and no one should have any doubts about it. It is easy for Dawkins to argue that scientists who are not Darwinians are "biased" and should not be taken seriously. But the fact is that there are many Darwinians and other evolutionists who acknowledge that Darwin's theory has not stood up to the test of modern science. One of the most prominent of these is Stephen Jay Gould.

The name Stephen Jay Gould first attracted my attention because so many scientists refer to him as an authority on fossils. When I read more about him and his diligent search for answers to the Darwinian interpretation of evolution, I was impressed and thought, "Here is an open-minded scientist who tries to face the facts honestly."

His credentials as an evolutionist are impressive indeed. Gould (1941–2002) was an American paleontologist, evolutionary biologist, and historian of science. He was also one of the most influential and widely read writers of *Popular Science* magazine of his generation. He served as president of the American Association for the Advancement of Science from 1999–2001. Earlier he was elected to the National Academy of Sciences, and served as president of the Paleontology Society. He also served as curator of invertebrate paleontology at Harvard University's Museum of Comparative Zoology. Stephen Jay Gould and Carl Sagan have been called "our current most popular scientific spokesmen for evolution."

Gould was clearly a scientist who recognized the problems with Darwinian evolution, and was actively seeking solutions to them. He recognized fully that the fossil record does not support Darwin's theory of gradual, incremental change from simple to very complex organisms. He faced the historical fact of the Cambrian Explosion and tried to find a new evolutionary theory to account for it. This theory, called the "Eldredge-Gould punctuated equilibrium" theory, has seemingly received very little support in the scientific community. It suggests that evolution has developed with long periods with almost no change, and also times of very fast development of new forms of life. One problem with the theory is that its supporters have not been able to explain either the cause or the process that might make such rapid evolutionary change possible.

A book published in 2003 gives the following assessment of this theory:

> Although . . . the newer punctuationalist model of evolutionary change appears more consonant with some aspects of the Cambrian/Precambrian fossil record, it, too, fails to account for the extreme absence of transitional intermediates, the top-down pattern of disparity preceding diversity, and the pattern of phylum first appearance. Furthermore, punctuated equilibrium lacks a

sufficient mechanism to explain the origin of the major body plans that appear in the Cambrian strata.[290]

It is interesting that Dawkins refers to Stephen Jay Gould three times in the book, but never in relation to the fossil record! Although Gould was an evolutionist, he clearly had *scientific* problems with Darwin's conclusion that the fossil record would someday support his view of evolution. Dawkins clearly has no problems accepting macroevolution and is out to sell it to whoever will listen. Dawkins argues that, even though there are gaps, "All the fossils that we have, and there are very very many indeed, occur, without a single authenticated exception, in the right temporal sequence." Also, "not a single solitary fossil has ever been found before it could have evolved . . . Evolution has passed this test with flying colours."[291] The scientific facts, of course, do not support such assertions.

B. Signature in the Cell

> Life does not consist of just matter and energy, but also information. Since matter and energy were around long before life, this third aspect of living systems has now taken center stage. At some point in the history of the universe, biological information came into existence. But how? Theories that claim to explain the origin of the first life must answer this question.[292]

In this book the philosopher of science Stephen Meyer sets about to show that the intelligent design theory is science. He obtained his doctorate in the philosophy of science at Cambridge University, and says he made a careful study of the historical sciences and their scientific underpinnings while at Cambridge.

Meyer seeks to explain and analyze the evidence more objectively than Dawkins does. He reports that Gould insisted that the "historical sciences," like geology, evolutionary biology, and paleontology had to use different methods than the "experimental sciences" like chemistry and physics did. Gould argued that historical scientific theories were testable, but not necessarily by experiments under controlled laboratory conditions. Instead, he emphasized that historical scientists tested their theories by evaluating their explanatory power.

Meyer says that "historical scientists and detectives alike must gather clues, study each one, and then form tentative hypotheses." They "must depend upon circumstantial evidence, the lingering indicators of what happened: the clues, the traces, the signs left behind." He gives three questions as important for doing this:

> "Is the proposed cause known to be adequate?"

> "Is there evidence that the cause may in fact be present?"

> "Which hypothesis best explains the evidence?"[293]

Is Intelligent Design Science?

We noted earlier that Dawkins agrees with this approach in general. Meyer argues that the intelligent design theory meets these three criteria better than Darwin's theory does.

Meyer sets about in his book to systematically analyze various possible explanations for events in the distant past. He attempts to eliminate competing explanations while developing a logical argument for intelligent design—in order to discover the best explanation for the biological and geological evidence. He explains it this way: "Contemporary philosophers of science call this the method of 'inference to the best explanation.' That is, when trying to explain the origin of an event or structure from the past, historical scientists compare various hypotheses to see which would, if true, best explain it. Then they provisionally affirm the hypothesis that best explains the data as the most likely to be true."[294]

Meyer relates that to him intelligent design means "the deliberate choice of a conscious, intelligent agent or person to affect a particular outcome, end, or objective." Intelligent agents have the power of choice and can affect the outcome of events.

In this book Meyer explains in detail the "functionally specified information in DNA, RNA, and proteins." He also describes what he calls the "functionally integrated information-processing sequences and systems in the cell." He says origin-of-life researchers have not been able to explain these existing without an intelligent source. Researchers have failed to explain the origin of the genetic code or information processing capability of the cell by any naturalistic or non-intelligent source or process. Neither chance nor necessity can explain them. He says there is clear evidence of an intelligent "signature in the cell."

Meyer likens the arrangements of bases in DNA and amino acids in protein in their functions to lines of computer code or phone numbers that ring a certain cell phone.

> This judgment is a direct consequence of the experimental verification of Francis Crick's sequence hypothesis. Indeed, since the confirmation of the sequence hypothesis in the early 1960s, biologists have known that the ability of the cell to build functional proteins depends upon the precise sequence arrangement of the bases in DNA.[295]

Meyer asserts that it is clear that intelligent design has demonstrated its capacity to produce the kind of information that exists in the cell.

> As I evaluated the causal adequacy of the other "suspects," I became progressively more intrigued with the design hypothesis. It eventually became clear to me that intelligent design stood as the only known cause of specified information-rich systems and, therefore, that ID provides the best, most causally adequate explanation for the origin of the information necessary to produce the first life. I came to this conclusion for three main reasons.[296]

He concluded that intelligent design theory is the best explanation for the existence of information in the cell for these three reasons: (1) There are no other causally adequate explanations; (2) Experimental evidence confirms the causal adequacy of intelligent design; (3) Intelligent design is the only known cause of specified information. He further argues,

> Undirected materialistic causes have not demonstrated the capacity to generate significant amounts of specified information. At the same time, conscious intelligence has repeatedly shown itself capable of producing such information. It follows that mind—conscious, rational intelligent agency—what philosophers call "agent causation," now stands as the only cause known to be capable of generating large amounts of specified information starting from a nonliving state.[297]

Meyer argues that "one can infer the past existence of a cause from its effect, when that cause is known to be *necessary* to produce the effect in question." Then he continues by saying that the only one known cause exists for the existence of the information discovered in DNA in the cell.

> For this reason, the specified information in the cell establishes the existence and past action of intelligent activity in the origin of life. Experience shows that large amounts of specified complexity or information *invariably* originate from an intelligent source—from a mind or a personal agent. Since intelligence is the only known cause of specified information (at least starting from a nonbiological source), the presence of specified information-rich sequences in even the simplest living systems points definitely to the past existence and activity of a designing intelligence.[298]

He emphasizes, "Intelligence is the *only known cause* of complex functionally integrated information-processing systems." And adds, "Since the intelligent-design hypothesis meets both the causal-adequacy and causal-existence criteria of a *best* explanation, and since no other competing explanation meets these conditions as well—or at all—it follows that the design hypothesis provides the best, most causally adequate explanation of the origin of the information necessary to produce the first life on earth."[299]

> Based upon the evidence presented in this book, I would contend that, whatever its classification, the design hypothesis provides a better explanation than any of its materialistic rivals for the design of the specified information necessary to produce the first life. *Reclassifying an argument* [as not scientific] *does not refute it.*[300]

A Strong Argument to Accept Intelligent Design as a True Scientific Theory

Meyer argues that intelligent design *is* a scientific theory in the same way that Darwinian evolution is a scientific theory. He insists that the design theory meets the same criteria for historical science as evolution does. He lists the following six points in following the established guidelines for historical investigation:

1. The case for ID (intelligent design) is based on empirical evidence.

    He maintains that "critics may disagree with the conclusions of these design arguments, but they cannot reasonably deny that they are based upon commonly accepted observations of the natural world."

2. Advocates of ID use established scientific methods for determining when observed evidence supports a hypothesis. Meyer says the principle method is "the method of multiple competing hypotheses." He believes the inference to intelligent design is the best explanation for the origin of biological information. He says there are only three competing types of explanations. These are chance, necessity, and design.

3. ID is a testable theory. He is convinced that intelligent design can explain the origin of biological information better than its materialistic competitors, and this shows that it has passed an important scientific test. He adds that "like other historical scientific theories, intelligent design makes claims about the cause of past events, thus making it testable against our knowledge of cause and effect."

4. The case for ID exemplifies historical scientific reasoning. Meyer argues that scientists use systematic methods to infer intelligent design. The specific methods they use are like the established patterns of inquiry in the historical sciences. This is true for each of the features of a historical science.

5. ID addresses a specific question in evolutionary biology. This is: "How did the appearance of design in living systems arise?" Meyer reports that "both Darwin and contemporary evolutionary biologists such as Francisco Ayala, Richard Dawkins, and Richard Lewontin acknowledge that biological organisms *appear* to have been designed."

    Then he asks,

    > Is the appearance of design in biology real or illusory? Clearly, there are two possible answers to this question. Neo-Darwinism and chemical evolutionary theory provide one answer, and the competing theories of intelligent design provide an opposite answer ... If one of these propositions is scientific, then it would seem that the other is scientific as well.[301]

6. ID is supported by peer-reviewed scientific literature. One of the criticisms opponents of intelligent design theory often raise is that the books and articles that promote it have not been reviewed by others in the same field to evaluate their scientific accuracy and value. Meyer states that this is false. He says an article of his was published in 2004 in *The Proceedings of the Biological Association of Washington*, and since then at least seven other scientific articles dealing with Intelligent Design have been published in main-stream peer-reviewed scientific journals.[302]

Meyer reports that Behe's *Darwin's Black Box* was reviewed by his peers. It sold over 250 thousand copies. Meyer says that this book by Behe almost single-handedly put the idea of intelligent design on the cultural and scientific map, and the controversy over Behe's argument from irreducible complexity has motivated specific new lines of scientific inquiry and new research questions. Behe's newer book, *The Edge of Evolution: The Search for the Limits of Darwinism*, was reviewed by several professors of biology, chemistry and medical genetics. This includes a member of the National Academy of Science. They have reviewed this book and attest to its scholarship and accuracy in reporting the current state of origin of life research today.

William Dembski's book *The Design Inference* (Cambridge University Press) was reviewed also. Meyers says it established a scientific method for distinguishing the effects of intelligence from the effects of undirected processes.

I believe Meyer has convincingly demonstrated that the evolutionists' arguments that intelligent design does not qualify as a true scientific theory are false. Then he goes on to suggest over a dozen scientific predictions based on the theory.

C. Chapter Summary

Denton and Behe, speaking from their personal knowledge of genetics, have concluded that certain biological structures could not have come into being in their present state of complexity and function in any gradual manner as required by Darwin's general theory of evolution. They conclude that these structures must have been designed by an intelligent designer. But neither of these biologists will, on the basis of their scientific discoveries, identify this designer.

Behe's argument that the cell could not have been developed by the Darwinian means of random mutation and natural selection is persuasive, and at this point my friends who are biologists say they have not heard of any valid counter-argument. The counter-attacks by Darwinians have been many, but at least to date do not appear to shake the arguments for design. The mass media has generally taken the side of the evolutionists. So have many recent court cases on this issue. However, the scientific evidence does not seem to support these actions.

Is intelligent design a *scientific* theory? Behe responds to this often-asked question by saying that it "relies heavily and exclusively on detailed scientific evidence, plus standard logic." Because of this he regards it "as a completely scientific conclusion."

About thirty years ago the argument for intelligent design became a focus of information theory in biology. Michael Behe focused his attention on a possible *pathway* to explain how Darwinian evolution could occur. He concludes that random mutation by natural selection cannot provide a *pathway* for evolution beyond a very limited degree. He says that random mutations simply accumulate as "genetic debris" and never produce the "new complex machinery" required for the cell to increase in complexity and develop more complex functions. Cells are irreducibly complex and could not have developed such complexity by small, incremental adaptive changes as Darwin theorized. Complexity is clearly *programmed* into the cell itself. In other words, the DNA in the cell contains information that determines how it develops and responds to its environment. The evidence suggests some kind of intelligent intervention is clearly needed at the point where a new biological class, phylum or kingdom can develop—if it has ever done so.

But Behe clings to the belief that common descent can account for large-scale evolution. His argument with Darwinism is only that random mutation and natural selection cannot possibly bring it about. Only "intelligent design" could have done it. Then how can intelligent design do this? He does not explain this.

Two books were published in 2009 that seem important for understanding the argument for the intelligent design theory today. They are *The Greatest Show on Earth: The Evidence For Evolution*, by Richard Dawkins, and *Signature in the Cell: DNA and the Evidence for Intelligent Design*, by Stephen Meyer.

Dawkins's book is considered a strong argument against intelligent design and for Darwin's theory that all life on earth developed from one or only a few simple organisms. It is strong on persuasion, but does not present the hard scientific data that challenge the theory on the basis of verified scientific facts—facts which are frankly acknowledged by leading evolutionists like Stephen Jay Gould as problems for evolution.

Meyer's book seems a more objective study of the hard scientific evidence that challenges Darwin's theory and supports the intelligent design theory. While at Cambridge University he says he made careful research into how the historical sciences differ in their methods of scientific verification from the operational sciences that can be tested and replicated today. He presents what seems to be convincing arguments that intelligent design does qualify as a truly scientific theory, and actually has more hard scientific evidence to support it than does Darwin's broad theory of evolution (macroevolution).

13

The Unwrapping of Evolution
Does Any Theory Satisfy Objective Analysis?

> Where did we come from and where are we going? Nearly every educated individual in the Western world believes Darwin and evolution science have shed light on these questions. But science has yet to show how, or even if, evolution on a grand scale has happened. Experiments, fossils, and molecular biology do not reveal the slow, gradual process Darwin postulated for connecting all life forms. Instead the evidence generally indicates species are separated by fixed boundaries. Further the new biology and geology paradigms credit *catastrophe*, not millions of years of mutations slowly accumulating through natural selection, as history's agent of evolution. Why then does science still rely so heavily on Darwin and his modern counterparts in describing a process of evolution?[303]
>
> — Harold Booher

Introduction

Much territory has been covered so far in the pages of this book. We have covered the disciplines of cosmology, biology and geology. We have considered how Darwinians, creationists, and others look at the same scientific discoveries and how they have drawn vastly different conclusions from them. We have tried to understand how experts in the scientific specialties understand and interpret the things they have discovered of the realities of nature.

Up to now we have sought to un-wrap the many layers of scientific discovery and theory. We have done so in order to have as objective an understanding as possible of nature as we "follow the truth wherever it leads." Now we will seek to evaluate what we have learned and try to put everything together.

Most books are written on science to report facts and interpretations based on the author's own worldview and his scientific orientation. They are written by Darwinians, evolutionists who question parts of Darwin's theory, advocates of intelligent design theory, and by Bible-believing creationists.

Books that seek to survey these vast areas from as neutral a prospective as is humanly possible seem extremely rare. But now we want to consider one of these, in order to evaluate better the scientific information we have studied. To do this, I looked for an author who has training in both the physical and social sciences. I looked for someone who can quickly discern between scientific fact and interpretation, and evaluate scientifically the material he is studying. I wanted someone who attempted to evaluate the competing claims of the "experts" as objectively and thoroughly as is humanly possible.

It appears that there is such a book. A book by Harold R. Booher surveys the vast literature on the issue of the origin and development of the universe and living things. It is titled *Origins, Icons, and Illusions* (1998).

Booher explains his unique qualifications for this task:

> My education and experience is interdisciplinary covering both physical and psychological sciences, with specialties in human factors and public policy. My field is particularly concerned with understanding the operations of complex systems involving people and technology.[304]

Booher has a BA from DePauw University, a BS in electrical engineering from Rose Hulman Institute of Technology, an MA in psychology from George Washington University, and a PhD in human factors from the Catholic University of America. When Booher was writing his book he was a Guest Scientist working at for the Army Research Laboratory and was a Standing Committee Member for on the National Research Council. He says his background is especially helpful to prepare him to write this book on origins science for the following reasons:

1. He is required in his work to look at information in very diverse scientific fields and study very carefully how systems perform. He believes that he can help others to better understand the complex issues in science. He does this by studying the processes needed for evolution to occur and evaluating the evidence to see if they actually support the theory. He does this across all the scientific disciplines, as an expert in systems analysis.

2. He has a high regard for the scientific method that is common to all fields of science. He says this method includes at least four things: (a) formulating hypotheses often derived from theory, (b) accumulating evidence to seek to validate hypotheses and (c) leaving a map for others to follow to see if they get the same

experimental results. Also, (d) theories must be tested and pass the tests before they can be considered verified.

He writes from his background in psychology as well as in science, as the subtitle of the book indicates: *Exploring the Science and Psychology of Creation and Evolution*. He faults the current state of development of *all* the theories of origins. He criticizes the theory of intelligent design as a field of investigative science, though he thinks that it fills the requirements for a viable philosophy of science better than its alternatives. But he is hardest on Darwinians. He does not mean all Darwinians; not even the vast majority of those who are doing research—but those who promote naturalism as an all-embracing philosophy. He calls it a "religion"—then backs off and says that it is not a religion because he feels all religions include a belief in a god or gods or some sort of transcendent power. He asserts it is being made a *substitute* for religion by its zealous advocates. He also claims that its advocates are using specific psychological tools to make their arguments appear more valid than they actually are.

Finally, Booher admits that the task was so great he found it necessary to seek the help of those who had more expertise in the various fields than he had. He writes that it took fifteen years to complete the research and verification of the contents of this book. He received critical comments from several scientists who have expertise in biological and medical research, as well as some with backgrounds in physics, and mathematics. He also had several philosophers of science review the book.

It seems Booher has done about as thorough a job as humanly practical. He says he continued his research until he could discover "whether Darwin or anyone since has provided—*through the scientific method*—substantive evidence to support either general or specific theories of evolutionary origins." He continues by asking,

> Do Darwin's critics have a case, *on scientific grounds*, that the favored theory of evolution is more modern dogma than science, or does the bulk of scientific evidence still support Darwin? And if the evidence now seems to rule against Darwinism, do competing theories have better scientific support?[305]

In the above quotations Booher emphasizes the phrases *through the scientific method* and *on scientific grounds* to separate theory and opinion from the science supporting it.

Booher is clearly concerned with the worldview of present-day Darwinian theorists. He does not seem to be concerned because of some pre-existing bias of his against the possibility of large-scale evolution. His concern, as a psychologist, is that *they* are biased, and only see the scientific facts so far discovered from their pre-suppositions that macroevolution *must* be true. He says his book "claims to be objective."

He quotes Julian Huxley as providing in 1959 an evolutionary vision for the future. This was said at the 100th anniversary of the publishing of Darwin's *Origin*. Huxley provided a vision of how Darwinians would advance from biological dominance to psychosocial dominance. Huxley said they will proceed by "breakthroughs to new dominant patterns of mental organization of knowledge, ideas and beliefs—ideological instead of physiological or biological organizations." Then Booher asks, "What does this mean if our world view, underpinned as it is, by science's weak proof of evolution, is warped?"[306]

Booher laments that few people understand that the many examples of small biological changes do not necessarily demonstrate evolution as a theory of origins. Asced theory. So Booher set about to explore all the theories about the origin of life to see if any could really be relied on as true.

Booher claims in his book that he is doing this investigation while trying to be free from bias. As a psychologist he is aware of how *unconscious bias* affects perceptions and our thinking processes. It seems that Booher tried to research the issue of origins without any presuppositions as to the final outcome of his research. I also have a background in psychology. I began the research reported in this book from the presupposition that the facts of science and the clear words of Scripture do not conflict. I did this because I believe that the God who established the laws of nature also guided in the writing of the Bible. But I have tried to remain conscious of my own bias, and keep it from affecting the scientific conclusions I make. I want this book to "follow the truth, wherever it leads."

Booher says, "It is important for us as individuals to know whether evolution is a proven scientific law, a scientific theory, or simply a philosophy believed by a majority of scientists to be true. Popularizers of evolution use it as a world view to instill a false sense of knowledge about ourselves and the universe."[308]

Reading this book carefully is a scientific education in itself! What is presented in it reminds us of much we have covered in earlier chapters, and I believe helps to reinforce the conclusions of Denton, Behe and the other authors I have introduced.

We will make use of Booher's extensive research to review and evaluate the issues we have addressed in this book. This is because he has attempted—honestly I believe—to explain the strengths and weaknesses of each position objectively.

First we will look at Darwinism as it is presented today—as an all-embracing philosophy of science rather than as a demonstrably proven scientific theory. Then we will look at Booher's observations in the field of cosmology. Then we will think about biology, geology, and human origins. We will also discuss his evaluation of intelligent design theory.

A. Darwin and His Influence on Our World

One hundred and fifty years ago Darwin published the *Origin of Species*. In it he formulated the theory that all life that exists today developed spontaneously from one or only a very few simple life forms. He argued that simple organisms developed over time into the marvelous complexity that exists today, solely by naturalistic means. He called these means random mutation and natural selection. Darwin was confident that research in biology and the study of fossil beds around the world would increasingly demonstrate the accuracy of his theory.

Booher reports the Scopes trial in 1925 in detail because he sees it as a turning point in American history on the issue. He shows how Darwinism took root in America as the dominant worldview based on a belief in its scientific proof—even though the scientific verification of the theory is lacking.

> This trial had ramifications beyond encouraging evolution as a way of looking at biological phenomena in nature. Since the Scopes trial evolution has continued nearly unopposed as a way of looking at nearly everything from economics and politics to the development of religion itself. It is so central a belief in modern society that it acts as the foundation for many of our educational, psychological, and social ideas. Consequently, it has come to be one of the most important elements in the development of our fundamental values.[309]

Darwinism seems clearly to be a central belief in many countries as well as in America. It seems to be presented everywhere in popular culture as a scientifically proven theory—even though many of even its most well-known advocates who are scientists admit that it is far from being that.

Darwin's ideas were soon applied to biology and genetics. They gradually extended to the social sciences, and recently even to moral philosophy. Social Darwinism, Freudian psychology, and even Marxist economic theory were influenced by Darwin's theory.

Basic to the theory is the idea that there is no purpose in nature and no goal toward which evolution is striving.

The Origin of Life on Earth

Darwin speculated that at one time there must have been "some warm little pond" that had the right mixture of chemicals and energy to produce a protein compound that was formed ready to undergo still further complex changes. Then, suddenly, life began! There has been much speculation since about what kind of conditions might have existed under which life could have begun.

Booher notes that one problem with this idea is that too much water can destroy "the precursors of life." But he adds,

> Unfortunately most other scenarios for the creation of life also have difficulties. For example the conditions needed to sustain life (like oxygen) are also hostile to many pre-life organic compounds. None of the evolutionary models, therefore, envisions an earth similar to the one we now live on. On the other hand, microfossils of the earliest and simplest bacteria and algae would indicate conditions not terribly unlike those of the present.[310]

Booher says that experiments have led scientists to give up any idea that spontaneous generation can take place in the present world. He surveys the literature reporting experiments to produce life from non-life by many scientists, and concludes that today only life begets life and there are no known exceptions.

Booher continues by saying that the experiments to demonstrate actual conditions on early earth when life began lack credibility. Then he adds that the problems of developing conditions that can produce DNA and RNA are much greater. He says there haven't been any experiments conceived on how to simulate this stage. Scientists can only speculate on how this might have happened.

Experiments to produce conditions that could foster life have all been discouraging. Booher emphasizes the following:

> No experiments have ever been conceived to test several of the chemical evolution model's critical assumptions; the necessary conditions for an early earth are not adequately modeled by the experiments that have been run, and neither geology nor chemistry support the product of viable macromolecules in nature.[311]

So we see that Booher's long and painstaking research has led him to the conclusion that no scientific experiments up to now have been able to simulate what circumstances were like when life began on earth. Not only that, but scientists have not conceived of any that might be tried in the future. But Booher quickly adds that the field of biochemical evolution and the study of life's origins is not without hope for future breakthroughs.

A biologist friend recently told me that "chemical evolution" has no scientific basis. It appears that he is right, judging from the research in the literature in this field by Booher and others quoted in this book.

Can Darwinism Explain Biological Complexity?

Booher thinks that the foremost scientific and philosophical question is how can Darwinism account for the complexity found throughout all life forms? He says the appearance of design is apparent throughout life forms. "In human endeavors, it is recognized that orderly processes and complex systems do not appear automatically; they are characteristics of something which has had intelligent planning and specification. And as a general rule, the greater the complexity of the system, the greater the intellectual investment of the designer."[312]

Then he refers to DNA:

> In a book on evolution, it would be a serious omission not to mention that DNA is the mechanism which insures "like begets like." It is the primary molecule to assure almost errorless, endless reproduction. If we press back to the primordial beginnings where life supposedly started to replicate itself, it is the one biological structure that would seem to be necessary in the first living organism if there was to be any assurance of a second.[313]

Booher thinks that theoretical biologists see a need for a radical new way of looking at science. He says they need to look at the whole organism to understand the origins of biological diversity. Just studying the genetic program and DNA is not sufficient to understand how living forms develop. He notes biologists point out that DNA does not replicate itself. It is the cell, which houses DNA, that reproduces.

DNA has its mysteries.

> It is an irony that must be somewhat humbling if we consider man's intelligence to be the ultimate result of the evolutionary process. How could it be that DNA, a random discovery by nature, has had from the beginning a computer characteristic which human design capability . . . can only dream of equaling—someday?[314]

He says that for an amino acid to produce one usable gene by chance would be an extremely rare possibility. Assuming the universe is no more than twenty billion years old, Booher estimates that there is *no chance* it could ever have happened! He doesn't think that a functional protein could have been produced by chance either.

> In summary, we can only surmise from studies of DNA and the cell that the cell came along with DNA for its own reproduction. DNA could not have made it alone. The first cell must have been a complex one, fabricated with enough detailed information to give rise to all the diversity in living nature.[315]

He goes on to state that this first cell made improvements in its basic genetic structure that was resistant to environmental influences.

What does all this mean for the concept of spontaneous evolution by naturalistic means? It seems to preclude the very possibility! All this adds up to increasing the wonder of how the life-forms we observe today could have been developed by any kind of evolution. If somehow a "first cell" did appear by natural means that began the development of life on earth, this would seem to me—judging by what science has discovered thus far—a greater miracle than what the Bible reports. The Bible reports that

God originated living things at more than one time, and that all the large categories of living things had separate beginnings.

Booher reminds us that

> Darwin did not conduct any experiments to prove his arguments and the few conducted since are generally non-supportive. Darwin assumed the variations he saw within species on a small scale could be extended on a grand scale. It would be left for future generations either by experiment or recovery of intermediate fossils to verify his assumption.[316]

Booher relates the failed experiments of two leading scientists. The famous geneticist Richard Goldschmidt, in 1933, reported on the twenty years he spent doing experiments to prove evolution. But he reported only negative results! Then N. Heriburt-Nilsson, director of the Botanical Institute at Lund University in Sweden reported in 1953 that he continued an experiment for more than forty years to demonstrate evolution, but he completely failed and concluded that the concept of evolution rests on pure belief.

Booher reports that Sir L. Harrison Matthews, a distinguished English biologist, wrote in 1971,

> The fact of evolution is the backbone of biology and biology is thus in the peculiar position of being a science founded on an unproved theory—is it then a science or a faith? Belief in the theory of evolution is thus exactly parallel to belief in special creation—both are concepts which believers know to be true but neither, up to the present, has been capable of proof.[317]

Booher points out that evolution can mean different things to different people:

> Since people can believe in the factual basis of what they each see and understand to be evolution the concept comes to mean just about anything anybody wants it to mean. This helps to make the world view of evolution so immensely stable. It is incapable of proof or disproof. Each person therefore can have his or her own personalized belief in evolution. As such it is unique to science.[318]

What Booher is saying here is that people have a *worldview* based on evolution. This individual worldview is a subjective one, and thus is not capable of being proved or disproved. He is not saying here that scientific evolution is incapable of being proved or disproved. He says later that the evolution reported in the mass media and in literature is really a worldview that is not fully supported by the scientific facts.

Darwinism Remains the Most Popular Theory of Evolution

> Although many criticisms can be made against Darwinism, it still remains the most widely accepted scientific theory for evolution because it is capable

PART 2—*What Do We Really Know about the World? The Science*

of falsification. Darwinism hypothesizes links which can be confirmed or rejected based on fossil and biochemical discoveries . . . Consequently, if it were not for the fossil and biological gaps, Darwinism would have few challenges from scientists. Phillip Johnson notes that one of the reasons Darwinism has remained so popular among evolutionary scientists is that without Darwin, "they would have no *theory* of evolution."[319]

He reports that scientists have come up with no other concept to explain the appearing of new genetic characteristics other than mutations.

But can Darwinism really explain biological change? Booher puts his finger on a central weakness of the theory here. Since it has been repeatedly observed in laboratories that most mutations are harmful to the organism, many biologists consider it impossible to explain how mutations can adapt organisms to change and adapt effectively.

He says,

> The neo-Darwinian explanation requires three levels of assumptions to evolve organisms from the simple beneficial mutation stage to the ultimate macro, the entirely-new-creature stage. The first level assumption is *mutations can have some beneficial effects on species*. The second is *natural selection acts on gene pools containing beneficial mutations thereby causing micro-evolution*. The third is that *over time, micro-evolution leads to macro-evolution*.[320]

But beneficial mutations are rare, and for any degree of evolution to occur they must be stored in a gene pool for recovery in the evolutionary process. Yet all the research so far indicates that this occurs to sustain evolution in only very small changes. Those mutations that are not harmful to the organism play more of a role in species survival than in evolution. Booher concludes that these rare beneficial changes probably play an insignificant role even in microevolution.

> Nearly all observable population fluctuations can be explained by a conservative natural selection model based on non-evolutionary assumptions. Random mutations and natural selection have not been shown even minimally plausible as the source and driver for macro-evolution.[321]

So Booher came to the clear conclusion that the three levels of assumptions mentioned above do not actually explain what happens in nature. They demonstrate that neo-Darwinism lacks scientific data to support it. Not only this, the data suggests that even the *possibility* of macroevolution is highly questionable.

Biochemist Michael Behe concluded after years of careful research that Darwin's theory cannot account for the existence of the cell. The cell is irreducibly complex, and could not have developed incrementally over time like Darwin thought. In his 2007 book *The Edge of Evolution*, Behe reported that biological evolution can progress naturally only so far. He says new varieties, new species and possibly even new genera can develop by the means Darwin suggested. But new families and kingdoms of biological organisms cannot. His conclusion was that Darwinian evolution is severely limited, like Michael Denton concluded earlier.

B. Cosmological Beliefs

The big bang theory is by far the most popular theory in cosmology today. Cosmologist Hugh Ross explains that the universe began as the size of a period, was extremely hot, and has been expanding ever since. This is estimated to have happened about 13.73 billion years ago. But big bang theorists have found it necessary to insist that the universe that is seen through modern high-power telescopes is actually only a small part of what is out there. To explain why the universe is expanding so rapidly they now believe that most of it is made up of dark matter and dark energy. These cannot be seen with our telescopes. But now in the twenty-first century they report that they can observe the effects of dark matter and dark energy on the visible part of the universe. Ross reports over thirty characteristics of the universe that reveal it is "fine-tuned" for life on earth.

Booher remarks that when we consider cosmology, "The impression of design is strong." He observed that

> more than any other scientific endeavor, the cosmological thinking of physicists tends to move the boundaries of science into territories previously claimed [as] the exclusive province of religion. No other area of science, not even biology, has the psychological impact of modern astrophysics in describing our cosmic status with God.[322]

Cosmologists, who look at the vast star-filled cosmos with its rotating perfection, usually do so with a sense of awe. Booher remarks, "The God of the cosmologists is not easily identifiable with the conventional God, however. Paul Davies, for instance, states that he does not 'subscribe to a conventional religion' but denies 'the universe is a purposeless accident.'" Booher continues by saying, "A considerable number of modern cosmologists seem to have a strong need for believing there is a purpose for why we are here, but, for them, the old Biblical God who had practical meaning in one's daily life has all but shrunk out of existence."[323]

In this way many scientists who are not religious have been drawn to the idea that a supernatural being must have created the cosmos. This is because they can conceive of no other explanation for the phenomena they observe when they study the universe.

Booher also investigated the level of scientific confirmation in cosmological research.

> Most of the ideas from physics today, used to describe the outer reaches of the cosmos and to explain how it came into existence, have yet to be verified

PART 2—*What Do We Really Know about the World?* The Science

> by adequate experimental observation. Unlike Newtonian and atomic physics with thousands of experimental confirmations, the new physics has made few predictions before the fact, and only a handful of observations are relied on for confirmation.[324]

Booher points out that the atomic particles essential to big bang and particle physics theories have not yet been detected. But the physics that led to the transistor and the atomic bomb relied on extensive experimentation. These practical effects demonstrated to some extent the accuracy of the theory. This is not true in astrophysics. His evaluation is that today's fundamental physics is primarily theoretical. This does seem to be a fair conclusion.

Did the Big Bang Happen?

> If the Big Bang really did happen; if it can be demonstrated with the confidence we can place in assuming the earth revolves around the sun, that galaxies and atoms exist, then the cosmologists are right and we should adjust our other beliefs to conform. But those astrophysicists who still subscribe to a higher value for observation over theory, the ones we seldom hear about in either the popular or scientific literature tell an entirely different story, one which gives adequate reason to reject the Big Bang in any form as an actual historical event.[325]

Booher lists four independent areas that are considered the most conclusive support for the big bang theory. The two strongest observations that support the big bang and an expanding universe are what are called the "red shift" of galaxies and the cosmological background radiation. But he says the theory does not satisfy mathematical equations unless other things are added.

> Big Bang faith requires unobservable matter made up of invisible particles because otherwise the Universe could not have formed in accordance with the Big Bang model. This cosmological tautology has developed because there is not enough matter in the Universe to exert the gravity required for early Big Bang fluctuations to have evolved into galaxy clusters.[326]

This problem forced theorists to assume the existence of dark matter and quarks as a solution. Booher likens this to the biologists search for missing fossils to support their theory of evolution. Ross and other cosmologists assume massive amounts of dark energy today. They theorize that over 95 percent of the universe is made up of invisible things like dark matter and dark energy that cannot be seen by their high-powered telescopes. They do this to understand the natural phenomena they can see. The two natural phenomena they explain by this are the speed of the expansion of the universe and the formation of galaxies.

In 2006 Ross reported that the science of astrophysics has made great strides in the last ten years. We noted in chapter 6 that he even enthused that "today, they have mapped out virtually all of its visible matter and most of its dark matter . . . Today, astronomers can say with confidence they have witnessed all of cosmic history."[327]

Booher would question Ross's confidence because it appears to be based on assumptions that are hard to verify scientifically.

Alternatives to the Big Bang Theory

Cosmologists and astrophysicists have come up with a few theories to take the place of the popular big bang theory. Most are considered extremely speculative, like the Multiverse theory of the existence of multiple universes besides our own.

Booher considers the only theory capable of replacing the big bang theory is the plasma universe theory. (Hugh Ross says that "plasma refers to high energy charged particles distributed in such a way that they form a gaseous medium.") But it appears that Ross and others, by assuming massive amounts of dark energy today, are really doing what the plasma theorists are doing when they assume massive amounts of plasma. The recent information on the universe has not caused them to abandon the big bang theory. It has only led them to refine it and increased their confidence that future discoveries will support the theory.

Are the Cosmological Concepts Testable?

Booher argues that almost all of the concepts in cosmology are not testable. They are built on ideas that are not supported by the old-fashioned standard of empirical evidence.

Speaking about ideas in cosmology like the big bang theory that Hugh Ross supports, Booher says,

> This belief requires extrapolating far beyond what we know to be scientific laws based on observation of nature. The very method, relied upon to give us assurance of how the laws of nature operate now, must be compromised in order to incorporate into our thinking a world view which integrates all the known forces (electromagnetic, weak and strong nuclear, and gravity) in the first millisecond of the Big Bang.[328]

Booher quotes Steven Weinberg, winner in 1979 of the Nobel Prize for physics. In writing about the first three minutes of the universe, Weinberg said, "I cannot deny a feeling of unreality in writing about the first three minutes as if we really know what we are talking about." Booher responds by saying, "It is therefore a testimony to his greatness as a scientist when he (Weinberg) distinguishes what cosmologists think about from what they actually observe in the Universe."

This criticism is certainly understandable. No one can possibly explain how the universe came into being by the laws of nature as we know them today. But whenever and however the universe came into being, it obviously didn't come about from the laws that now govern the universe. Matter and energy cannot be produced from nothing today. It must have come into existence through a force or forces that were outside the universe. No known forces are observed today that could have created the universe and the laws that now govern it.

Booher asks, "What is out there—really?"

Of course astronomers cannot conduct the experiments scientists do with nature today. As one astronomer noted, everything we can know about a star, must be determined from its spectrum and its color, for all we receive from the stars is radiation.

What cosmologists see through their telescopes suggests the universe is actually disintegrating. This observation seems to support the second law of thermodynamics, because it appears the sun and stars are slowly but surely dying out. Everywhere in the cosmos heat is turning to cold, matter is dissolving into radiation, and energy is being dissipated into empty space. Booher says all these observations of what is happening today only suggest the enormous complexity that existed in the past. They do not show the processes that led from the creation of space, time and matter to the complexity observed today.

Then Booher says that visual and semantic illusions are used in popular cosmological science presentations. They give the illusion of realities in the cosmos and how it developed that are not supported by hard scientific evidence. Of course illusions are used to do this also in the fields of biology and paleontology.

Summary on Cosmology

> Astronomy provides only three key observations to hold up the Big Bang. None of these observations, Hubble's red shift relationship with galaxy velocity, the abundance of helium, and the cosmic background radiation, requires a Big Bang interpretation.[329]

Booher acknowledges that the concept of the big bang is widely accepted by Christians as scientific validation for the creation account in Genesis. Astrophysicist Robert Jastrow and dozens of other cosmologists believe it is. But Booher cautions that it would probably be wise for both theologians and scientists to tread cautiously in making a link between cosmology and biblical theology.

Booher does not criticize only the big bang theory. He believes that no cosmological theory is adequate to explain important things.

Booher and Ross are not in complete disagreement. Booher refers to the evidences of fine-tuning in Ross's 1993 book, and seems to recognize their validity. He reports in his book the thirty-five "design parameters" that allowed the appearance of life on earth, and cites as a source *The Creator and the Cosmos* by Ross. But he is far less convinced than Ross is that actual observations of the universe demonstrate that the universe began in space and time as a tiny object and has been expanding ever since. It is hard to imagine, from our observations today, exactly how this tiny object exploded into the vast universe that our telescopes reveal. In this Booher seems on solid ground. But arguing that the natural processes that we observe today cannot explain such a phenomenon does not prove that it did not in fact happen.

C. The New Biology

Booher says that biology appears to be moving away from neo-Darwinism. The "new biology" is still based on evolutionary concepts, but seeks to adjust to the realities of the fossil record as it is understood today. Booher notes that the new biological paradigm accepts the record as telling us that macroevolution actually proceeded in steps or "jerks" and occasionally by "leaps."

The new theories accept catastrophes to explain the appearance of new species in the fossil evidence, and relatively rapid extinctions. Three groups of new biological evolutionary theories are competing for attention. What seems the most popular was started by Niles Eldredge and Stephen Jay Gould. Booher calls this a *"higher" Darwinism*. They see the fossils as revealing long periods of little change followed by sudden appearances of new life forms. They move away from the classical gradualism of Darwin and acknowledge the effect of catastrophes. Gould emphasizes that biological history is unpredictable. He and others think there were catastrophic extinctions that resulted in some species becoming the most fit and others becoming extinct. They say that in the long period of geological history abrupt changes could take only a few thousand years.

Booher says that Gould and Eldredge have made an important contribution to scientific investigation. They have done so by effectively returning paleontology to looking carefully at the fossil record to understand what actually happened. They acknowledge that the "missing" transitional forms that many scientists insist must exist may actually have never existed.

Booher points out two large problems still facing the approach of Gould and Eldredge. The first is the absence of fossil data to support it. The second is that it and all the other new theories still have no real support from genetics to show how macroevolution can take place.

He says the main new rule of the new biology is to recognize that random genetic mutations are not the primary source of evolutionary variation.

Booher quotes Francis Hitching as saying that neo-Darwinism assumes you can deduce the organism from the genes, but nobody has ever done so. Booher says it appears that genes only function in secondary roles to basic patterns that are already established in embryological development.

> One of the greatest mysteries of development is the radically different forms and functions that cells take on in the same organism. Each cell from an individual whether taken from the skin, the heart, the nervous system, the bones or anywhere have identical DNA. "How," Hitching asks, "do cells 'decide' what their future job in the organism is to be, when each has the same set of genetic instructions?"[330]

This mystery is related to "information theory," which Stephen Meyer and others have studied extensively. It appears to be a problem that evolutionists of all schools of thought have not found a solution to.

Besides the "higher biology" we have been describing, Booher explains two other trends in biology today. These are chaos theory and complexity theory. Chaos theorists believe that small errors in a system can lead in time to large and unpredictable consequences. Complexity theorists see all dynamic changes to be a result of spontaneous organization from simpler systems to more complicated ones.

The new biology assumes some goal-directed attributes are written into the code of life and guide evolution.

What can we conclude from these new trends that can shed light on the science of origins? First, many leading scientists today have looked at our state of knowledge in science today and concluded that the Darwinism taught in classrooms and used for philosophical arguments is not supported by scientific observation. In fact, they acknowledge that some data in biology and in geology actually contradict classical Darwinism. Thus they do not ignore the fossil gaps and biological gaps that so many Darwinists do. Second, they remain evolutionists in their worldview and seem to be honestly seeking theories that do fit what they have learned about the natural world. Third, they appear to face the evidence of "purpose" or "teleology" in biological structures. They acknowledge the many evidences of "design" in the universe. But most avoid any theological implications related to this acknowledgement.

The scientific establishment continually promotes Darwinism in all their meetings and publications. Booher says they reward those who do research that seems to support macroevolution. They work to shut out evolutionary scientists who have concluded from their research that Darwinism needs drastic revision in order to account for the results of their scientific experiments. There are few exemptions to this. One was Stephen Jay Gould. As we have seen, Gould acknowledged that the fossil record does not support Darwinism. Yet he was also a member of the National Academy of Sciences, and served as president of the Paleontology Society.

These scientists have come to embrace the reality of catastrophic events in earth's history and their importance in understanding the story of the development of life on earth. But they do so without accepting the interpretations of catastrophes made by young earth scientists like Henry Morris.

The discussion of catastrophes in earth's history is intimately connected with geology. Indeed, much of the discussion on trends in biology is very much related to how to interpret the geological record. So next we turn to Booher's research in this field.

D. The New Geology

To understand modern advances in geology we need to consider four areas. These are catastrophism, the fossil record, historical time, and human origins.

1. Catastrophism

> Catastrophe is clearly a long neglected alternative to the study of geology. It is likely to once again become a more productive model for describing and understanding the geological formations of the past. There seems no better explanation for wholesale extinction of species. Unfortunately it has little to offer at this point in showing how new species evolve.[331]

Darwin assumed the slow, gradual build up of geological strata and the fossils they contain. But later discoveries of actual geological formations have not supported this assumption.

In chapter 8 we read that Davis Young defined methodological uniformitarianism as a belief that the laws of nature are invariant in time and space and the earth processes of the past behave in accord with those laws just as they do now. He says this does not mean that change has always occurred at a uniform rate throughout geologic history. But it does mean that all geological features were formed by processes that still operate today. Catastrophes occur, but follow the same laws of nature then as now. This is consistent with what Booher calls the "new geology."

Booher agrees with Davis Young that various local catastrophes like volcanic explosions can quickly kill and cover over millions of organisms. Then he says, "Professor Young is entirely correct, but local catastrophe examples do not help uniformitarian assumptions for geology in general." Booher adds,

> The problem is, once formations routinely become interpreted as the result of catastrophe, what happens to the geological perception of great ages? . . . The old view would have been that a particularly thick sedimentary rock with fossils encased required maybe millions of years to form . . . The new view, whether regional or global, is that the entire layer with organic life entombed was laid down in a period of perhaps days or even minutes rather than millions of years.[332]

Then he says, "The 75 feet high formation resulting from the Mount St. Helens eruption appears very much like a small model of Grand Canyon . . . Already after a few years, the walls [the laminated layers produced from the mud flows] have become rock."[333]

Booher notes that the result of the eruption of Mount St. Helens is not proof that the Grand Canyon layers formed in a day. But he considers it solid evidence that such a catastrophe can produce the distinctive layering effect visible in the Grand Canyon. His argument here is similar to that of Gary Parker in our chapter 10. Booher says that close up pictures of Mount St. Helens are strikingly similar to those of the Grand Canyon.

Booher points out that the scientific credibility of historical geology was based on the strength of the assumption that change in the past is similar to that of today. But, he says, "Unfortunately, under the new non-uniformity geological paradigm, this would-be strength has now been shown to be based on false assumptions. The inference that what is happening now is no different from what happened in the past is no longer acceptable."[334]

Booher gives several examples of geological features that do not correlate with geological phenomena we can observe today. These include mountain chains uplifting from the sea floor, or earthquakes that can account for the mountain chains that exist today. They include glaciers assumed to account for certain existing rock formations that are far greater than glaciers existing today. Volcanoes are still active today, but none that can account for the great lava fields and other formations assumed caused by volcanic activity. Even the coral reefs and salt domes are not like those being produced today.

Booher's research into recent developments in geology led him to a clear conclusion. This is that scientists have become increasingly aware that the physical evidence cannot be adequately explained without the acknowledgement that catastrophic events played a large part in earth's history.

Evolution by Catastrophe?

Booher says that Velikovsky is credited with reviving catastrophism in the twentieth century. Immanuel Velikovsky (1895–1979) studied some of the dramatic natural events reported in the Old Testament like the thick darkness in all the land of Egypt. He found traditions among American Indians of darkness for five days that seem to describe the same event. He theorized a cosmic event where Jupiter "belched out Venus" about four thousand years ago. Booher says that Velikovsky predicted certain general characteristics about Venus, Mars, the moon and Jupiter that were later found to be true, even though considered heretical to scientists at the time he made them.

Booher quotes Velikovsky as saying, "We are led to the belief that evolution is a process initiated in catastrophes. Numerous catastrophes or bursts of effective radiation must have taken place in the geological past in order to change so radically the living forms on Earth, as the record of fossils embedded in lava and sediment bear witness."[335]

Velikovsky's description of disturbances to the earth's terrain during the last four thousand years indicated they were much greater than the recent explosion of Mount St. Helen. His ideas had a great impact on geological theory, and Carl Sagan spent much time trying in vain to refute him. Booher agrees with the many geologists who say that the earth's crust abounds with evidence of catastrophic upheavals.

Booher is describing a wide issue in current geology theorizing. Geologists are moving away from uniformitarianism because of the need to explain the clear evidences of catastrophic events all over the world. Part of this need is to explain the large extinctions of animal life as evidenced in the fossil beds. Booher cited human reports of catastrophic events and also geological and paleontological evidence that support catastrophes on a worldwide scale. He concluded from his research that these events actually happened.

But Booher sees two "major roadblocks" to replacing the Darwinian model with the catastrophe model. He says the first and the most obvious one is that unlike neo-Darwinism, catastrophism does not provide an observable biological mechanism for studying evolutionary changes. The second roadblock is actually demonstrating in the fossil record that new species really coincide with catastrophic events.

2. The Story of the Fossils

The greatest challenge in geology to Darwin's theory is what is commonly known as the "Cambrian Explosion." Paleontologists today believe that *almost all* of the world's forty phyla—the largest classification of the animal kingdom—appeared suddenly during the Cambrian period.

Darwin anticipated that the "gaps" in the fossil record would eventually be filled. But the Cambrian Explosion has shown that there were really no fossils of complex organisms before that time, and in a very short period of time amazingly complex organisms appeared.

Booher explains the importance of the fossils to evolution in this way:

> Given that the primary proof of evolution through direct experimentation has been unsuccessful, the study of life forms in an evolutionary context has had to progress primarily through an understanding of the fossils ... Fossils, then, are the primary evidence necessary to convince us of the fact of evolution.[336]

Booher's emphasis above on the importance of fossils is remarkably similar to that of Henry Morris in chapter 10. Morris wrote that the modern view of evolution has as its main factual basis the discovery of various fossils, their relative position in the geological strata, and their assumed age. When the order and assumed age of the fossils are questioned on scientific grounds, this becomes a major challenge to Darwin's theory.

Booher argues that for the scientific community to continue to insist on evolution as a concept validated by science the fossils should demonstrate transitions clearly linking everything throughout the geological strata. This is because the only current scientific model there is to show evolution is the neo-Darwinian one.

But Booher concludes as follows about fossil remains in general:

> Surprisingly, a smooth progression of fossil forms, as suggested by the Darwinian model, (which clearly would strengthen the claim that evolution is fact) is nowhere to be found in the fossil data—plant or animal. *The required links are essentially nonexistent*. In fact the start of the fossil record is in itself amazingly abrupt. Labeled the *Cambrian explosion*, over 95 percent of all phyla appear within a range of only 10–40 million years within the Cambrian period. The phyla start separately and remain so with no new major life forms added for the rest of the geological record.[337]

He says Stephen Jay Gould's problems with the evolutionary concept of gradualism are still valid. Gould wrote the following:

> My own favorite target is the belief in slow and steady evolutionary change preached by most paleontologists (and encouraged, admittedly, by Darwin's own preferences). The fossil record does not support it; mass extinction and abrupt origination reign . . . [Paleontologists] have purchased their gradualistic orthodoxy at the exorbitant price of admitting that the fossil record almost never displays the very phenomenon they wish to study.[338]

The paleontologist Gould showed that the gradual change predicted by Darwin is rare. Gould said that change occurs suddenly with the appearance of new, well-differentiated species.

Gould and Eldredge theorize, based on these sudden appearances, a *punctuated equilibrium* theory. *Punctuated equilibrium* is the idea that evolution has developed with long pauses, followed by times when new forms of life develop quickly. This idea of growth followed by no growth is repeated over and over. There are "gaps" when no growth in evolution is seen, and they think some "gaps" are hundreds of millions of years long. This is very different from Darwin's idea that evolution was a continual development of very gradual changes in organisms. This new theory is an attempt to retain the concept of macroevolution while honestly facing the actual realities seen in the fossil beds.

Since no known biological mechanism has ever been discovered that could bring about such sudden changes, researchers in laboratories are said to be very skeptical of Gould's idea.

Booher says that if the "punctuated equilibrium theory" becomes popular, "the primary historical data, fossils, will provide support for evolution in the same manner that they always have for special creation."[339] In other words, this theory of evolution would explain the fossil record no better than does the concept of special creation to scientists.

3. Historical Time

> *Geology looks old.* If one stands before the Grand Canyon on a quiet day it appears that it would take eons of such days to pile up the hundreds of earth layers before gorging out such an enormous trench. The fact that geology looks old does not necessarily mean such a perception is accurate... Geology looks old primarily because of the enormous sizes of formations which would take a considerable period of time to create by the processes we experience personally. The perception of old age in geology may be accurate or it could be an illusion like the movement of the sun. Catastrophic processes... are known which could create similar formations literally overnight.[340]

Obviously scientists need to consider more than how old geological formations look to conclude that they actually are old. They have devised and tested various methods in trying to understand their age. Various "radiometric measures" have been devised to do this. In this regard Booher discusses an important issue, whether radiometric measures are absolute. These measures are considered the main method to accurately judge time in the prehistoric past.

Booher reports that evolutionists like Tim Berra state that over the last thirty to forty years over one hundred thousand radiometric ages have been reported in the scientific literature using radiometric measures. Booher considers this assessment accurate. Berra concludes that these all support the conclusion that the earth and moon are about 4.5 billion years old. Booher responds that despite his research he didn't find any methods which could justifiably claim such accuracy." He explains that most natural historians acknowledge that one of the best techniques for directly measuring the ages of fossils is with Carbon 14, but this can be used only for fossils laid down during the last fifty thousand years.

Then Booher points out that geology professors like William Stansfield of the University of California at Santa Barbara say that there is no absolutely reliable long-term radiological "clock." Stansfield said this because he was aware that age estimates on a given geological stratum by different radiometric methods often vary by hundreds of millions of years.

Booher found that radiometric methods are highly reliable when tested in the laboratory. He says that the problem lies with the assumptions underlying radiometric dating techniques. He discovered that the three most important assumptions that this dating is based on are:

1. "The rate of decay has been uniform throughout the time being measured; i.e., the isotopes half-lives have been constant."

PART 2—*What Do We Really Know about the World?* The Science

2. "The initial conditions when the rock was formed are known."

3. "There has been no contamination of the sample during the decay period."[341]

Booher concludes from his research that only this first assumption appears valid. The other two assumptions cannot realistically be made. Even some prominent geochemists like Melvin Cook have come to a similar conclusion. Henry Morris also made a similar assessment. This sheds serious doubt on the reliability of these measures of historical time.

Booher concludes his report on the reliability of radiometric measures as follows:

> In view of so many problems with the assumptions underlying all the long term age dating methods, there is little reason beyond convention to believe radiometric clocks can date absolutely very far back into the past. On the other hand almost no radioactive short time measure exists. One notable exception is an indicator which strangely, has a half-life on the order of minutes, not thousands of years.[342]

His conclusion about the techniques of radiometric time measurement is that these techniques and the constant speed of light support an old universe and earth, but there are no absolute measures of geological time.

The clearest conclusion as to how reliable the methods are for dating geological structures and fossils seems to be "no one really knows"!

4. Human Origins

The issues of fossils and the development and age of mankind are interwoven. Because of this scientists the world over are seeking fossils that show an evolutionary development from the lower animals to humans. Every few months we hear of a new find of fossil bones that appear to have some characteristics of human skeletons.

Booher states his objective in writing about the history of mankind was not to support the arguments of the creationists or the evolutionists. Rather it was to make an attempt to give a fair assessment of the quality of the scientific evidence from wherever it is found. He argues that even though many do not agree with the religious beliefs of the creationists, their influence has brought about a more careful study of the claims of the Darwinian evolutionary worldview.

Booher discusses what he calls the "Ape-Human Chain":

> The fossil collection . . . does reveal some rather unique features sufficiently different from modern humans and living apes to pique our curiosity. But any truly credible discovery of a transitional link seems destined to be continually a future event. Nearly every past claim for discovery of an ape-human link can be disregarded because it turns out to be (a) indistinguishable from apes,

or (b) indistinguishable from humans, or (c) outright fake. There is no better example of the latter than the famous Piltdown Man.[343]

He says there are many individual fossils which seem to be *Homo erectus*. He believes that, more than any other hominid, *Homo erectus* appears to be either fully man or a true transitional link. Booher makes the important point that "the most plausible true intermediate" between apes and humans—*Homo erectus*—may actually be "fully man."

Was Lucy a hominid? Booher says the actual *hard* evidence that allows Lucy to be classified hominid at all is apparently at most only five bones. Only two suggest a possible human link, and they were found several miles from the other bones.

Booher emphasizes the reluctance of scientists to recognize the weakness of the fossil record to support Darwinian evolution. His book is titled *Origins, Icons, and Illusions*. He believes it is an *illusion* that the fossil record shows connectedness. He says that the need for meaning encourages many prehistory scientists to unconsciously extend every possible fossil fragment found to fit into the evolutionary model. He adds that the scientific community depends very heavily on this *illusion* to reassure itself that science supports Darwinism, and considers it treason when some of its most prominent members start to question the scientific support for it.

He mentions three famous anatomists who applied their skills objectively to try to construct facial models of "hominids" from the fossil remains. They all had vastly different results. And their interpretations were rejected by the scientific community, who were hoping for more humanlike features. Then Booher adds, speaking about paleontology in general,

> However, when we take a critical, non-evolutionary assuming, look at the massive amount of data which has been compiled with the fossils and at the geological models of the strata which contain them, we see no clear evolutionary connections between the species. If Darwinian gradualism ever happened, paleontology is yet to show it. A continuous evolutionary column, family trees, and a chain of being exist solely upon the faith of our modern prophets.[344]

This is a remarkable conclusion from one who attempts to evaluate the issues objectively! The scientific establishment continues to support macroevolution as a continuous process. Booher is as much as saying here that the acclaimed scientific data from geology used to support this is really not there, and the theory is based solely on faith!

5. Evolution and the Test of Science

Booher has concluded that there is a basic contradiction to the theory of evolution as it is believed and taught today. He says that evolution is a theory that so far does not have adequate support in actual scientific observation and experimentation. It also appears to directly contradict the natural law that is considered as having the most support of any natural law known to mankind. This is the process of thermodynamics. He writes as follows:

> For complexity to increase with time requires a special directing force acting from the creation of the cosmos to the appearance of human consciousness. This unique force is theorized to be the process of evolution. But evolution still has not been reconciled with the better documented processes of thermodynamics and information flow.[345]

The contradiction here is really quite simple. For evolution to cause increasing complexity as time passes it must move in a direction *contrary* to what is called "time's arrow." This term in classical thermodynamics means that everything points in a declining direction—toward a universal state of equality or diffusion. High energy systems move toward lower energy, and complexity moves toward a less complex state. But evolution assumes the exact opposite.

Albert Einstein wrote of the law of thermodynamics, "It is the only physical theory of universal content concerning which I am convinced that, within the framework of applicability of its basic concepts, it will never be overthrown."[346] Since the laws of thermodynamics are considered the best verified physical laws known to science, this is a real problem for the general theory of evolution. This observation is not new. Henry Morris made the same point about thirty years earlier.

Booher points out that "time's arrow" represents the irreversibility of some of the energy lost as a result of energy transformations, being driven downward by *entropy*. He states that today we know of no exceptions to this law. General evolution is based on the idea that life evolved from the simple to the complex. The question arises as to how any form of evolution can produce complexity when entropy works so strongly in the opposite direction.

6. Summary on Geology

It now seems clear that Henry Morris and Gary Parker were right to emphasize the vital role that catastrophes have had in earth's geological past. The "new geology" shows that many geologists are now more open to consider not only the importance of catastrophes but also evidence that the earth may not be as old as once thought. However, very few geologists would consider the possibility that the earth is as young as Morris believed.

Booher concluded from his research that there is no complete geological column anywhere on earth. The Grand Canyon is over one mile deep, but he says it represents only 1 percent of the hypothetical geological column. Several hypothetical periods are completely missing. This can lead to questions about how strong the geological evidence really is that supports the belief that the earth is billions of years old.

The new emphasis on catastrophes has led to a consideration of the possibility of "evolution by catastrophe." This is the idea that something other than "survival of the fittest" best fits the fossil record of mass extinction and the subsequent rapid development of new forms of life. A problem here is that this new theory does not seem any better than classical Darwinism to explain progressive biological changes.

Radiometric methods seem to fare no better. Laboratory experiments support the accuracy of some of the measures used to compute geologic time, but laboratory conditions are far different from actual conditions in nature. Assessing natural conditions in the distant past seems to demand a lot of guesswork. The initial conditions of the rock formations cannot be discovered now, and how many changes occurred later in the formations is difficult to assess.

Booher makes a general conclusion that most fossils are found in sedimentary rock, and these rarely contain radiological materials. Although radiometric data are commonly cited as indicating the earth is about 4.5 billion years old, Booher's conclusion is that the evidence does not support this as something we can confidently rely on. He says there are no absolute measures of geological time. There are also indications of a young earth, but he says more evidence seems to support a much older earth.

Scientists and scientific theorists don't really know much about the "when" and "how" these geologic columns and fossil fields developed. Different experts interpret differently how the geologic columns must have been formed. The fossil record is also read differently by geologists of different persuasions.

How old is the earth and life on earth? The best answer seems to be, "No one really knows on the basis of science alone."

E. Is Intelligent Design Science?

Booher notes that before the twentieth century nearly all the major contributions to science came from those who had a belief in a supreme intelligence. He mentions Isaac Newton and Michael Faraday.

He says that lawyer Phillip Johnson presents one of the clearest arguments against a completely naturalistic definition of science. Johnson calls naturalism a definition of science which does not allow any role for intelligent design in formal scientific activities. Johnson shows that such a restrictive definition of science is only a sociopolitical agenda for a favored philosophy. That is, it promotes a philosophical program of scientific naturalism rather than the scientific method of inquiry. Johnson says this limits science, and fails to even begin to understand our actual ignorance of nature.

Booher explains it is important to understand that Johnson's argument does not require some form of creation science. It simply does not exclude this as a possibility. Naturalistic philosophy should not be allowed to draw a line around "science" and thereby use semantics to exclude all other contenders. Booher says science so restricted is not really doing its job. This is because "science should be willing to invalidate the Darwinian hypothesis and admit ignorance in this area of knowledge when that is what the evidence shows." He continues,

> There is nothing in the scientific method which dictates that intelligent design can have no role as an original source or integrating force in scientific theorizing. But somewhere along the line naturalism as a philosophy which drives all scientific theories has become dominant.[347]

Doesn't the pursuit of wisdom and true knowledge begin with an acknowledgement of our ignorance? Then an acknowledgement of what is not yet known can clarify our vision and direction so we can pursue that knowledge—if it is attainable.

Booher makes an important point about the distant past:

> It is fair to claim the Darwinian hypothesis is on a different scientific level from hypotheses which must invoke miracles. The former is falsifiable while miracles are not. On the other hand, just because a statement is testable in the here and now does not mean it can tell us what happened with events that can not be repeated or observed today.[348]

Booher goes on to say that Darwin's hypotheses *have* been falsified, so it is fair to ask why they have not been rejected. They have been demonstrated to be false, but still have not been abandoned. They have been shown false in the sense that the fossil record does not support Darwin's claims, and also in that random mutations do not produce true evolution of species, as Booher makes clear in his book. His argument is that, since Darwin's hypotheses have been falsified, they should not be considered on a higher scientific level than intelligent design just because the latter isn't falsifiable.

He says it is also important to understand something else. Since events that happened in the distant past cannot be observed and tested, a somewhat lower level of theorizing is required. This is because falsification is not possible for events that cannot be repeated. In such cases, models and logical analogies are appropriate, in order to choose between alternative explanations. He says, "Reliance upon analytical method is the situation with all studies of origins where the choice seems to be a logical one between evolutionary and intelligent design propositions."[349] Booher argues that one analogy model can't be called more scientific than another, because "naturalistic speculations" are not any more scientific than "intelligent creator speculations." Therefore it is as proper for science to test assumptions based on intelligent design theory as it is to test hypotheses based on naturalism. In this he agrees with Stephen Meyer, who says scientists are advised to accept tentatively "the best causal explanation of the evidence in question."

Booher explains that the information in one single-celled organism is as great as that in a volume of an encyclopedia. He also quotes a scholar who says that this information follows the same kind of pattern as combinations of letters by humans to convey information, and nothing remotely resembling this "code of life" is found elsewhere in nature. He asks how we can avoid the conclusion that it took intelligence to produce a living organism? He says,

> We do not have to define the specific nature of the designer to infer something higher than human intelligence. Although human intelligence may be the highest intelligence we have experienced, this does not preclude a still higher intelligence being able to produce things intelligible to us.[350]

This reminds me of Behe's point when he said we cannot know the nature of the designer simply by studying his design.

Booher insists that the Darwinian "natural mechanism" is not a plausible source for the complexities seen in biology unless two things can be demonstrated. One is to answer how the small experiments of Darwinians are related to the huge gaps between life and non-life and animal and human consciousness. The other is being able to reproduce the level of intelligence displayed in even the simplest living organisms by random evolutionary means. He is convinced that science is nowhere close to doing either of these.

> Based on what we now know about Darwinian biological mechanisms and the fossil record, a plausible natural mechanism does not yet exist for creating the intelligence seen in DNA, the cell, and living organisms. On the other hand intelligent models created by human intelligence come the closest of anything within our experience to modeling the information possessed by biological entities. It is certainly a worthwhile consideration at this state of our scientific knowledge that an intelligence higher than ours had a hand in their origins.[351]

Booher points out that a major argument used against intelligent design is the imperfections found from time to time. This is also a favorite theme of debaters for evolution and against intelligent design. I have heard evolutionists use this argument in some of the debates I attended.

Then Booher brings up an important scientific issue. Can intelligent design make predictions that are scientifically testable? Can its advocates pose legitimate scientific hypotheses? Can its theory be falsified? He asks,

> Is it any more scientific to assert an *unknown, inconceivable* natural cause started the big bang, if it happened, than to say if it happened, God initiated it? Can either be replicated, tested or observed better than the other simply by asserting one or the other as a personal preference?[352]

Booher says that naturalists are so stymied at this point they need a miracle as much as creationists do. He maintains that this is true of both the beginning of the universe and the creation of human consciousness.

Booher thinks creationists and advocates of intelligent design should be able to come up with testable hypotheses as well as evolutionists can. (As a matter of fact, Hugh Ross, in his 2006 book *Creation as Science*, suggests several testable hypotheses for future investigation.) Booher argues that intelligent design can function as a positive science by proposing specific concepts in both comparative biology and catastrophe geology. He emphasizes that both naturalism and intelligent design need to formulate testable hypotheses and test them in order to come up with hard scientific data to support their theories.

He concludes his argument in this way:

> When the fair-minded apolitical scientist looks at all the contributions to scientific knowledge throughout history that have come from believers in intelligent design, he or she should find it incredulous that it can be defined to have no role in modern science or questions of origins.[353]

Summary and Conclusion

> It is important that the confidence of any discovery claimed to be found using the scientific method be as objective as possible regardless of the scientist's personal philosophy or religion.[354]

Booher surveyed the scientific and philosophical literature on the origin of the universe and life on earth. He did this from his background in the physical and psychological sciences, and as an expert on how complex systems perform. He says he evaluated "the evidence and theoretical processes considered to support evolution as an operational system . . . across the scientific disciplines." He also did so by testing the conclusions of each theorist by the scientific method. He found that the old paradigms and frameworks for investigative science have been discarded by many leading scientists. They have done this because they found the old "traditional" theories were unable to adequately explain the realities of nature that they actually observed.

Booher calls the new paradigms the new physics, the new biology and the new geology. He also explains how the scientific establishment prevents most of the science behind the new paradigms from being published in scientific journals. The popular press, on its part, dutifully parrots the interpretations of the scientific establishment whenever they report new discoveries.

He began his book by investigating why Darwinism is such a dominant mindset in America, and why it seems to be unquestioned by the media and even in scientific journals—even though the scientific basis for it is weak. He found the roots in the famous Scopes "Monkey" Trial in 1925.

He explored the field of cosmology, and found the actual scientific evidence for the popular big bang theory to be astonishingly weak. It is based on the assumption that when the universe began it was tiny, has been expanding ever since, and will continue to expand indefinitely. In order to shore up the big bang theory its proponents theorize that the visible material in the universe is only a relatively small part of the universe.

Booher's general conclusion on the field of cosmology is that all the theories proposed go far beyond observation and what is known about scientific laws that can be tested today. He says much of the thinking about the new physics is based on illusion and mathematical speculation.

In reviewing current research and thinking in biology, Booher concludes that no one knows how life began or the processes that divide life into "separate self-replicating entities." All attempts to simulate in laboratories conditions on earth when life began—and produce the precursors of life—have failed completely. He says that the Darwinian model is the only one so far that is a testable hypothesis for microevolution, but that macroevolution is seen neither in the fossil record nor in the laws of biology known today. Biological change is limited, and organisms adapt and change only within their "basic types." This conclusion is supported by evidences of catastrophe that accompany the sudden appearance of more complex organisms in the fossil record.

Booher does not believe that Darwinism can explain real biological change on the macro level. It cannot explain the complexity of life today. And it certainly cannot explain the origin of life on earth. Yet, in spite of this, it remains the most popular theory of evolution. A better explanation seems to be the new biology, because it accepts catastrophism as the explanation for the sudden appearance of complexity in the fossil record. But this is a catastrophe theory with a geological history much longer than that argued by young earth creationists.

An even greater departure from Darwinism in the new biology is the acknowledgement of *design* and *purpose* in biological organisms and biological change. This rejection of random genetic mutations as the primary source of biological change is a fundamental departure from Darwin's theory. The new biology assumes some goal-directed attributes are written into the code of life and guide evolution. Booher says the source of "intelligence" found in DNA remains a mystery.

The new geology has embraced catastrophism as necessary to explain geological history. But this is "evolution by catastrophe" and not the young earth creationism of Henry Morris and Gary Parker. Immanuel Velikovsky says he found many events in the cosmos to explain the dramatic catastrophes recorded in the histories of various ancient people. Booher notes that the effects of catastrophe have been observed in our day in the geological changes in Mount St. Helens. The Grand Canyon also exhibits features similar to what happened to Mount St. Helens.

But Booher sees two problems with replacing Darwinism with this theory. One is that catastrophism provides no biological mechanism for studying evolutionary change. The other is the inability to show in the fossil record how species actually change by catastrophe.

Another issue related to geology is the scientific measurement of historical time. Radiometric dating techniques can be tested and found reliable in controlled scientific experiments. The problem is that no one knows what else besides time might have affected the fossils. Experts say it is unreasonable to assume what the initial conditions were or the absence of later contamination. This means that we cannot be sure that any dating method is reliable.

The findings related to fossils and the human race are especially important in origins research. Booher argues that if evolution can be proved historically the evidence must come from the fossil record. Booher surveyed the scientific literature on "the ape-human chain" and found it to be unconvincing. He says he did so while attempting to "present a fair assessment of the quality of the 'scientific' evidence from whatever quarter it comes."

He explains why even famous paleontologist Stephen Jay Gould rejects the slow and steady evolutionary change theorized by most paleontologists. Gould's "punctuated equilibrium theory" tries to take into account the mass extinctions and abrupt origination of species that the fossil record abundantly illustrates. But it lacks widespread scientific support, and does not explain how evolutionary changes can occur.

Booher concludes that the small changes seen in organisms when new species develop naturally seem incapable of contributing to large-scale evolution. Biochemist Michael Behe, molecular biologist Michael Denton, creationist Gary Parker, and many other scientists have come to a similar conclusion.

Of all the so-called links between the higher primates and man Booher considers *Homo erectus* to be the most plausible true intermediate. But he thinks that *Homo erectus* may eventually be found to be the same as modern man. Booher points out that paleontologists disagree on how to interpret the findings regarding *Lucy*, and there is no clear proof that she is in the "ape-human chain." As a psychologist, Booher believes the scientific community depends quite heavily on illusion to reassure itself that in case after case the facts do support Darwinism. They usually reject evidence that contradicts their beliefs.

He concludes that if we look critically and without assuming evolution at the massive amount of data on the fossils and the geological models of the strata which contain them, we find no clear evolutionary connections between the species. Paleontology is yet to show that Darwinian gradualism ever happened.

In "Entropy and Time" Booher returns to the laws of thermodynamics in some detail. Evolution is a theory that so far does not have adequate support in actual scientific observation or experimentation. It appears to directly contradict thermodynamics. But thermodynamics is considered to have the most supporting proof of any natural law known to mankind. He asks how evolution of complexity can happen in the face of the counter tendency of entropy? He says this basic problem has never been solved by any mechanism of evolution.

Booher analyzed the many attempts by science to explain how macroevolution is possible in spite of its obvious conceptual conflict with thermodynamics. He summarizes this by saying that time's arrow is apparently not routinely reversed with biological systems any more than they are with natural physical systems. He says living systems have processes that can temporarily and locally slowing down the effects of the second law of thermodynamics, but it seems this is all they can do.

It is true that special creation seems to violate the first law, which says matter and energy cannot be created or destroyed. Evolution appears to violate the second law with every macro change. This is because this law dictates that complexity disintegrates into simplicity, while evolution assumes the gradual increase of complexity. Creationists have no trouble with the first law, because the Bible clearly says that the universe and all in it were created by a transcendent God. But evolutionists apparently have no answer to the problem of the second law.

This is a major problem for all theories of macroevolution, and evolutionists can only get around it by seeking to discover a yet-unknown counter force or rationalize by claiming the second law does not apply to biological organisms.

An alternative is offered by creationists and advocates of intelligent design. They believe that macroevolution by chance does not happen. Some do accept evolution, but believe a higher intelligence has provided periodic bursts of creativity in the past that overrode the second law.

The final result of Booher's many years of exhaustive research into the theories and research regarding origins is his conclusion that no theory has adequate scientific support. Scientists tend to interpret all of science on the basis of their unproved and mostly unprovable theories. The scientific contributions of creationists and intelligent design theorists are systematically shut out of the professional journals and underreported in the mass media. He thinks that of all the theories in vogue today, intelligent design theory has the most scientific support. But he says science has a long way to go to come up with a theory of origins that is consistent with all that we now know about nature.

Booher returns repeatedly to the psychological aspect. A major theme in Booher's book is that the general theory of evolution (macroevolution) is based more on illusion than on solid scientific proof.

Booher points out two things to explain why inconsistencies between scientific observation and the dominant evolutionary view are not acknowledged by the dominant scientific community. One is that they oversimplify their discoveries in order to strengthen their arguments. The other is that they systematically block minority viewpoints from being published in the professional journals and even in the popular press. They thus present the appearance of consensus and solid scientific support that does not in fact exist.

Booher insists that *all good science should be considered*. He insists that the scientific discoveries and interpretations of creationists and others who are ignored by the dominant scientific community should be considered—if these ignored scientists are properly trained and honest, and their data collection methods meet the highest scientific standards.

PART 2—*What Do We Really Know about the World?* The Science

I don't think there is a better way to conclude this chapter than to repeat the words of Booher at the beginning of this chapter.

> Where did we come from and where are we going? Nearly every educated individual in the Western world believes Darwin and evolution science have shed light on these questions. But science has yet to show how, or even if, evolution on a grand scale has happened. Experiments, fossils, and molecular biology do not reveal the slow, gradual process Darwin postulated for connecting all life forms. Instead the evidence generally indicates species are separated by fixed boundaries. Further the new biology and geology paradigms credit *catastrophe*, not millions of years of mutations slowly accumulating through natural selection, as history's agent of evolution. Why then does science still rely so heavily on Darwin and his modern counterparts in describing a process of evolution?[355]

14

Summary of the Science
When All Is Said and Done What Do We *Really* Know about the Natural World?

> Our physics seems inadequate to explain the early times in a way that is consistent with the conditions existing today. This is a crucial requirement of science—no gaps should exist in the cause and effect chain linking two moments in a physical history. If our physics fails, understanding on the fundamental level weakens; *we have a crisis in science.*[356]
>
> —Steven Hawking

Introduction

I spent over eight years studying the science that illustrates and defends Darwin's theory of evolution and modern Darwinism. I also studied the science and arguments that challenge the adequacy of his theory. Then I heard of Booher's book *Origins, Icons, and Illusions*. Booher seems to have the unique training and experience to look at all the scientific information and analyze how much the facts of nature as we know them today support the theories that are presented. As a psychologist, Booher was aware of the various *illusions* that are used today to promote a false sense of confidence in the scientific basis for Darwin's theory.

I was struck with how closely Booher's findings and conclusions mirror my own. He was able to study scientific research in more detail than I did, and his education and experience in both the physical and psychological sciences is much greater than mine. He said he also had experts in the various fields of science review his book for accuracy.

Put simply, what do we really *know* about the kind of large-scale evolution that Darwin proposed 150 years ago? What have the many thousands of dedicated scientists discovered that shed clear light on the possibility—and the actuality—of Darwinism?

The mass media and the scientific establishment almost always present "evolution" as an established scientific fact. But I have found that the reality is quite different. There are eminent scientists who claim to be evolutionists who have concluded that the form of evolution presented by the popular press did not happen. One of these was Stephen Jay Gould, former president of the Paleontology Society. Many leading evolutionary scientists frankly recognize that some of the realities of nature falsify (contradict or disprove) Darwin's theory of broad evolutionary development and are seeking new modifications to his theory. Some reject Darwin's theory altogether and are seeking viable alternatives. But these scientists are seldom mentioned in the popular media.

Any honest and open report on what we know about nature must frankly admit that there are important things that are not known to us today. Often the most fair and impartial conclusion about some issues in science is that nobody knows. Michael Behe discovered this in his original research on the origin of the cell and in his careful survey of the scientific literature. Booher also discovered this in his studies of the various fields of science.

Humility and a sense that we still have much to learn about nature should characterize the true scholar and seeker after truth. This reminds me of Donald Rumsfeld's famous words in 2003 when he was the US Secretary of Defense. "There are known knowns. These are things we know that we know. There are known unknowns. That is to say, there are things that we know that we don't know. But there are also unknown unknowns. These are the things we don't know we don't know."[357]

I discovered that Booher arrived at the same general conclusions as I have through my own study of cosmology, biology, geology and paleontology, and also of human origins and the nature of man.

This chapter is primarily a report of what I have discovered and concluded when researching the science related to evolution with as open and unbiased a mind as I can have as a human. It is based primarily on the information presented in chapters 4–13. We all have our own biases, rooted in our own personal view of the world, and our own psychological and social needs. But my purpose has been to "follow the truth, wherever it leads." To do this I need to be aware of my biases and try to not let them limit my understanding and openness to truth.

As a fruit of my years of research I believe a few clear and definite conclusions can be made about origins research and about evolution in general. This chapter is written to explain these.

What Can We Learn From Science?

1. From Cosmology

The universe has always existed, or it had a beginning. Cosmologists have discovered that matter, energy, space and time appeared "suddenly." They call this the "big bang." They confidently assert that this happened about 13.73 billion years ago. Even secular cosmologists like Sean M. Carroll, who advocates the theory of multiple universes ("multiverse") in place of the big bang, are confident that our universe is about fourteen billion years old. But this confidence is based on how they interpret what they see through their telescopes and have learned of the laws of physics.

The first law of thermodynamics states that matter and energy can neither be created nor destroyed. This is a problem for big bang theorists and all scientists who recognize the fact that the universe had a beginning.

If the universe had a beginning, it had a "cause." The cause had to exist before the universe was created. It had to "transcend" the universe—exist outside of it. It had to be immensely powerful. This points to a "cause" or designer that is transcendent. He exists separate from the universe and from before the universe existed—because it is his creation. He exists outside of our space-time dimensions.

2. From Biology

Cosmologists and biologists assume that simple, primitive life developed spontaneously on earth long ago. Darwin imagined "some warm little pond" where this happened. But scientists have never been able to explain how life could have developed on earth by natural means. And the scientists who have tried the hardest and longest to simulate the atmosphere of primitive earth in the laboratory and develop simple life admit that they have failed miserably. Stanley Miller's experiment was one of the "icons of evolution" that convinced Strobel of the fact of Darwinian evolution. But scientists today are in agreement that Miller's experiment did not in fact simulate the early atmosphere of earth like he assumed.

Booher's long and painstaking research also led him to the conclusion that no scientific experiments up to now have been able to simulate what circumstances were like when life began on earth. Not only that, he says scientists have not conceived of any that might be tried in the future.

Darwin based his theory on several assumptions. He knew that random mutations do change plants and animals in small ways. He assumed that natural selection would result in beneficial mutations gradually developing more complex organisms. But Darwin also declared,

> If it could be demonstrated that any complex organ existed which could not possibly have been formed by numerous, successive, slight modifications, my theory would absolutely break down. But I can find no such case.[358]

What can we say for sure about the testimony on the complexity of biological organisms? Biochemist Michael Behe pointed out that for organisms to evolve from simple to complex forms—no matter how much time it would take—a biological *pathway* is needed. He concluded that the cell is so irreducibly complex that no conceivable pathway exists in nature. This means that evolution from simple organisms to very complex ones could not have developed incrementally. Random mutations are real, but they do not evolve into more complex organisms that are capable of thriving and reproducing. Natural selection also does occur—frequently. But this again does not produce offspring that increase in biological complexity from generation to generation.

Behe says this point is crucial and points out that: "If there is not a smooth, gradually rising, easily found evolutionary pathway leading to a more complex biological system within a reasonable time, Darwinism won't work."[359]

Behe says that the result of his research since 1996, as reported in *The Edge of Evolution*, only reinforce his conclusions in *Darwin's Black Box*. Speaking of the latter book, he argues that irreducibly complex structures could not have evolved by random mutation and natural selection. Numerous scientists like Richard Dawkins and Kenneth Miller have written about how this does happen, but it seems what they have written lacks scientific validation.

I think we can conclude, based on what biologists have discovered about the laws of nature, that evolution from simple forms of life to the complexity of life on earth today by Darwinian means could not have happened. No pathway exists to allow for this. This means that, in Darwin's words, his theory does "absolutely break down." Behe argues convincingly that the only explanation for the existence of the cell is that it had an intelligent designer.

Beneficial mutations are rare, and for any degree of evolution to occur they must be stored in a gene pool for recovery in the evolutionary process. Yet all the research so far indicates that this occurs to sustain evolution in only very small changes. It indicates that mutations that are not harmful to the organism play more of a role in species survival than in evolution. Booher concludes that these rare beneficial changes probably play an insignificant role even in microevolution. We referred to Booher earlier as saying that "random mutations and natural selection have not been shown even minimally plausible as the source and driver for macro-evolution." Genetics may be able to explain microevolution, but cannot explain how to turn a reptile into a mammal, or a fish into an amphibian. In chapter 10 we quoted geneticist Richard Lewontin, a geneticist, as saying that the essential operation of natural selection is to enable organisms to maintain their state of adaptation rather than to improve it.

I believe we can safely conclude that an intelligent designer created the laws of nature so that *limited* evolution can and does occur. Behe and many other believers in intelligent design theory are also convinced that the designer continues to work in nature.

3. From the Fossils

Darwin was well aware of the rich fossil remains from the Cambrian period. He wrote,

> The several difficulties here discussed, namely—that, though we find in our geological formations many links between the species which now exist and which formerly existed, we do not find infinitely numerous fine transitional forms closely joining them all together;—the sudden manner in which several groups of species first appear in our European formations;—the almost entire absence, as at present known, of formations rich in fossils beneath the Cambrian strata,—are all undoubtedly of the most serious nature.[360]

But Darwin was confident that eventually fossil beds would be found that reveal the development of organisms through time from the original simple organisms to the complex organs that exist everywhere in the world today.

What can we say for sure about the testimony of the fossil record? Geologists in general believe that they have already discovered all or almost all of the large fossil beds on earth. Harold Booher says he surveyed the literature of historic geology quite extensively. He informs us that even world-famous paleontologists like Stephen Jay Gould concluded that no fossil beds will ever be discovered that show the slow incremental development of evolution that is needed to support Darwin's theory.

Discoveries since Booher and Gould published their findings have only confirmed their conclusions. Recent discoveries in fossil deposits near Chengjiang, China, are considered the most complete fossil record found anywhere in the world. These deposits reveal the sudden origin of major body plans. They show sudden diversity and the absence of plausible transitional organisms. This discovery and how it is a powerful refutation of the neo-Darwinian assumptions is discussed in a book by Stephen Meyer and others published in 2003. The authors evaluate the discoveries in China as follows.

> Neither the pace nor the mode of evolutionary change match neo-Darwinian expectations. Indeed, the neo-Darwinian mechanism cannot explain the geologically sudden origin of the major body plans to which the term "the Cambrian explosion" principally refers. Further, the absence of plausible transitional organisms, the pattern of disparity preceding diversity, and the pattern of phyla first appearance all run counter to the neo-Darwinian predictions or expectations. Only the overall increase in complexity from the Precambrian to the Cambrian conforms to neo-Darwinian expectations.[361]

PART 2—*What Do We Really Know about the World?* The Science

I see no hope that fossil deposits will ever be found that show the gradual, incremental evolution from simple to complex life that Darwin expected—especially to the extent it would demonstrate the Darwinian evolution happened. But what if what was totally unanticipated by such well-known paleontologists like Gould actually happened—that somewhere in the world a fossil bed was found that showed a long progression from one biological class to another? Would that change the situation? Yes and no. It would lend support to the argument that Darwinian evolution and theistic evolution could have happened. But it would not answer the problem that the fossil record around the world overwhelmingly demonstrates the sudden and unexplainable appearance of complex organisms.

There are two issues to face regarding the fossil record. The first is whether modern science has supported Darwin's expectation that it would eventually demonstrate the historical fact—or even the feasibility—of gradual large-scale evolution. The second is the reliability of the methods used to estimate the age of the fossils.

What has the advancement of science in the 150 years since the publication of *Origin of Species* discovered regarding the fossil record? Simply put, the fossil record has failed to support Darwin's theory. It has not demonstrated that large-scale evolution from very simple organisms to the complexity that exists today has ever occurred anywhere in the world.

In the meantime the popular press regularly reports new discoveries of transitional creatures that are said to prove macroevolution. In chapter 10, I reported the recent discovery of a fossil of a large "fishapod" that was estimated to have lived some 375 million years ago and was a "missing link" between sea and land animals. It was called a "walking fish," but no evidence was given that it walked or even could walk—only speculation based on bone fragments that it may have had a fish's scales and gills, a suggestion of lungs, and what looked like a primitive hand with "five fingerlike bones." This "transitional creature" was cited as one example of many proofs that the fossil record does support Darwin's theory. However, no evidence was provided that it was anything more than a malformed fish that left no descendants.

I think it is also worth remembering that many scientists, both evolutionists and creationists, argue that current dating methods in geology suffer from serious problems. Some of these methods are found to be highly reliable under controlled laboratory conditions—but only one of the three assumptions underlying radiometric dating techniques may be valid. This assumption is that the rate of decay has been uniform throughout the time being measured. The other two essentials are that the initial conditions when the rock was formed are known, and there has been no contamination of the sample during the decay period. Several of the scientists I have quoted in this book have concluded these two essentials cannot be assumed when dealing with the distant past. This sheds serious doubt on the reliability of these measures of historical time.

It seems clear now that the findings of modern biology and geology are consistent in suggesting that Darwinian style macroevolution has failed the test of scientific investigation.

4. From the Nature of Mankind

Darwin said the following in *Descent of Man*: "We thus learn that man is descended from a hairy, tailed quadruped, probably arboreal in its habits, and an inhabitant of the Old World."[362] But scientific research in the 150 years since does not support this assertion. Not only is the fossil record fragmented, the similarities between higher primates and humans in their DNA have other possible explanations than common descent. The basic question has never been answered: How can human intelligence possibly have developed through time by unintelligent processes?

Darwin himself stated,

> A moral being is one who is capable of reflecting on his past actions and their motives—of approving of some and disapproving of others: and the fact that man is the one being who certainly deserves this designation, is the greatest of all distinctions between him and the lower animals.[363]

No one has been able to explain the existence of human personality and human intelligence without assuming an "intelligent designer."

Mankind is clearly "intelligently designed." But there are flaws in his body and in his nature. The eye is not designed as effectively as it could be. Sicknesses abound. Men kill men and do many cruel things. Social problems abound everywhere. The common response to this is that a designer is under no obligation to make his design as perfect as he is able to. Behe and others point out that designers who are clearly capable of designing things better often choose not to. But I am convinced that a better answer is the Bible explanation that mankind has sinned against God and as a result degenerated from his original physical and moral condition.

Darwin concluded *Descent* with the following argument:

> We must, however, acknowledge, as it seems to me, that man with all his noble qualities, with sympathy which feels for the most debased, with benevolence which extends not only to other men but to the humblest living creature, with his god-like intellect which has penetrated into the movements and constitution of the solar system—with all these exalted powers—Man still bears in his bodily frame the indelible stamp of his lowly origin.[364]

Yes, man shares many characteristics with the higher animals, even to sharing much DNA. But he also shows in his spiritual nature "the indelible stamp" of his Creator. Anthropology professor William Kornfield wrote that "only man has culture, which for a number of anthropologists constitutes a difference in kind rather than degree from the animal world. It would seem that God made Adam separate from the primate world with all his physical, mental, moral, and spiritual characteristics present at the same time."[365]

5. On the Other Side of the Argument is Homology

This refers to the similarities of some bone structures in very diverse organisms, and it is commonly argued that this is proof of macroevolution and common descent. Scientists sometimes argue that this suggests similarity in design only, and is not a proof of common descent. Jonathan Wells explained to Strobel that homology in vertebrate limbs are said to point toward a common archetype or design, and not toward descent with modification. He pointed out that the similarities could theoretically be either common descent or common design, but no mechanism has ever been discovered to show how it can be a result of common ancestry. "A designer might very well decide to use common building materials to create different organisms, just as builders use the same materials . . . to build different bridges that end up looking very dissimilar from one another."[366] (It has even been said that 99 percent of all human genes are similar—but not identical—to those of a mouse!)

But why do biologists who acknowledge the biological evidence against macroevolution and acknowledge the clear evidences of "intelligent design," still cling to the belief in common descent? Some seem to because of "homology." It appears to be an indisputable fact that some bone structures in the whale, in birds, and in mammals have very similar bone structures.

It is common for engineers and architects to construct things that have different purposes but are similar in many ways. Could God have done the same in designing and creating separately whales and birds and mammals? Certainly He could have! Would He have done so? What reason would He have? We can speculate on this, but the reality is that we don't know why He would or would not do so. "Oh, the depth of the riches both of the wisdom and knowledge of God! How unsearchable are His judgments and His ways past finding out! For who has known the mind of the Lord?" (Rom 11:33–34a).

What Do We Really Know about the Realities of Nature Today?

First, we can be sure that life did not begin on earth by any natural process. This is so obvious even to many evolutionary scientists that they are now hoping to discover life on a different planet or moon. Since this earth is so uniquely suited to complex life, it is doubtful this will happen. But cosmologist Hugh Ross believes that simple life probably is on the moon or a nearby asteroid or planet because it was blasted there when a meteor or something larger hit the earth in the distant past. This does not seem farfetched when we read that meteors have landed on earth that scientists are convinced came from Mars.

It may well be true that simple life does exist outside our planet. Microbes have been found recently in places on earth where scientists have long assumed no life could exist. But this is far different from finding complex organisms in such places. Even if simple life is found elsewhere in the universe there would be no way to know if it developed there by purely natural processes.

Second, that the slow, incremental development of simple organisms to complex ones by random mutation and natural selection is greatly limited. Evolution on a limited scale seems to have occurred. There is abundant hard scientific evidence that it has. But there are natural limits, beyond which evolutionary change by random mutation and natural selection cannot occur.

Biochemist Michael Behe has done extensive research in this field, and has concluded that there is limited evolution, but it is only possible up to the development of new species or slightly beyond. There is no way that dogs and cats, for example, could have evolved from the same ancestors by natural processes. While acknowledging the limitations of science to make absolute statements, I believe we can confidently conclude that all living organisms today could not have developed naturally from one or only a few ancient simple organisms as Darwin assumed.

In 2009, in *Signature in the Cell*, Stephen Meyer made a strong scientific case that reaffirmed Behe's conclusion that the cell could not have come into being by any natural process.

Third, we can conclude from the fossil record also that the large-scale evolution from very simple to complex organisms did not happen. Darwin himself said that if the fossils didn't eventually show this progression it would undermine his theory. But large-scale evolution is not seen in the fossil record, and it seems clear from the careful research of Behe and other biologists that it cannot happen.

Fourth, we can conclude that human and cellular intelligence could not and did not evolve by natural biological means. Behe, Meyer and Johnson and many other scientists and scholars have shown that the intelligence in DNA is so wonderful and complex that it could not have developed by any non-intelligent process. The existence of human intelligence is also a marvel that cannot be explained by naturalistic scientific processes.

Fifth, astrophysicists tell us that the universe is slowly disintegrating—deteriorating from order and complexity to disorder and diffusion. This is expected from the second law of thermodynamics, but is contrary to the evolutionary theory of complexity developing from simplicity.

It is clear that there is a huge *gap* between what the mass media often present as "the facts of evolution" and what the scientists doing the research have discovered. There are scientific theories which contradict the words of the Bible. I have so far discovered no *confirmed scientific facts* that are incompatible with what the Bible clearly says about the natural world.

PART 2—*What Do We Really Know about the World?* The Science

What can we learn from a careful study of the verified facts of science and of the theoretical frameworks—the theories—that scientists construct to explain and interpret them? First, that *we humans tend to become prisoners of our own theories!* I think it is a sign of greatness when a scientist or a philosopher of science can stand outside his "world of assumptions" and honestly face the "realities of nature." This is true in all fields of endeavor.

The spokesmen and women of the mass media are clearly not trained to do this. In graduate school I was taught to look at a research report and try to discern the interpretations from the facts presented. I believe this is a helpful thing to learn. As Rumsfeld wisely said, there "are the things we don't know we don't know."

It is usually not the things we know, nor the things we realize we don't know, that lead us to wrong conclusions and actions. It is the things "we don't know we don't know."

We really don't know with any scientific assurance how old the universe is. We don't really know the age of the earth either, nor how long there has been life on it. We cannot learn from science how life first came into existence. But I don't think this knowledge or lack of it need affect how we live our lives in any way.

In spite of the presentations of the mass media and some popular scientists that the issue is settled and we are the product of natural processes, the hard science is not there to support these assertions. The *illusion* is presented of connectedness of primitive organisms to present day complex organisms, and the fact of large-scale evolution—but the science is simply not there to support it.

I believe the main conclusion to our study of what is actually known in science is this: *the progress of modern scientific discovery has failed to demonstrate that Darwinian evolution, macroevolution, did happen or even could have happened.*

All the findings of science today are *compatible* with creation. But I think it is going too far to say all the findings *prove* creationism. Stephen Meyer explains that historical scientists compare various hypotheses to see which would, if true, best explain the biological and geological evidence. Then they provisionally accept that hypothesis.

It seems clear that intelligent design is a better theory than Darwinism, because it allows for intervention from intelligence that is outside—transcends—our space-time continuum at all points where evolution fails to explain natural events. It seems true that intelligent design cannot be falsified in the sense that Behe, Meyer, Booher and other scientists say that Darwinism has. Darwinism has been falsified in that Darwinism does not explain how the intelligence in the cell and other indications of "design" came into being. But intelligent design seems clearly to meet the criteria needed to be considered a true "scientific theory."

In part 3 we will turn to what the Bible reveals about nature. We will consider whether theism provides a better explanation for the realities of nature. We will also try to discern what is clearly revealed in the Bible about the realities of nature from what is inferred by various scholars.

PART 3

What Do We Really Know about the World?

The Bible

Preface to Part 3

> Science is an attempt to interpret the facts of nature. Christian theology is an attempt to interpret the words of the Bible. Since, according to that theology, God created the universe and is also responsible for the words of the Bible, and since He does not lie or deceive, there can be no contradiction between the words of the Bible and the facts of nature. Any conflict between science and Christian theology must be attributed to human misunderstanding.[367]
>
> —Hugh Ross

Naturalism and most forms of evolution share a worldview centered on nature as all there is to existence. Beginning with Genesis 1:1, the Bible presents a worldview centered in God as Creator: "In the beginning God (*Eloheem*—the supreme God) created the Heavens and the Earth." The universe and all in it are depicted as *derived* from the will and activity of God.

I see no reason why God could not have created the universe by a "big bang." It is true that this does not fit some Christians' theology. But which came first, the modern study of systematic theology or the revealed "truth" from God? There is an argument that, since God is omnipotent, He would show His mighty power by creating an "already perfect universe" at the beginning. He would not create it incomplete and then change it from a temporary form into the magnificently ordered cosmos we observe today. But this argument has no special persuasive power to me. Certainly God *could* have! But this argument does not lead logically to the conclusion that He would choose to do so.

I will illustrate the central issue another way. We can argue all we want that a holy God would not permit man, whom He created after His own image, to fall into sin. *But He did*. We know this because He told us so. It is God's revealed truth that should determine our theology.

Preface to Part 3

"No, God! I don't want Adam and Eve to sin! That would mess up the whole human race!" Wishful thinking, isn't it? God tells us they did sin, and this knowledge helps us understand much truth about ourselves, human nature and the way things really are today. There is much about the universe we don't need to know in order to know God and His will for us today. The Bible is clear on what we need to know; to believe; and to obey; in order to reap His eternal rewards.

There is no reason to reject the big bang theory because it does not match our views about God. It is entirely possible that God did it this way. It is also entirely possible that God did it in a different way, and that cosmology will someday repudiate or redefine the big bang theory when new discoveries cause cosmologists to revise present-day scientific knowledge.

Has God spoken clearly? *Yes!* He has spoken clearly about the things we really need to know. These include the following: God created all things. He created us. He holds us accountable for our actions. In spite of our failures, if we respond to Him in faith He will provide a way for our final success. These are important truths in the book of Genesis. They are clearly more important to the main messages of the first three chapters than *how* and *when* He created the heavens and the earth. We must remember the oft-told truth that the Bible is first and foremost a revelation to mankind of God and His way of salvation, and was not intended to be a textbook on science.

> The account of creation is unique in ancient literature. It undoubtedly reflects an advanced monotheistic point of view, with a sequence of creative phases so rational that modern science cannot improve on it, given the same language and the same range of ideas in which to state its conclusions. In fact, modern scientific cosmogonies show a disconcerting tendency to be short lived and it may be seriously doubted whether science has yet caught up with the Biblical story.[368]

The above words by W. F. Albright (1891–1971), whom Bernard Ramm called "one of the world's greatest living archaeologists," remind us of a great wonder—the wonder of the fact that the Genesis account of creation was written at a time when there were in various cultural groups many mythical accounts of our universe and its beginning. None of those ancient accounts stand up to rational scrutiny—let alone scientific investigation. But the Bible does. Neither the concepts of modern science nor the vocabulary needed to express them were yet in existence, yet the words of Scripture ring true today!

Albright's words also remind us that science has more then once been forced to reject prevailing paradigms for different ones as new facts about the universe and living things became known. Scientific paradigms come and go, but the written Word of God remains unchanged.

We have now come to the second half of the book; half-way in our quest.

Preface to Part 3

In part 3 we will consider the ideas of several prominent men-of-science who believe that the creation account in Genesis was revealed to Moses by God some 3,400 years ago. These authors can be divided into two main groups. One is those who accept almost all current thinking in their scientific disciplines and seek to reconcile the words of Scripture with the known "scientific facts of the day," like Hugh Ross and Davis Young. The others are those like Henry Morris who begin with their interpretation of Scripture and seek to reconcile the discoveries of science with it.

But we must continually remind ourselves that both our understanding of the facts of science and our understanding of certain verses of Scripture may need correction or refinement as new information becomes available. We know enough about our physical universe and our world to be confident of the existence of certain natural laws, like the second law of thermodynamics, and how natural laws worked in the past to bring about the realities of nature that we observe today. But there are still "mysteries of nature" that scientists do not yet understand, like what the basic building blocks of all matter are, and the exact nature of "dark matter" and "dark energy."

There needs to be room in our imperfect understanding of present physical realities for new interpretations to develop, and philosophers of science remind us that this will always be the case. They argue that science will never develop beyond the point where knowledge is considered tentative and new paradigm shifts can no longer be anticipated. We do *know* some things, but even our understanding of what is well known is subject to further scrutiny and refinement.

A similar thing can be said about our knowledge of the universe and physical realities gained by a study of Scripture. Scripture follows the laws of language, and can be properly interpreted with confidence when we understand and follow these laws. But certain questions remain that directly relate to our understanding of the Bible account of creation. Among these are how narrowly or broadly the word "day" (*yom*) should be interpreted in Genesis 1.

We do *know* some things clearly, things that directly relate to the main theme of the Bible, like how God provided for and applies eternal salvation. But some words and meanings of Scripture, as they relate to the timing and conditions of our earth's creation, are subject to more than one honest interpretation.

The Order of Development of Part 3

The conflict between a belief in creation by the God of the Bible and belief in a Darwinian explanation of origins is a conflict of faith systems. As such it is clearly an issue of two competing *philosophies of science*. This is one reason I have chosen to begin part 3 with a discussion of *The Christian View of Science and Scripture*, by Bernard Ramm.

It would be difficult to fault Ramm for a shallow understanding of science. Although not a scientist by profession, he obtained a doctorate in the philosophy of science at the University of Southern California, read widely on the subject, and taught

Preface to Part 3

university courses in this area. (Like myself, he readily admitted that he had to rely on scientists for the technical details of the several sciences he refers to in his book.)

I am also convinced that Ramm cannot be faulted for his lack of knowledge of the Bible and how to interpret it. This was one of his specialties, and I remember using his book *Protestant Biblical Interpretation* when studying this subject at Columbia International University over fifty years ago. Ramm was highly regarded as a scholar who knew the Bible and was at the same time well acquainted with the scientific thought of his day.

I want, in general, to introduce authors in the chronological order of when they wrote. A chronological order is helpful when we remember that some of the discoveries in cosmology and in biology were not known when Ramm wrote, but are taken into account by later authors like Ross. Ramm wrote *The Christian View of Science and Scripture* in 1956, and Whitcomb and Morris say responding to it was a reason for their publishing *The Genesis Flood* in 1962. From this we can see that Ramm's book is important for an adequate study of our subject.

After considering the philosophy of science as it relates to understanding the Bible, we will discuss how to reconcile modern science and Genesis. The first view we will consider is that of Harry Rimmer, who represented what I call the "traditional view of creation." Next, in historical order, is Bernard Ramm.

Henry Morris's writings on how to interpret the creation account in Genesis follow. Then we will return to Davis Young's ideas, as they in turn are said to be a response to *The Genesis Flood*. Young wrote *Creation and the Flood* in 1977. Then we will return to Hugh Ross, whose books on cosmology were introduced in chapter 6. Subsequently we will introduce a few others' ideas on how to reconcile Scripture with science.

Some Differences Are to Be Accepted, and Perhaps Welcome

We will not find full agreement and identical answers in the following pages. Just as there are many disagreements in the scientific community on how to interpret the "facts of science," there are disagreements on how to harmonize Scripture with modern scientific discoveries. All the authors we include in part 3 accept the Bible as God's revealed Word and acknowledge its authority, but they differ in important ways in how they seek to harmonize them with scientific discoveries. And some express strong criticism about what the others have written.

It is not necessary to our task to decide which author is most correct in his views. Our task is far humbler. It is to honestly evaluate each of the views presented, seek to discover what can be concluded with confidence, and admit our lack of full understanding when we face questions that do not yet have a final answer. I believe we will accomplish our goal of reconciling science and Scripture more adequately and honestly and satisfactorily if we recognize our human tendency to pretend that we know more than we do. I believe we can make some conclusions with confidence, and we can trust that the problems which seem insolvable today actually do have satisfactory solutions. Our quest will not lead us to answer all the riddles of science and revelation. But I trust they will lead us to conclude that the facts of nature and the words of Scripture are nowhere in essential conflict.

Science and the Bible are widely considered to be in fundamental disagreement—so much so that many feel they cannot accept both without sacrificing rational thinking. But can we find a "middle ground" that seeks to accept all that is *verifiable* in modern science (not of course all that is *theorized*) and all that is clear in Scripture? I trust we can!

I hope a careful consideration of these things will lead us to think about how the Bible, written so much earlier than modern science, so clearly demonstrates its divine inspiration, and encourages us to also seek its guidance in how to live life to the fullest here on earth.

15

A Philosophy of Science for Harmonizing Science and Scripture

Introduction

> From medieval universities with faculties composed completely of Bible believers we have now reached the point where very few modern universities have Bible believers on the staff. The battle to keep the Bible as a respected book among the learned scholars and the academic world was fought and lost in the nineteenth century.[369]

In the above quotation, Bernard Ramm lamented the fact that the influence of modern philosophy and science since Darwin's day has crowded a study of the Bible out of the curricula of public universities in the United States and in Europe. He claims, "Orthodoxy did not have a well-developed philosophy of science or philosophy of biology. The *big* problems of science and biology must be argued in terms of a broad philosophy of science."[370]

Ramm insists that evangelicals cannot have their voices heard just by challenging the claims of evolutionists in their interpretations of details such as the authenticity of a bone or geological formation. In the past, he says, "It was impossible to settle the complex problems of Bible and science, theological and empirical fact, without a well-developed Christian theism and philosophy of science."[371]

Our "philosophy of science"—how we fit our knowledge of science into our worldview—greatly affects our thinking in all areas of life. Earlier we looked briefly at how Hugh Ross and John Klotz expressed their philosophies of science. Now, in this chapter, we will consider this field in its broader aspects. Doing this should help us prepare to consider the chapters that follow.

Ramm acknowledged that the discipline of philosophy of science only developed in recent years. Evangelicals lost "the battle of the Bible and science" in the nineteenth century. He says the tragic result was that physics, astronomy, chemistry, zoology, botany, geology, psychology, medicine and the rest of the sciences are taught without any influence from the teachings of the Bible or the Christian perspective.

Bernard Ramm (1916–1992) seems well-qualified to address the need to bring the teachings of the Bible back into the mainstream of academic studies. He studied theology at Eastern Baptist Theological Seminary and philosophy at the University of Southern California. His master's thesis and doctoral dissertation were on the philosophy of science, and he taught university classes on this subject. He also authored several books on bible-related issues. He thus appears to have a unique background to grapple with the issue of reconciling science and biblical revelation.

Ramm's book *The Christian View of Science and Scripture* seeks to analyze the conflict between theology and science, and explore the ways to reconcile the two. It includes chapters dealing with astronomy, geology, biology, and anthropology. His book is important to our study because of his contributions to a biblical philosophy of science. Also because of his perspectives on how scientific discoveries can be reconciled with the words of Scripture. He developed what he called a "Christian view of nature."

I. Philosophy of Science

> Our premise [is] that the Holy Spirit conveyed infallibly true theological doctrines in the cultural mold and terms of the days of the Bible writers, and did not give to the writers the secrets of modern science. It is a misunderstanding of the nature of inspiration to seek such secrets in various verses of the Bible. However, contrary to liberalism, we affirm that the theological does at times overlap the scientific.[372]

Ramm gives five examples of how the theological overlaps the scientific. These are "matter is not eternal but created; the simple preceded the complex in the order of life; man is the latest and highest creation of God; Jesus was actually born of a virgin"; and "the universe will have a demise and make way for the new heavens and the new earth."[373]

Ramm believed that the development of a Christian philosophy of nature was essential for any intelligent reconciliation between Christianity and science. Under the need for a philosophy of nature, Ramm listed three points.

First, the biblical approach to nature is essentially *religious* and *theological*. Although the Bible clearly tells us *that* God created, it is silent as to *how* He created. What the Bible says about nature is sufficient for man's religious and spiritual needs. But it was never intended to teach systematically about nature.

Second, science is rich in findings based on human experience, but it is unable to deal with the large problems of philosophy and theology. It is strong on the *how* of nature, but weak on the *why*.

Third, a philosophy of nature is needed to act as an umpire to deal with the quarrels between the domains of the Bible and science. Such a Christian philosophy will include what the Bible says about God and creation. It will acquaint itself with the study of the philosophy of science in a broad way. It will also learn all it can about the findings of science.[374]

Ramm points out that Hebrew people in Bible times saw God's activity in all of nature, and had no understanding of chains of causality and secondary causes. Everything that happened was directly dependent upon God in their thinking. He quoted Aquinas as follows: "Therefore God is in all things by his power, inasmuch as all things are subject to his power; He is by his presence in all things, as all things are bare and open to his eyes; He is in all things by his essence, inasmuch as He is present to all as the cause of their being."[375]

Throughout the book the foundation of Ramm's philosophy of science is clearly seen. This is that there is no separation of the spiritual and the material to Ramm. This is because God is working constantly in all of nature by His Spirit. He refers to famous men like Augustine and Thomas Aquinas as holding this same view. A philosophy of science, therefore, cannot separate the two. Science can only study the natural; theology studies the spiritual, and the philosophy of science must learn from both. A study of science is needed to tell us the *what* and the *how*; Scripture alone can tell us the *why*.

II. How to Interpret a Bible Passage

> Genuine relevant thinking cannot be accomplished in the realm of Bible and science until the nature of Biblical language has been deeply probed ... The language of the Bible is that of ancient Palestine and Greece. To correlate the two languages requires an understanding of both of them.[376]

Since Bernard Ramm taught biblical interpretation and also the philosophy of science, it should be helpful to consider his views on how to understand the way the Bible deals with scientific issues.

He faults many fundamentalists for assuming that if a record is inspired its meaning is always obvious, and also for assuming that we should ignore the literary form and any special subtleties of the Bible text. Ramm responds by saying, "No real grappling with the issues is possible till one has worked out his own theory as to the nature of Biblical statements about natural matters." Then Ramm presents the following points:

PART 3—*What Do We Really Know about the World?* The Bible

1. *"The language of the Bible with reference to natural matters is popular, not scientific."* He refers to popular language as the language in which people converse. His point is clear here. An example is to note that the language a scientist uses when writing for a technical journal is very different from that he uses when chatting with his neighbor. "The Bible is a book for all peoples of all ages. Its terms with reference to Nature must be popular."[377]

2. *"The language of the Bible is phenomenal.* By phenomenal we mean 'pertaining to appearances.'" Ramm refers to words like rise, set, fall, ascend and descend as relative terms, that do not speak of absolute movements of the sun or the earth, but rather to the appearances of natural things. Again this is easy to understand. Even a cosmologist will speak of the time when "the sun came up this morning."

3. *"The Bible does not theorize as to the actual nature of things."*

   "The Bible is singularly lacking in any definite *theorizing* about astronomy, geology, physics, chemistry, zoology, and botany."[378] These matters are dealt with according to popular and phenomenal terms that are free from scientific assumptions.

   Ramm says he agrees with W. B. Dawson when he wrote, "A remarkable point in Biblical references to nature, is that we find no definite *explanation* anywhere of natural things . . . The writers of the Bible show more than severe self-control, and must indeed have been divinely guided, in thus keeping to description and avoiding theoretical explanations of natural things."[379]

4. *"The language of the Bible employs the culture of the times in which it was written as the medium of revelation."*

   This means the Holy Spirit guided the writers to use the words of their ancient cultures, but worked to preserve them from the errors of their day. The Bible came to us in human languages using familiar human concepts and symbols. "If God spoke through Hebrew-speaking prophets and Greek-speaking Jews, what He had to say was to a degree colored by the natures of the Hebrew and Greek languages."[380]

   The revelation is *understandable* and *meaningful*, because it can be understood by prescientific people. He emphasizes that it is *prescientific* but not *anti-scientific*—to the human eye the sun rises and sets, and men of science even today speak in these popular terms. "Because the Bible uses prescientific terms it is a Bible for all ages and is adapted to all stages of human progress."[381] He says that because the Bible is non-scientific, even though scientific thinking has changed greatly in the past, it is as true for our time as it was true for its own time, and is likely to remain true for all time to come.

Ramm claims that "there is a very distinctive Biblical attitude towards Nature." He presents the following guidelines:

a. "The first feature of the Biblical view of Nature is that it is a very frank *creationism. God is the Almighty Creator of heaven and earth.* Therefore Nature exists fundamentally for spiritual purposes, and is capable of *teleological* explanation."[382] Teleology is the study of design or purpose in natural phenomena. So a teleological explanation means that nature can be understood when one knows the purpose of natural phenomena. Ramm says that there is a purpose in nature because nature is ordered by a Personal Intelligence.

He adds that the same God Who created His creation preserves it. The God of providence is the same God Who gave us the moral law. And the God Who gave us the moral law is the God of redemption. Because God is the reality behind nature, morality, redemption, and reason, all these are established to accomplish His purposes.

b. "The Biblical view of Nature also clearly maintains that *the universe is maintained by the Providence of God.* Biblical theism is unfriendly to deism and pantheism. It refuses to identify God with His works and it refuses to bar God from His works."[383]

> *God's providence is His working all things to their destined goal.* In cooperation with the redemption of God it forms the basis of the Christian philosophy of history. In that God works through the natural and the human, providence applies to nature. The possibility of miracles, and the possibility of answered prayer, are deeply involved in the Biblical doctrine of providence . . . The God of the Bible is not manacled to the causal laws, nor is He a prisoner in His own creation . . . The providence of God demands the freedom of God in Nature. The Bible is just as frank about its providence as it is about its creationism.[384]

c. "In keeping with the consistent creationism of the Bible is the constant prohibition of the Bible of *any worship of any part of the creation* . . . The worship of the created is forbidden in the Ten Commandments which prohibits any material representation of God."[385]

d. "The Bible clearly teaches that the *regularity of Nature is the constancy of God, and the Laws of Nature are the laws of God . . . The uniformity of Nature* is a Biblical notion and not the sole creation of modern science."[386] "Uniformity" here includes order and regularity. The "laws of nature" that science discovers reveal the "order" that God built into nature. Ramm says that "in Genesis 8:22 God promises the regularity of seed time and harvest, cold and heat, summer and winter, day and night, i.e. all such shall be regular and in order."[387]

PART 3—*What Do We Really Know about the World? The Bible*

Ramm goes beyond this to maintain that we err to make distinction between the natural and the supernatural. God controls and superintends the natural order just as He does what is supernatural. He contends that the Bible attributes directly to God what we consider the working of natural law, just as it does answers to prayers and miracles.

e. *The Biblical outlook on Nature is that Nature is temporal.* Ramm says this is an assertion of both fact and value. "It is *factual* in that it asserts that the universe was created by God in the past, and will be concluded by God in the future."[388] In other words, the universe had a beginning in time, and will have an ending also. He exclaims that "In the glorious theism of the Bible only God is eternal self-caused Being and Reality." The universe exists to accomplish God's purposes.

f. "*The Scriptures consider Nature a realm of probation and judgment.*"

Ramm points out that God said that creation was good, not perfect. He notes that diseases, storms, and many other imperfections exist on earth. Animals kill other animals in order to live. People are capable of doing both good and evil. We must study diseases and chemistry to learn how to cure our ills. We must study the properties of metals to have modern industry. We are in this way given tasks to learn how to accomplish, and problems to learn how to solve by studying nature.

> The entire system of nature involves tigers and lions, storms and high tides, diseases and parasites. It is part of our probation to learn how to capture or control the tiger and the lion, to learn how to protect ourselves from storms and tornadoes, to learn the mysteries of chemicals and bacteria for the healing of the body. If we fail in this probation innocent and sinful suffer alike.[389]

This is not the common theological explanation of the "evils" in the world today. Ramm does not believe that these "evils" are a result of God's curse on nature after mankind sinned. He rather emphasizes that nature, in its present imperfect state, exists "for probation" (to test mankind) and for "judgment" (to reward man for wisely learning to understand and benefit from nature, and to punish him when he fails to do so). This argument points out the importance of scientific investigation. It implies that natural science is vital to carry out the "agenda" God left man by allowing these problems in the natural world. This is a valid point, whether or not the problems remain from creation or are a result of the fall of mankind into sin.

III. Summary and Conclusion

Ramm emphasized the necessity of having a clear philosophy of science or "philosophy of nature." This is needed before one can attempt to harmonize what the Bible says about creation and nature with modern scientific discoveries. He says such a philosophy must recognize three things: (1) The Bible is essentially religious and theological; it tells us it was God Who created, but is silent on *how* He did it. (2) Science is unable to handle problems of philosophy and theology; it emphasizes the how but is inadequate to explain the *why*. (3) A philosophy of nature is needed to act as an "umpire" to settle the conflicts between science and Scripture in order to harmonize the two.

He says such a philosophy of nature must recognize that God is the reality behind all phenomena in nature, and that He maintains all natural things by His Spirit. God maintains all natural things, from the order of the cosmos to the continuing existence of life on earth.

He says the biblical view of nature includes the following features:

1. God is Creator, and the universe exists in order to accomplish His purposes.

2. God maintains the universe actively by His providence. In doing so He may work miracles and answer prayer. He may work by His natural laws or act contrary to them.

3. God forbids us to worship any part of His creation.

4. The laws of nature are the laws of God.

5. Nature and the universe are temporal; they had a beginning and will have an ending. This tells us that the spiritual is more important then the natural, because the spiritual is eternal.

Ramm adds a sixth point—that God did not create the earth without "evil." There is sickness and natural disasters and dangers. These are God's "tasks" for humans to deal with on earth. I think this is the only point here that will be rejected by most Bible scholars.

Ramm cautions that a clear understanding of the differences between the *language* of the Bible and that of science is essential to harmonizing the two. The Bible is written in the language of common people, using words reflecting the culture of the day. It is written to express appearances of natural phenomena and not their essence. It does not theorize about astronomy or any other science. And it is written in a language that expresses an ancient culture. In all these ways the language of the Bible is very different from the technical language of the sciences. It is not *anti-scientific*, but it is *prescientific* and *non-scientific*.

He also says that "Scripture is committed to no theory of the solar system nor the structure of matter, etc." Then he adds, "Christianity is a religion and not a science...

Therefore, if a scientist comes to God he must come the same way as any person comes to God . . . He must repent; he must confess his sin to God; he must believe in Jesus Christ with all his heart."[390]

It would be good to keep in mind these differences as we consider how our several writers seek to reconcile the findings of modern science with the account of creation in Genesis.

# 16

The Traditional View of Creation

When a new Christian, I was taught that Genesis 1:1 refers to the original creation, and Genesis 1:2 as referring to a "state of chaos" that occurred later. I learned that the reason for this chaos was the rebellion of "Lucifer," the chief angel, who led a rebellion against God and was cast down onto the earth. Later he appeared as a serpent to corrupt mankind in Genesis 3. (Lucifer is also called Satan and "the devil.")

This teaching is *inferred* from Scripture. That Satan did rebel against God, and is now in conflict with the purposes and plans of God is clear from the Bible. But that his fall was the "cause" of a fall of earth into chaos—the "without form and void" of Genesis 1:2—is not clearly taught in the Bible.

The teaching I am referring to is explained in *Modern Science and the Genesis Record,* published in 1937, by Harry Rimmer (1890–1952). Rimmer had a doctorate in divinity and another doctorate in science.

He took the position that Isaiah 5:18 states that God created the earth and "did not create it a waste place." It became a waste at the time referred to in Genesis 1:2. He believed that the reason earth became "waste and chaotic" is because it was earth that was the site of Lucifer's rebellion. Ezekiel 28:11–17 refer to a beautiful *mineral* garden of Eden; which Rimmer said was the condition when Satan fell, before the botanical garden of Eden depicted in Genesis 2:8.[391]

Other writers take the position that Satan was cast from heaven to earth when he rebelled. With either position earth is seen as a place without sickness or decay or death, until Adam and Eve fell into sin as recorded in Genesis 3.

Christian cosmologists today generally take a different view. Their studies have led them to believe that there was a lengthy period when the earth remained in an unformed state. These two views are similar regarding belief in an original world condition very unlike what we have today. But they differ greatly in the assumed cause of the "chaotic" state in Genesis 1:2. Which view is true cannot be decided conclusively, but there are still Bible-believers today who believe and teach that the earth "*became* waste and void" because of Lucifer's rebellion.

The Gap Theory

The idea that Genesis 1:1 is divided from 1:2 by a lengthy time-period like Rimmer expressed is a view of creation that was common among Bible-believers when he wrote. It was one that attempted to reconcile the known facts of geology with Scripture. This is commonly called the "gap theory." It was first introduced by the Scotsman Thomas Chalmers (1780–1847) and popularized in the United States with the *Scofield Reference Bible* in 1909

In the Scofield Bible, the footnote explaining "created" in Genesis 1:1 is as follows:

> But three *creative* acts of God are recorded in this chapter: (1) the heavens and the earth, v. 1; (2) animal life, v. 21; and (3) human life, vs. 26, 27. The first creative act refers to the dateless past, and gives scope for all the geologic ages.[392]

In explaining the words "without form, and void" in Genesis 1:2 the Scofield Bible speaks of many indications of "a cataclysmic change as the result of a divine judgment" on the face of the earth. It also claims that "There are not wanting intimations which connect it with a previous testing and fall of angels."[393]

The gap theory includes three beliefs.

1. A belief in a literal interpretation of Genesis.

2. A belief in a very old but undefined age of the earth.

3. A belief that there was a time period and events between Genesis 1:1 and 1:2 that can explain the origin of the older fossils and geologic strata. Many who hold this view believe that during this time period Lucifer the chief angel sinned and was cast down to earth and this caused chaos on earth and made it "waste and void."

As a rule, large-scale (Darwinian) evolution is not considered acceptable to those who hold this view.

The gap theory usually does not assume that the six days of creation were twenty-four-hour periods. It does not seek to date either how long ago the creation during those days occurred or the age of the earth. It is of course rejected by young earth advocates.

Various objections have been raised against this theory. Young earth creationists point out that Exodus 20:11 states, "For in six days the LORD made the heavens and the earth, the sea and all that is in them." They infer that this means that all of creation was created in six twenty-four-hour days reported in Genesis 1. Perhaps the answer to this objection is that the first use of *day* is in Genesis 1:5. This suggests that what

The Traditional View of Creation

are now called *days* did not exist until "God separated the light from the darkness" in Genesis 1:4, and that the long period prior to this is not excluded from the meaning of Exodus 20.11, as quoted above.

Rimmer also referred to the length of the days in Genesis 1.

> *Are the Days of Genesis literal days of twenty-four hours each, or are they periods of time?* To that question we can only reply, "We do not know"; and then set forth the evidence that shows also why we *cannot know*. This word "day" is one of the mysteries of Scripture. There is no scholar living so erudite that he can be dogmatic here.[394]

Rimmer explained all the times *yom* (often translated as "day") appears in the Bible, reported the eight different words used when they were translated into English, and concluded as follows. "It is thus absolutely impossible to take any one meaning of the many that are permissible and say 'Yom' must be translated thus, and thus only, in every possible case."[395] He argued that, even when translated as "day"—which is usually the case—*yom* can have different meanings just like "day" does in English. ("In my grandfather's day," is a common usage both in English and in Hebrew.)

Rimmer's views here reflect those of Hebrew scholars who explain the various ways *yom* is used in Hebrew literature.

Rimmer insisted that any view that Scripture is in conflict with the scientific facts is shallow.

> Is it a literal truth that science is at variance with Moses? Indeed it is not; rather the contrary is true: there is such a magnificent and complete agreement between the established facts of physical science and the first chapter of Genesis that no human explanation of this strange phenomenon is possible. How is it to be explained on any natural basis? Here is a chapter of a book written in the fifteenth century before Christ, in a day of ignorance and superstition.
>
> Yet when we examine that book in the light of modern scientific discoveries, it contains the most recent facts of physics, botany, astronomy and kindred sciences; and has maintained this marvelous harmony with scientific truth for ages before these sciences were born.[396]

Then later he returned to this theme.

> How did Moses know these great truths . . . so as to write in perfect harmony with modern science? *We do not claim that Moses knew these things; but rather that when he wrote, he wrote by the inspiration of the Holy Spirit, Who guided him.*[397]

Rimmer, speaking of Genesis 1:9–13, observed that "three marvelous events, in just one hundred and twenty words, are here recounted in the Divine record . . . In simple as well as concise language this tale is told, for in the one hundred and twenty words of this clear-cut statement one hundred of them are words of one syllable!

Human wisdom is not thus expressed." He remarks that only supernatural intelligence "can express such mighty deeds in language so simple that a child can grasp it all."[398]

He gave a simple argument that the universe could not exist without a designer. "There are no machines without inventors, and mechanics cannot be separated from the intelligence that produced them."

> See, then, how the world of botany praises God! The sea, the dry land, the green grass, and the tallest trees, they all cry aloud that Moses was right, when he said that they are the results of the creative power of an omnipotent God.[399]

> Every creature which lives in the sea, the air, or on the land, as well as every creature which ever did live, and which has left its fossil bones to tell the tale of the past, cries aloud the fact of design in its creation.[400]

Rimmer pointed out that the theory of theistic evolution sees man as starting low in the scale of evolution and progressing, but specific creation starts with man *higher than he is now*. "It has him in the hour of his origin a sinless, perfect being, dwelling in instant communion with his Maker."[401] He said that the history of humanity is not described by evolution or developing civilization, or by human accomplishment. It is rather described by creation, degeneration, and re-creation.

17

Ramm's View of Creation

> God is almighty Creator and . . . all is because He made it. He is Nature's Preserver and Sustainer and Provider. The laws of Nature are his laws, and the regularity of Nature is a reflection of God's faithfulness. This strong creationism and theism of the Bible must then be imported at this point into our considerations of the geological record. *All life, all forms, all geologic changes, all geological laws* ARE OF GOD.[402]
>
> —Bernard Ramm

Introduction

Harry Rimmer taught that Genesis 1:1 refers to the original creation, and Genesis 1:2 to a "state of chaos" that occurred later. This was a result of Satan leading a rebellion against God and being cast down onto the earth. This interpretation attempts to show that God's original creation was "perfect" but later became "waste and void." It also provides for geologic ages where various forms of life could exist and produce the ancient fossils that have been discovered throughout the earth. This is commonly called the gap theory.

Bernard Ramm reconciles modern science and the teachings of the Bible in a different way than Rimmer. His theory of creation includes two central concepts, which he calls "progressive creationism" and "moderate concordism."

Although Ramm reviews the scientific fields of astronomy, geology, biology and anthropology in his book, he gave far more space to geology.

He was concerned that scientists may not consider what the Bible says because of five misunderstandings of what evangelical Christians believe. He gave five "denials" here.

1. It is not true that all evangelicals believe that the world was created 4004 BC, but to the contrary, evangelicals in large numbers believe that the universe and the earth are as old as the reliable evidences of science say they are.

2. It is not true that all evangelicals believe that man appeared 4004 BC. Many evangelicals will push the date of man's origin back to the time of the earliest civilization (say, 10,000 BC), whereas others are willing to admit that man is hundreds of thousands of years old.

3. It is not true that evangelicals believe that the earth is flat or that the earth is the center of the solar system. Neither of these is the biblical position . . .

4. It is not true that all evangelicals believe that evolution is contrary to the Faith.

5. It is not true that evangelicals believe that the last word on specific details of physics, astronomy, chemistry, geology, biology, and psychology is to be found in the Bible. Evangelicals believe that the great metaphysical backdrop and historical setting is given in the Bible for the sciences. But by so asserting this evangelicalism does not seek to stifle all reason, all research, nor does it seek to dogmatize beyond the facts nor to have theologians dictating to scientists.[403]

A. Progressive Creationism

> Conservative Christianity is caught between the embarrassments of simple fiat creationism which is indigestible to modern science, and evolutionism which is indigestible to much of Fundamentalism. It is the conviction of at least this one evangelical [Ramm] that the only way out of the impasse is through some form of progressive creationism which we have imperfectly sketched here.[404]

This view recognizes two types of creation: creation *ex nihilo* (from nothing) by nothing more than God speaking (by fiat); and a process over a considerable length of time that followed the laws of nature God established in order to bring about the forms of nature he preordained.

We explained in chapter 15 Ramm's "philosophy of nature." This includes the work of God's Spirit in nature. Ramm explained this as follows:

> God is world ground. Nature depends upon Him for her origin, character, and movement to her destined ends. The Spirit of God is God's *innermost* touch on Nature seeing that it complies with His will, and imparting to Nature the spiritual energies the material world needs for its preservation. Progressive creation is the means whereby God as world ground and the Spirit of God . . . bring to pass the divine will in Nature.[405]

Ramm gives three points to summarize his position here. First God had a *concept* in mind—a purpose and a "blueprint" like a construction engineer uses. This is followed by a sovereign act of creation where nothing existed before. Then Genesis 1:2 refers to an assembling of raw materials to create something that never existed before. He also believed there were several acts of fiat creation in the history of the earth. In other words, God acted to create several times during the process.

After this comes the *process* of creation. God spoke *outside* of nature to begin the task of creation, and now turns the process over to the Holy Spirit Who is *inside* nature. The Spirit follows the divine blueprint to guide a process from the original vacant condition to realize the divinely intended form in the natural world. To make dry land appear, the Spirit sets those laws of geology to work which will produce dry land. He initiates the necessary conditions for the seas to swarm with fish. Ramm says, "The laws of Nature, under the direction of the Holy Spirit, actualize over a period of time and through process, the plan of God."[406]

Put simply, Ramm is saying three things: (1) That God had a blueprint in His mind for the universe. (2) Following this blueprint He brought the universe into existence without the use of anything already in existence—but in a very incomplete and transient form. (3) He developed the universe and this world over a very long period of time by His Holy Spirit working within nature. The Holy Spirit did this while following the blueprint and natural laws that God established at the beginning. This combines the "creation from nothing" revealed in Genesis with the long processes discovered or assumed by modern geology and biology.

Later Ramm clarified how this view is very different from theistic evolution.

> In theistic evolution there is a continuous line from the original cells on the prehistoric waters to man . . . But progressive creation teaches the transcendental activity of God. There is no continuum of life from amoeba to man, but the great phyla and families come into being only by the creative act of God. To equate this theory with theistic evolution is not proper.[407]

Ramm acknowledged that the large divisions in the animal and plant worlds were created separately. The fact that this appears to be true is demonstrated by the fossil record, as we have seen in part 2. He added that "progressive creation is the belief that Nature is permeated with the divine activity but not in any pantheistic sense."[408]

So Ramm is saying that Genesis 1:2 reports a "blank and void" state; and that the "six days of creation" were a very long process of development to the condition of the earth that we observe today.

> Almighty God is Creator . . . and Omnipotent Sustainer. In His mind the entire plan of creation was formed with man as the climax. Over the millions of years of geologic history the earth is prepared for man's dwelling . . . The vast forests grew and decayed for his coal, that coal might appear a natural product and not an artificial insertion in Nature. The millions of sea life were born and

perished for his oil. The surface of the earth was weathered for his forests and valleys. From time to time the great creative acts . . . took place. The complexity of animal forms increased.[409]

Finally, after a process that extended over many geological ages, the earth was prepared and *man* was created. This is how Ramm explains a very long process of creation that included several introductions of new forms of life. This process was preparing for the final act of creation—the appearance of human beings on earth, "in whom alone is the breath of God."

So we see Ramm differed from Rimmer's view that the earth was created originally in a completed form and *became* "waste and chaotic." It also differs from evolutionary theory in that Ramm acknowledges that each major division of plant and animal life was created separately and did not have one common source.

> *Creation and development are both indispensable categories in the understanding of geology and biology.* The fiat creationist can be embarrassed by a thousand examples of development. Progression cannot be denied geology and biology. The chasms in the order of life can only be bridged by creation. Biology cannot be rendered totally meaningful solely in terms of progression. Both Genesis and biology start with the null and void, both proceed from the simple to the complex, and both climax with man.[410]

B. Pictorial Day and Moderate Concordism

Then Ramm presented his view and calls it *Pictorial Day and Moderate Concordism*. He gave the following five points to explain this theory:

1. "The main purpose of Genesis is theological and religious."

   The *who did it* of creation is revealed, but not the *how*.

   > Finally, it is God Who does, God Who makes, God Who forms, God Who acts, God Who creates. The *how* is so plastic to the divine that it loses its relevance to the divine. Only by the ponderous methods of science followed through centuries of time do we commence to unravel the *how* of the universe.[411]

2. "With reference to the six days of creation, we reject the literal interpretation because by no means can the history of the earth be dated at 4000 B.C. or even 40,000 B.C."[412]

   Ramm also rejected the gap theory and parts of the age-day theory, which we will explain later.

He said that the language of Genesis is "*phenomenal* and *popular*, not *scientific* and *causal*." Therefore it is a mistake to look for scientific data in Genesis. He also argued that the meaning of "day" in the Hebrew text (*yom*) has many uses and it cannot be argued that it must be a literal twenty-four-hour day or an epoch or age. His argument here is similar to Rimmer's.

The six days refer to a snapshot of what happened, a revelation of an aspect of creation each day.

> The creation record is part topical and part chronological to convey to man: (i) some sense of the order in creation; (ii) that God made everything so nothing may be worshipped. Man as the last in order is highest in importance, and for that truth the order is necessary.[413]

3. "God *told* man the story of creation"—Ramm said it is wrong to think as some do that it was communicated only through *visions*. Ramm's use of the word "pictorial" refers to word-pictures and not to visual images.

4. "By *moderate concordism* we mean that geology and Genesis tell in broad outline the same story. Both agree that the earth was once in what may be called a chaotic condition. Both agree that certain cosmic conditions had to be realized before life could begin, e.g., the need for light, dry land, separation of waters and atmosphere. Both agree that the simple is first and the complex later. Both agree that the higher animals and man were the last to appear."[414]

5. "The truth about the geological record can only be settled with the combination of *geology* and *theology*. If Genesis is completely silent about secondary causes, and if geology is ignorant about first causes, then it is only as we bring the first causes and secondary causes together that we will get the truth for the full understanding of the geologic record."[415]

It is clear from the above explanation that Ramm's reconciliation of science and the Bible record is generally similar to Ross's.

Interesting Ideas of Ramm

Ramm rejected the idea that *all* death was a result of man's sin. He argued that "*the Bible ascribes death from sin to man alone*. Plant life had to die even in pristine Eden. To insist that all carnivora were originally vegetarian is another preposterous proposition. Why such huge teeth and sharp claws?"[416] He insisted that animals ate other animals in order to live even before man sinned.

Ramm believed that God created the earth in such a way that there were "imperfections" like diseases, and storms from the beginning—even before man sinned. When man was created he was given the task of learning how to cure diseases and protect himself from inclement weather. He also had to learn to protect himself from lions, tigers and parasites.

Conclusion

Ramm said he rejects the gap theory and some forms of the age-day theory. He called his view "progressive creationism" and "moderate concordism."

Progressive creationism is three things to Ramm: God had a blueprint in His mind for the universe. God followed this blueprint when bringing the universe into existence "from nothing" into a very incomplete and transient form. He then developed the universe and world over a long period of time by His Holy Spirit working within nature.

Moderate concordism and what he called "pictorial day" is explained by five things. The main purpose of Genesis is theological and religious, not to explain the how and when of creation. The language of Genesis is "*phenomenal* and *popular*, not *scientific* and causal." The text is meant to communicate the great fact that *God is Creator*. God told man the story of creation in word-pictures. Moderate concordism means that geology and Genesis tell in broad outline the same story. This is that the earth was once in what may be called a chaotic condition, that certain cosmic conditions had to be realized before life could begin, that the simple came first and the complex later, and that the higher animals and man were the last to appear. The truth about the geological record can only be settled with the combination of *geology* and *theology*. Genesis reveals the first cause of our universe, and geology reveals the secondary causes. We must bring these two causes together to gain a full understanding of the geologic record.

In this way Ramm plots a course between two concepts. One is an evolutionary theory that sees God as beginning the process and then leaving it to develop without divine intervention. The other is a literal interpretation of Genesis that believes that God spoke from heaven to create the separate kinds of living things in the mature form we observe today. He argues that creation is progressive, and God is directly involved in the process through the Holy Spirit, and continues to be involved even today. Ramm developed these ideas to harmonize the biblical account of creation with the findings of modern science.

18

Young Earth Creationism

> There are really only *two* basic philosophies or religions among mankind. The one is oriented primarily with respect to God, the Creator, of Whom and by Whom and for Whom are all things. Man is a creature of God, among the highest of His creatures but nevertheless utterly dependent upon and responsible to Him... The other basic philosophy is oriented primarily with respect to man.[417]

Introduction

We look next at *The Genesis Flood* because John Whitcomb and Henry Morris wrote that a major reason for writing this book was their desire to respond to *The Christian View of Science and Scripture* by Bernard Ramm.

They declare complete confidence in the divine inspiration and "transparency" of the Bible, "believing that a true exegesis thereof yields determinative Truth in all matters with which it deals." They add that "the instructed Christian knows that the evidences for full divine inspiration of Scripture are far weightier than the evidences for any fact of science."[418] In a later book Morris says, "No geologic difficulties, real or imagined, can be allowed to take precedence over the clear statements and necessary inferences of Scripture."[419]

From the quotations above it is crystal clear that Whitcomb and Morris take the Bible as the revealed truth from God and that it is to be taken very literally in all its references to creation. The Bible and not science is considered infallible and without error. They are convinced that the book of Genesis is a literal historical account.

As its name implies, *The Genesis Flood* deals mainly with the question of the universality and effects of the flood of Noah's day. In it they explain systematically only three of the six days of creation in Genesis 1 and 2. But for many years this book was

PART 3—*What Do We Really Know about the World?* The Bible

considered the most authoritative work on young earth creationism in circulation. It is still held in high regard by many Christians.

In this book the authors base their arguments for a young earth on a few beliefs. They believe that the universe is around ten thousand years old, that God in six literal twenty-four-hour days created the earth, plant and animal life, and created mankind on the sixth day. This is basically the position introduced in America by George McCready Price in the 1920s.

How they reconcile the discoveries of modern science with the Bible is fairly simple. They do so by suggesting that the earth was created "with the appearance of age," and that there was no such thing as death anywhere in the world until Adam and Eve sinned. They argue that plants, animals, and humans were created in their present "mature" condition, and there has been no advancement in their condition as described by the theory of evolution. They say that the original creation had "the appearance of age," in that fully mature organisms were created at the beginning.

They also argue that the flood in Noah's day was worldwide in scope and accounts for the geologic record that geologists point to when arguing the earth is billions of years old. This includes the existence of fossil beds and the geologic columns. A major theme of their book is a response to Ramm's arguments that the flood of Noah's day was only local and not worldwide. They also object to Ramm's claim that the earth and life on it are very old.

They go into great detail in their attempts to reconcile discoveries in the field of geology with their young earth position. This was covered in chapter 10.

This chapter will report on how Whitcomb and Morris explain (1) their outline of geological history; (2) how long life has been on earth; (3) how long they believe mankind has been in existence; (4) the effects of Adam's sin; and (5) the results of a worldwide flood. A survey of these will introduce us to how they reconcile science and Scripture.

1. Outline of Geological History

In chapter 6 of *The Genesis Flood* the authors say they present "a Scriptural framework for historical geology."

> The uniformitarian geologists of the nineteenth century, rejecting the Biblical testimony of deterioration and catastrophe and all the geological implications thereof and accepting instead the philosophy of evolutionary naturalism, built their system of historical geology upon a foundation of sand.[420]

Whitcomb and Morris report that the Bible implies that there have been at least five great epochs of history. These are as follows.

First Was the Initial Creation Itself

> The initial act of creation quite evidently included the structure and materials of at least the earth's core and some sort of crust and surface materials . . . It seems reasonable that, even if the earth's creation was accomplished as an instantaneous act, its internal heat and the waters on its face would immediately have begun to perform works of profound geological significance.[421]

Next Was the Work of the Six Days of Creation

> Especially on the third day was a tremendous amount of geological work accomplished. On that day, the Genesis account tells us that dry land was made to appear above the surface of the waters. This can only mean a great orogeny, as the rocks and other materials of the primitive earth were uplifted above the waters.[422]

> The fourth day witnessed the establishment of the sun and moon in their functions with respect to the earth. Since the sun now provides all the energy received by the earth for its geological processes, this event also has profound geological implications.[423]

Note here that they do not say these heavenly bodies were *created* on the fourth day, but that this is when they began to function as light-givers for earth.

Then the Antediluvian Period

The creation story says that during the original creation God "made a firmament" that divided water above it from water below it. "Firmament" here refers to the lower atmosphere of the earth. The waters below seem to refer to the oceans and the waters above to a thick canopy of water-laden clouds.

> The waters "above the firmament" seem to imply more than our present clouds and atmospheric water vapor, especially since Genesis 2:5 implies that during this time rainfall was not experienced on the earth. These upper waters were therefore placed in that position by divine creativity, not by the normal processes of the hydrologic cycle of the present day. The upper waters did not, however, obscure the light from the heavenly bodies and so must have been in the form of invisible water vapor.[424]

Then the Deluge

Whitcomb and Morris believe that the flood in Noah's day is clearly the greatest physical convulsion that has ever occurred on earth since the creation of life on it, and in fact all but obliterated everything on the face of the earth!

Then the Modern Post-Deluge Period

> In general, uniform processes of nature would henceforth prevail; thus the geological dogma of uniformity can, with certain limitations, be applied to the study of this period.[425]

The authors argue that how creation began can only be known by revelation, not by science. They say, "It is not amenable to the scientific method, which implies reproducibility of experimental results. It was a once-for-all event, never repeated and not observed by man. Therefore the only real knowledge of the mode of origin must be by means of divine revelation."[426] And God's revelation does not tell us how he did it.

2. How Long Has Life Been on Earth?

In answering this question the authors dwell on two main beliefs. First, that living things were created in a mature condition without any process. Second, that counting the genealogies from Adam on to the birth of Christ may not be an accurate way to determine the true age of the human race. They believe that life has been on earth no more than a few thousand years.

Appearance of Age

This means several things. When God created the earth he provided soil and nutrients ready to nourish plant life, with no need for a long period to degrade rock until soil is formed. Plants were created in a mature state, capable of producing seeds to reproduce themselves. Animals were also created in a mature state, ready to produce offspring. Their view is that all the preparation needed to sustain life was present during the original creation as reported in Genesis 1. They argue that an extended period between the creation of the earth and the creation of plant and animal life was not needed and did not occur.

Whitcomb and Morris realized that a long period of time is needed now to prepare soil before it can support plant growth. Speaking of the third day of creation, they conclude, "But here it must have been created essentially instantaneously, with all the necessary chemical constituents, rather than gradually developed over centuries of rock weathering, alluvial deposition, etc."[427]

An interesting idea of theirs is that the universe could have been created instantaneously with the photons of light at some distance from stars, rather than exploding from one central location. They think this can explain why it *appears* that light has traveled for billions of years until picked up recently by human telescopes—though the universe may actually be much younger than that.

3. How Long Has Man Been on Earth?

Here they discuss how to interpret the genealogies.

> Evangelical scholars who feel the necessity of bringing Genesis 11 into conformity with current paleo-anthropological timetables should realize the full implications of such harmonization efforts. It would seem to us that even the allowance of 5,000 years between the flood and Abraham stretches Genesis 11 almost to the breaking point.[428]

They remark that it is "highly significant that no truly verified archaeological datings antedate the time of about 3,000 B.C."[429]

They argue from various details in the genealogies in Genesis and in the New Testament that it isn't necessary to press the numerical data in these chapters into a strict chronology. They believe the chronologies in Genesis 5 and 11 were not for the purpose of presenting a strict chronology by which one can infer how many years had passed, but rather had other purposes—such as presenting a testimony of how faithfully God preserved the Messianic line. This is also widely believed by evangelicals and most Old Testament scholars.

They thus acknowledge that the Old Testament genealogies cannot be used, as Ussher did, to show how old the human race is. But they also take issue with those who argue that the human race from Adam could have started hundreds of thousands of years ago.

4. The Effects of Adam's Sin

> With the fall of man, a new order of things ensued, not only in God's spiritual economy with respect to man but also with respect to the earth itself, which was "cursed for man's sake" (Genesis 3:17; 5:29). The whole of creation was delivered into the bondage of corruption (i.e., "decay"), groaning and travailing in pain together (Romans 8:21, 22).[430]

In *The Genesis Flood* Whitcomb and Morris insist on two things. First, the death God pronounced on Adam in Genesis 3 was *physical* as well as spiritual. They say, "In the face of such clear passages as Romans 5:12–21 and 1 Corinthians 15:21–22, few who accept the Bible as the Word of God will deny that Adam's sin and fall introduced *spiritual* and *physical* death into the human race."[431]

Whitcomb and Morris go even further. They see *all* death and decay as an expression of the curse that came as a result of Adam's fall. They believe no plant or animal died before this, so all fossils are a result of death that occurred after the fall. Therefore, since Adam existed only a few thousand years ago, all the rocks that contain fossils could not have been formed many millions of years ago as geologists believe. They say *all* fossiliferous rocks were formed after the fall. This position is of course necessary to support their belief that death did not occur anywhere on earth before Adam sinned and the ground was cursed.

Second, they say God's judgment on Adam's sin and that of his descendants brought about great and permanent changes in the laws of nature.

> To put the issue into its sharpest delineations, a literal interpretation of the Fall demands as its corollary a thorough-going Biblical catastrophism; and the doctrine of the Flood can be fully understood only in the light of the Fall and the Edenic curse.[432]

They argue that an acceptance of the fall of the human race when Adam and Eve sinned as a historical fact *demands* the acceptance of the flood of Noah's day as a worldwide historical event. Bernard Ramm, Davis Young, and many others would of course challenge this.

They say the following about fossils that are considered by paleontologists to be of humans that lived long ago:

> We say, on the basis of overwhelming Biblical evidence, that every fossil man that has ever been discovered, or ever will be discovered, is a *descendant* of the *supernaturally created* Adam and Eve. This is absolutely essential to the entire edifice of Christian theology, and there can simply be no Christianity without it.[433]

Here they seem to be on solid theological grounds. We will return to this in chapter 23.

The authors point out that Romans 8:19–22 says that "the creation was subject to vanity," and that creation will someday "be delivered from the bondage of corruption." Also "the whole creation groans and travails until now." They say, "This passage teaches very clearly that some tremendous transformations took place in the realm of nature at the time of the Edenic curse."[434]

Their point is well taken that the natural realm on earth was "cursed," and became in some way or ways different than it was before Adam and Eve sinned. But to argue, as Whitcomb and Morris do, that there was no decay, no entropy, no death at all in the world does not follow inevitably from this argument.

Morris, in *Biblical Cosmology and Modern Science,* places great stress on his belief that there was no entropy in the universe before sin entered the world. (This is because he believes that the first and second laws of thermodynamics did not come into effect until man sinned.) He says,

> All natural processes operate within the constraining framework of the two Laws [of Thermodynamics], and this means that every process, in addition to being a conservative process, is also fundamentally a decaying process.[435]

Morris believes that, after the original creation, matter and energy can neither be created nor destroyed. Various forms of energy are simply interchangeable. The energy of the universe is continuously being converted into heat. It is thus running down.

Since the cosmos is still far from dead, it must have had a beginning, and that beginning could not have been accomplished by the present, conservative, processes of the cosmos.

Morris believes that God created a perfect universe at first, one without any form of deterioration—a universe that would have remained virtually unchanged if man had not sinned. The laws that govern the universe today, especially the laws of thermodynamics (which are the most fundamental natural laws that we understand today) were established by God after sin entered the world. Then, second, he is arguing that the very existence of the laws of thermodynamics is conclusive proof that the cosmos had a special creation at a finite time in the past, and that these creative processes are now not operating.

5. The Flood of Noah's Day

> A careful study of the Biblical evidence leads us to the conclusion that the Flood may have occurred as much as three to five thousand years before Abraham.[436]

This is saying that the flood probably occurred from seven to nine thousand years ago. They add that "there is no record of a tree, or any other living thing, being older than any reasonable date for the Deluge."[437]

Whitcomb and Morris believed that a worldwide flood, the flood of Noah's day, accounts for most of the geologic features that appear to be of great antiquity. What appear to be strata of rocks throughout the world that were laid down in succession during many long geologic ages were actually piled on top of each other in no more than one year by a gigantic flood. They cite as supporting this position the fact that in some parts of the world the strata are not found in the same order as are the strata uniformitarian geologists point to as supporting their arguments for an old earth.

Morris also argues that few fossils are produced in modern times. He says that the large fossil beds support his arguments that a catastrophic flood caused them and also caused them to pile up in geologic columns on the continents of the world.

Conclusion

Whitcomb and Morris maintain that Scripture must interpret science, and not the other way around. They also insist that the universe and earth were created at the same time, and cannot be much older than ten thousand years. The earth, plant life, animal life, and the human race were all created in a span of six twenty-four-hour days. The earth and life on it appear much older because they were created in a mature state, "with the appearance of age."

Three great events in nature recorded in Genesis can explain why the earth appears much older than it is. The first was the great land movements that formed the continents and seas during the first three days of creation. The second was the "Edenic curse" on nature resulting from Adam's sin. The third was a worldwide flood in Noah's day.

They argue that the age of the earth, the vegetable kingdom, the animal kingdom, and the human race are all about the same age, as they were created only days apart. The geologic columns used to date the earth were not formed over millions of years. They were all built up within a period of a year by the waters of the flood.

They directly challenged Bernard Ramm's view that the world was already populated by thorns and thistles before Adam and Eve were thrust out of the garden of Eden. This is of course not surprising, because they totally reject the thought that death and deterioration in the plant and animal kingdoms existed before the fall. They believe that since God created everything "good" thorns and thistles and sickness did not exist until sin entered the world.

How successful were they?

Do Whitcomb and Morris succeed in reconciling the Genesis account of creation with science—particularly with geological science? Considering the information presented in chapter 10 and here in chapter 18, can we conclude that no "irreconcilable differences" remain between the two?

The first question here is, *Do they interpret the Bible in a way that is consistent with sound exegesis?* Here I believe they did well. They consistently take key words and phrases in Genesis and interpreted them in a very "safe" and literal way. If modern science supported their very literal interpretation of Genesis then the debate, at least among Bible-believers, could be considered over. (Perhaps it should be noted here that Whitcomb and Morris seem to have chosen a "narrow" interpretation of some verses over what appear to be equally acceptable alternate interpretations that can be also considered "literal." An example is their putting Genesis 1:1 and 1:2 with verses 3 to 5 and calling this period "one day." They give as a reason for this Genesis 2:4, which says, "This is the history of the heavens and the earth when they were created, in the day that the Lord God made the earth and the heavens.")

But the debate is far from over! The second question we must answer is, *Do they interpret the realm of nature in a way that is consistent with currently accepted facts of science?* Here they have many challengers, who also accept the Bible as the inerrant Word of God but interpret it differently.

Davis Young is one of these. We proceed to consider his position in the next chapter.

More details on how they interpret the first three days of creation are reported in the appendix.

19

Geology and Old Earth Creationism

> Although I oppose an idea [that the earth is only a few thousand years old] that is common and on the surface sounds biblical, the reader must not draw the conclusion that I am opposing Christianity or attacking the Bible or even opposing the idea that the Flood was global in nature. I write as one who is firmly committed to the infallibility and inerrancy of Scripture and in full agreement with historic Christianity. I simply believe that the young earth view is unscientific and not necessarily biblical. I believe that continued promotion of such ideas will in the long run do damage to the credibility of Christianity . . . I am a creationist, and I believe in the biblical record of creation.[438]
>
> —Davis Young

Introduction

I introduced Davis Young in chapter 9. He accepts the current geological theory of methodological uniformitarianism. He believes the laws of nature that affected geological changes in the distant past are the same as those at work today. But he acknowledges that changes have not occurred at a constant rate throughout geologic history, and catastrophes have occurred in the past.

In this chapter I will discuss Young's attempt to reconcile the teachings of Genesis with the present-day scientific knowledge gained from geology, as he understood it. Because Young wrote in response to reading *The Genesis Flood*, I have chosen to explain his thinking on this issue right after my chapter 18 on *The Genesis Flood*. Young refers to *The Genesis Flood* on at least thirty-three pages of *Creation and the Flood*, and on seven pages of *Science, Scripture and the Young Earth*.

Young is in agreement with Whitcomb and Morris that the Bible is without error, but strongly disagrees with their interpretation of historical geology.

He refers to the apparent clash between modern geological thought and Genesis.

> The established interpretation of Genesis suggested that the earth had been created in six days; geology now indicated that it had developed over a period of millions of years. The established interpretation of Genesis said that the earth was very young; geology said it was very old. Scripture recorded the occurrence of a gigantic flood; geology could find no evidence for such a flood. The current interpretation of Scripture suggested that life had been created directly by God; geology said that life had evolved from more primitive life and ultimately from non-life. Scripture said that man was created by God in His own image; geology said that man was an advanced animal and was related to the apes.[439]

He says this conflict of "facts" shook the faith of many in the inerrancy of the Bible. But he maintains that "there is no compelling reason why belief in the authority and infallibility of the Bible should ever have been undermined as a result of this debate."[440]

He insists that if there is an apparent contradiction between Scripture and "some group of facts from any realm of human thought," it is because of one of three possible causes:

> (1) the exact human interpretation of Scripture on that point is incorrect or not clear; (2) the interpretation of the group of facts is incorrect; or (3) the "facts" are really not facts. But the words of Scripture are *never* incorrect. *The words and real teachings of Scripture are never in ultimate conflict with the real facts of science or history.*[441]

Young insists that "the reader will find that the charge that there are geological errors in the Bible cannot be substantiated."

In *Creation and the Flood*, Young attempts to explain the verses in Genesis that relate to geological issues in such a way as to reconcile them with current scientific opinion—"current" being around 1977 when he wrote the book (*The Genesis Flood* was written in 1961).

This chapter will present Young's view of creation, including the way he interprets Genesis 1, and his criticisms of *The Genesis Flood*.

Young's View: Old Earth Creationism

Young does not quarrel with the dominant theories of geology, but sought to reconcile his interpretation of Genesis 1 with them. In setting his view apart from both theistic evolution and young earth creationism, he makes two main points.

1. The Six Days of Creation Are Historical Days of Indeterminate Length

Young insists that these are a *figurative* week made up of *figurative* days.

First, they are *historical*. Young faults many theistic evolutionists—those who believe in God but also accept Darwin's general theory more or less completely—for not regarding the first chapters of Genesis as historical. He insists that the creation week is historical, not allegorical or mythological or symbolical. He says, "The day-structure is figurative only in the sense that these days are not identical to our days, but rather indeterminate stretches of *real, historical time*."[442]

Second, they are *of Indeterminate Length*. Young criticizes what he calls "biblical catastrophism." This is the view that God created the world quite recently with "the appearance of age." It also argues that God used a worldwide flood in Noah's day to make the geological formations and fossils that appear to be ancient. He says that this view causes serious problems with geological theory and practice. If the six days of creation are of indeterminate length, and not days of twenty-four hours each, then he believes there is no problem with reconciling the Genesis account with an old earth.

In other words, he believes that the Bible is the verbally inspired Word of God, but believes it can be interpreted as allowing for long days of creation without being inconsistent with proper scriptural interpretation.

> Indeed, if it can be demonstrated beyond all doubt that Scripture demands a 24-hour view of the days, then the Christian scientist must accept that . . . Scripture, *not* science, must determine our beliefs of the world and it's history. Science may help us ask new questions of Scripture, but Scripture still provides the answers.[443]

2. Evolution of Plant and Animal Life Is Possible, and Did Occur

In *Creation and the Flood* Young repeatedly explains his belief that animals and plants *evolved* during several long periods called days, but man was directly created by God—as one adult male and one adult female.

Young further argues that it is not necessary to interpret the six days of creation "throughout as a purely divine miracle that is . . . totally devoid of process . . . There is considerable evidence that, after the formation of the initial stuff of the universe, the laws and processes of the present time may very well have been in operation during creation week."[444]

In this way Young argues that Genesis 1 reports historical events that occurred in the realm of nature. These occurred in an essentially chronological order over a period of six consecutive days of unknown length. The account of each day describes genuine physical events that occurred in space and time either on the earth or with respect to the earth. The six days of creation therefore deal with events of *geological, biological,* and *astronomical* significance.

Young's Interpretation of Genesis 1

> Scripture fully permits the interpretation that the creation week of Genesis 1 is a figurative week consisting of figurative days. According to this view the account is not concerned with the question of the duration of the creation period, and the Bible does not tell us the age of the earth.[445]

Young concludes that if the total length of the creation period is unclear in Scripture then believing the earth is 4.6 billion years old is not incompatible with the Bible.

Then Young states his belief that the events reported in Genesis 1 are consistent in their sequence with that postulated by geological science. But he says that the Bible is speaking in terms of very broad, large-scale phenomena and not in terms of precise, scientific, technically describable phenomena. "The account is concerned with major events, the highlights of creation."

> Keep in mind, however, the brevity and conciseness of Moses' account. Because of his economy of expression, the mistake of believing that only that which is recorded in the chapter occurred during the creation must be avoided. Moreover, Genesis 1 is a generalized description of major events.[446]

Young argues that it is "the brevity of the account" which makes it look like God acted instantaneously in Genesis 1: "Let there be—and it was so." Moses had a religious purpose for writing. "Moses was concerned to show that God—the sovereign, personal God—has created the universe so that His creatures might behold His power and worship and adore Him!"[447]

He further argues that, just as we might list only the major activities we did on a given day in a report, many other things might have happened on a given day that Moses did not mention.

> Perhaps a few beasts were actually made on the fourth day. Some of the activity of the sixth day may overlap to earlier days. All we are compelled to believe is that in comparison with other activities the sixth day was pre-eminently the day of creation of beasts and man and that the great bulk of beasts and creeping things were formed on the sixth day. It is a great mistake to insist that not a single animal could possibly have been created before all the plants were created simply because the third day reports on plants and the fifth and sixth days report on animals.[448]

Young acknowledges that there are several difficulties in reconciling Genesis and the geological record. Rather than leading one to discredit the biblical account, they should lead us to seek to understand both the Bible and science better. He says that internal evidence makes it clear that Genesis 1 is meant to be a historical account. It clearly claims to report real space-time events. But internal evidence does not clarify whether the events took place in 144 hours or over a longer period of time.

> Internal Scriptural arguments can be given which lend support to either view. The view a Christian holds would seem to be a matter of individual conscience. The 144-hour view, even though a Biblically possible interpretation, presents extremely difficult practical problems for the geologist. It makes the practice of geology virtually impossible. The practicing Christian geologist then is likely to opt for the long-day interpretation of Genesis 1, which likewise has Biblical support and is therefore a legitimate option.[449]

In other words, Young believes in the modern geological explanation for the fossils and geologic columns. But if it can be conclusively demonstrated that the Bible means that no processes of death operated before Adam sinned, then Young will accept this as an article of faith—no matter how inconsistent this is with present-day scientific observation.

Young says that the facts of geology should serve as a *stimulus* for a new and deeper analysis of what the Bible says, in order to obtain a more accurate understanding of the meaning of Scripture.

> Extrabiblical evidence may be of immense assistance in increasing our grasp and appreciation of Scripture, but by all means the conclusion must be avoided that the extrabiblical evidence is *identical* to what Scripture says or that exegesis is based on extrabiblical evidence as much as it is on the text of Scripture.[450]

Conclusion

Balancing his faith in science, Young testifies to his faith in the inerrancy of the Bible. He does not believe that any discoveries of science challenge the authority or infallibility of Scripture.

He does have a strong concern that the young earth position expressed in *The Genesis Flood* will discourage its young adherents from entering geology as a field of study. He believes that this would be unfortunate, because he is convinced that a clear understanding of what the Bible actually says can be reconciled with sound geology.

Young agrees with Whitcomb and Morris that the six days of creation are historical. But he believes that they report a process of creation that in its order is consistent with scientific discoveries, and that the days of creation are of indeterminate length. In other words they are historical, but the term "day" is used in a figurative sense. He is willing to accept the idea that plant and animal life developed by processes of evolution over a very long period of time. But he seems to believe that major categories

Geology and Old Earth Creationism

of plant and animal life were created separately, and says that the first humans were created directly by God in essentially the same state of development as exists now.

As a geologist, he says he is a "methodological uniformitarian," and seeks to interpret Genesis in a way that is consistent with the dominant theories of geology today. This includes the belief that all ancient geologic formations seen today can be explained by natural forces that operate today.

Does Davis Young succeed in reconciling what he knows of modern geology with Scripture, and does he do so in a way that can be supported by the rules of biblical interpretation?

His way of reconciling a long period of evolution with the account in Genesis 1 and 2 is to assume that the creation days were periods of unknown length that covered many years. What was reported as being created on a given day expressed very briefly only the main kinds introduced on that day. Not all kinds are listed. Although beasts are listed as created on the sixth day this only means that most of them were made then. A few may have been created earlier. As Young believes plants and animals evolved, he thinks that their main evolution was completed on the day mentioned for their creation.

This is one way to interpret Scripture in a way that makes it generally consistent with modern geological thought. It avoids the need to assume, in order to explain the older fossils, that plant and animal life existed before the six days of creation. It allows for the existence of fossils of simple marine animals much earlier than land life even though marine life is not mentioned until the fifth day.

He emphasizes that

> the truths and data of Scripture are independent of the opinion of science. The Christian must always be careful to remember that science is susceptible to change as new discoveries are made . . . at some point in the future geology may want to change its opinion regarding the relative sequence of events of earth history.[451]

He continues this thought by insisting that "the Christian has no right at any point in time to expect or insist upon complete agreement between Scripture and the available facts of science. It is unreasonable to expect complete harmony until all the facts are in."

Young believes that both scriptural interpretation and scientific discovery are subject to improvement over time, and that the Christian can be confident that progress in both will continue to demonstrate a marvelous agreement between God's revelation and His creation. This faith that future scientific discoveries will increasingly support the Genesis account of creation is shared by Hugh Ross. We will return to his contribution in the next chapter.

I think it is important to note that Young wrote his books before *Evolution: A Theory in Crisis* by Denton and *Darwin's Black Box* by Behe were published. I get the impression from Young's writings that, when he wrote *Creation and the Flood,* he assumed the scientific proof for the validity of macroevolution was much greater than it is now known to be. He may be yielding much more to current scientific opinion on evolution than the scientific facts warrant.

In sum, Young is seeking to reconcile a modern scientific view of historical geology with a Bible account that he considers historical and without factual error. He interprets Genesis 1 as reporting a process of creation of indeterminate length that in its order is consistent with scientific discoveries which suggest the order that living things appeared on earth. His interpretation does not seem to violate the rules of biblical interpretation as they are recognized by biblical scholars today.

Some of Young's detailed comments on Genesis 1 are reported in the appendix.

20

The Cosmos and Old Earth Creationism

Cosmologist Hugh Ross emphasizes that the universe had a beginning and will have an ending and that this is in contrast with God Who is eternal.

> The Bible says in unequivocal terms that the "heavens and the earth" began, that they exist for finite time only, and that God exists and acts inside, outside, and before the universe's space-and-time boundaries. He alone is everywhere present and always existing.[452]

He says that *The Bible's Emphasis is that Only GOD Could have Created Everything.*

> The Hebrew verb for "created" is *bara*. This verb, or predicate, appears in the Bible with only one subject: God. Its usage suggests a kind of creating that only God, and no one else, can do. Hebrew linguists define it as "bringing into existence something new: something that did not exist before." While it can refer to creation *ex nihilo* (that is, out of nothing), its use is less restrictive.[453]

He gives a concise statement summarizing his belief of how God created.

> The God of the Bible generated the universe transcendently, that is, independent of matter, energy, and the dimensions of length, width, height, and time. He personally designed and built the universe and our solar system so that life could flourish on Earth. Though the Bible does not identify the specific means by which God produced the lower life-forms, it does state that He specially created through fiat, miraculous means birds, mammals, and human beings.[454]

In chapter 6 we showed how Ross presented recent discoveries in cosmology and astrophysics and showed how they demonstrate that our cosmos had a wise and powerful Creator.

Ross apparently accepts all the current theories that govern the research in cosmology and astrophysics today. He believes Einstein's general theory of relativity has been supported by recent discoveries in the cosmos. He also believes the big bang theory is a valid basis for understanding the universe and its creation.

If *all* assumptions of the current big bang theory are correct, the universe is apparently between ten and twenty billion years old.

Ross believes that the earth is 4.6 billion years old, and that there have been several times when life became extinct and only fossils remain. Different kinds of life existed when the earth was rotating faster and when more of earth's surface was covered by water. These died out when conditions on earth changed. He sees no conflict between these ideas and Scripture.

Now we will look at how Ross reconciles these ideas with the creation account in Genesis. We will also look at how Ross reconciles the Genesis account of the creation of Adam and Eve with the apparent existence on earth of humanlike creatures that are said to have existed on earth in antiquity.

Ross says that the Hebrew words translated "the heavens and the earth" "consistently refers to the totality of the physical universe: all of the matter and energy and whatever else it contains."

> All of the stars, galaxies, planets, dust, gas, fundamental particles, background radiation, black holes, physical space-time dimensions, and voids of the universe—however mysterious to the ancient writer—would be included in this term.[455]

The Point of View of Genesis 1

Ross suggests we begin by accurately applying Galileo's rule, which is: "Begin by establishing [not assuming] the point of view." Ross points out that "Genesis 1 precisely and clearly identifies the point of view for the creation account: 'Darkness was over the surface of the deep, and the Spirit of God was hovering over the waters.' (Genesis 1:2)."

Ross says this verse "suggests that the reader interpret the events of creation from the perspective of an observer on the surface of the earth." By this he means that the point of view for the creation events revealed are from the perspective of an observer on the surface of the ocean, and underneath the layer of clouds.[456]

How Ross Interprets Genesis 1

In Genesis 1:16 the verb translated "made" ("God made two great lights ... He also made the stars") is *asah*, which connotes completed action. This refers back to what God did in verse 1, because "heavens and earth" in 1:1 includes the galaxies, stars and planets.

Ross explains several apparent contradictions between scientific discovery and the order of creation in Genesis 1 by explaining more exactly the meanings of the Hebrew words used.

Ross gives the order of creation events, based on Genesis 1, as follows:

1. "Creation by God's fiat miracle of the entire physical universe." He means here our four-dimensional universe, including space (length, width, height), time, matter and energy. All the galaxies, stars, and planets are included.

He says that at this stage earth is empty of life and unfit for it. The reason is that "Earth's primordial atmosphere and the solar system's interplanetary debris prevent the light of the sun, moon and stars from reaching the surface of the earth's ocean."

2. "Clearing of the interplanetary debris and partial transformation of the earth's atmosphere so that light from the heavenly bodies now penetrates to the surface of the earth's ocean." There is a partial transformation of the earth's atmosphere to allow some light to enter, but it cannot yet enter as freely as it does today.

3. "Formation of water vapor in the troposphere under conditions that establish a stable water cycle." (The "troposphere" is all that portion of the atmosphere below the stratosphere.) This is the area where clouds form.

4. "Formation of continental land masses together with ocean basins."

5. "Production of plants on the continental land masses."

6. "Transformation of the atmosphere from a translucent condition to one that is at least occasionally transparent." In other words, there is a change in the atmosphere around the earth to allow more light to enter. Genesis 1:14 speaks of placing the sun, moon and stars in the "firmament" ("expanse of the heavens") on the fourth day. Since they were obviously created earlier in Genesis 1:1, it must mean these bodies only became *visible* on the fourth day. This is because Ross says that we must interpret Genesis 1 from the point of view of being on the surface of the earth.

Thus light was not created after the earth was.

> Light was not created on the first creation day. On that day the light already created "in the beginning" suddenly broke through to the earth's surface. This breakthrough required the transformation of the atmosphere (plus the interplanetary medium) from opaque to translucent. On the fourth creation day we see yet another atmospheric transformation, this time from translucent to transparent. Through that transformation, the sun, moon, and stars became visible for the first time on earth's surface. It's not that God made (or created) them on the fourth day; He simply made them visible on that day.[457]

Ross denies that this view is unique with him; he named six others with advanced degrees in science, in theology, or in both who have published similar interpretations of Genesis 1.

7. "Production of swarms of small sea animals."

8. "Creation by God's fiat miracles of sea animals and birds."

This seems to be God's second direct miracle in this process. Ross evidently believes that all that happened after point 1 above through point 7 were processes set into motion by the first miracle. This is suggested by the fact that the Hebrew word for "create" is not used between verses 2 and 21.

9. "Creation by God of land mammals capable of interacting with the future human race." Ross refers here (Genesis 1:20–25) to animals that are *nephesh* in Hebrew. He says this means "soulish creatures, creatures that can relate to humans; creatures with qualities of mind, will and emotion. These can only be birds and mammals."

Ross says that "creeps on the ground" in 1:25 and "creeping things" in 1:24 refer to "short-legged mammals such as rodents and hares."

(Ross uses the word "creation" in point 9 because verse 21 does even though verses 24 and 25 do not.)

10. "Creation by God's fiat miracle of the human species."

It is noteworthy that Ross refers to points 1, 8, and 10 above as being created by "God's fiat miracles." These correspond to the three times "create" is used in Genesis 1. For the other steps in the process Genesis says that "God said," or "God made" and not "God created." This distinction seems quite important to Ross.[458]

Then Ross returns to his "point of view" argument, to show there is no contradiction between the above ten points and either the findings of science or the words of Genesis.

> With the point of view and initial conditions correctly identified, the sequence of Genesis creation events no longer seems difficult to harmonize with the record of astronomy, paleontology, geology, and biology. The few purported conflicts with the fossil record stem from inaccurate interpretations of some Hebrew nouns for various plant and animal species.[459]

In reference to Genesis 1:11–13, when God made land vegetation on the third day, Ross says,

> Scientific evidence for ocean life predating land life poses no threat either. The Spirit of God "brooded" over the face of the waters (Genesis 1:2), possibly creating life in the oceans before the events of the six creation days begin.[460]

The Long Days of Creation

> Genesis 1 states that within six "days" God miraculously transformed "formless and void" earth into a suitable habitat for humanity, and then created human beings from dust. The meaning of the word *day*, of course, is the focal point of the creation time-scale controversy.[461]

All the authors who seek to reconcile Genesis 1 with current scientific thought deal with this issue. Ross goes into the meaning of the Hebrew word "day" in great detail, but many of his arguments are already treated in this book as they are similar to what others have written.

Ross gives nine arguments for considering them "days" of millions of years. Among these are the following four points:

As God exists in a dimension separate from time, the length of God's days are not the same as ours. Psalm 90:4 refers to one thousand years being like one day to God. Ross points out that the author of this psalm is Moses—the one who also wrote Genesis.

The Hebrew word used (*yom*) and translated as "day," includes different meanings than the English word. (*Yom* is frequently used for a long period, and frequently for a season or time when something special happens. It is similar to the English use in "my grandfather's day" and "the day of the dinosaurs.")

The uniqueness of the seventh day. The wording in Genesis for days one through six imply each had a beginning and an end. But this isn't true of the seventh day. Ross refers to Psalm 95:7–11 and Hebrews 4:4–11. "From these passages we gather that the seventh day of Genesis 1 and 2 represents a minimum of several thousand years and a maximum that is open ended (but finite). It seems reasonable to conclude, then, given the parallelism of the Genesis creation account, that the first six days may also have been long time periods."[462]

So much is recorded as happening on the sixth day that it could not have taken place in one twenty-four-hour day. He says, "Altogether, many weeks', months', or even years' worth of activities took place in this latter portion of the sixth day." He notes that "the fact that the Bible does consider the antiquity of the founding of the earth a suitable metaphor for God's eternity suggest the biblical view of an ancient earth."

Ross adds that "interpreting the Genesis creation days as tens of millions or even hundreds of millions of Earth years in no way lends support to evolutionism. These time frames would be too brief by countless orders of magnitude for simple life to arise and become complex by natural processes."

But Ross does acknowledge that God established the laws of nature in such a way that only minor biological changes have occurred since.

> Since the time these animal kinds were created by God, they have been subject to minor changes in accordance with the laws of nature, which God

established. However, the Bible clearly denies that any of these species descended from lower forms of life. Human beings are distinct from all other animals, including the bipedal primates that preceded them, in that humans alone possess body, soul, and spirit.[463]

Creation of Mankind

Ross takes great pains to reconcile the Genesis account of the creation of Adam and Eve with the existence on earth of humanlike creatures that apparently existed on earth in antiquity.

Ross explains the difference between humans and hominids by saying that the latter do not have spirits. God made man, and only man, in His image.

> What makes humans different is a quality called "spirit." None of the rest of Earth's creatures possesses it. By "spirit" the Bible means awareness of God and capacity to form a relationship with Him. Worship is the key evidence of the spiritual quality of the human race, and the universality of worship is evidenced in altars, temples, and religious relics of all kinds. Burial of dead, use of tools, or even painting do not qualify as evidence of the spirit, for non-spirit beings such as bower birds, elephants, and chimpanzees engage in such activities to a limited extent.[464]

He emphasizes that "by the biblical definition, these hominids may have been intelligent mammals, but they were not humans. Nor did Adam and Eve physically descend from them."[465]

Ross admits that some differences between the Bible and secular anthropology remain. But he believes that advances in the study of anthropology are narrowing these. As an example, he refers to the following:

> New evidence indicates that the various hominid species may have gone extinct before, or as a result of, the appearance of modern humans. At the very least, abrupt transitions between hominid species is widely acknowledged.[466]

The Reliability and Trustworthiness of Scripture

Although it is not yet possible to reconcile *all* the differences between science and the Bible account of creation, Ross believes that the marvelous agreements clearly attest to the accuracy of the biblical record.

> Obviously, no author writing more than 3,400 years ago, as Moses did, could have so accurately described and sequenced these events, plus the initial conditions, without divine assistance. And if God could guide the words of Moses to scientific and historical precision in this most complex report of divine

activity, we have reason to believe we can trust Him to communicate with perfection through all the other Bible writers as well.[467]

This is an important observation. The Bible was not written for the purpose of revealing the history or order of natural things, but to reveal the existence and nature of God and reveal spiritual realities, including the existence of the afterlife and the importance of having a right relationship with God. Ross is saying that the accuracy of the Bible regarding natural things is a demonstration of its divine inspiration and veracity. It demonstrates that the Bible throughout can be trusted to reveal things that are *true*.

Belief in a Young Earth

Ross says that "nearly half the adults in the United States believe that God created the universe within the last 10,000 years." They think this is what the Bible says. Then Ross says the following: "Caltech physicist and Nobel Laureate Murray Gell-Mann said (in testimony to the Supreme Court around 1980) it would be easier to believe in a flat earth than to believe the universe is 6,000 years old, or anything other than about 15 billion years old."[468]

Ross adds, "There are many different models of origins under the banner of old-universe creationism. All go against at least some of the tenants of evolutionary biology. But young universe creationism challenges virtually all of science."[469] (We will report the young earth response to this later.)

In *Creation and Time* Ross reports that young earth scientists like Walter Brown, Henry Morris, and Edward Blick argue that there are from fifty to eighty separate scientific evidences that indicate the universe and earth are only thousands of years old. Ross says he has examined these carefully and concluded that they involve one or more of the following four problems: These are faulty assumptions, faulty data, misapplication of principles and equations, and the failure to consider opposing evidence.

In this way Ross accepts all the information gained by cosmic research, and challenges the claims of "young earth" proponents. He says that he has had many discussions with advocates of recent creationism about the inaccuracies of the evidence they use to support their view and they admitted to him in private that they really do not have solid evidence for it.

Conclusion

Ross says that God created the universe from nothing over ten billion years ago and then fashioned the earth by natural processes beginning about 4.6 million years ago. The point of view of the creation days in Genesis 1:2–27 is that of one standing on the earth, which at first was covered with water. This explains why the shining of light on the earth comes after the creation of the sun, moon and stars. At first no light pierced through earth's atmosphere, then it did so dimly. Finally, on the fourth creation day, the atmosphere changed from translucent to transparent and the light from these heavenly bodies was clearly seen on earth.

Ross agrees with many other scholars that the Hebrew word for "day" can be interpreted as a long, indeterminate, but not infinite, period of time.

He argues that, based on these explanations, there is no conflict at all between the account of creation in Genesis 1 and the scientific facts.

Ross differs from Young in believing that biological evolution is very limited in scale. Here he is closer to Morris and Whitcomb and the other young earth creationists. He differs from the young earth creationists primarily in his belief in the age of the universe, planet earth, and the beginnings of life on earth.

An explanation of how Ross explains some of the key verses in Genesis 1 can be found in the appendix.

21

Other Views on Interpreting the Bible

Introduction

We have now seen what I consider some of the most important viewpoints on our topic. In this chapter we will look briefly at a few other views.

Richard Carlson, in his book *Science and Christianity: Four Views*, says that he is convinced that there is no one single Christian viewpoint when it comes to the relationship between natural science and the Christian faith. He presents four positions, and calls these: (1) Creationism, (2) Independence, (3) Partnership, and (4) Qualified agreement.

(1) Creationism

> Creationism could be simply defined as a belief system that places principal or final authority in the Bible, many times (but not always) in terms of a literal reading of the Bible that is regarded as inerrant or infallible. In any conflict arising between scientific and theological conclusions, the science is taken to be defective or incomplete or inadequate or at least suspect, for the Bible is seen to be free from any error and is the final authority in all matters concerning faith.[470]

Then Carlson says that, though there are differing beliefs among creationists about the age of the universe and the earth and whether or not some evolution has occurred, no creationist believes that humankind has descended from some earlier subhuman form. He calls intelligent design a variety of creationism distinct from what he calls the "young earth, anti-Darwin and literal interpretation" of some creationists. Carlson chose two science professors to represent this view.

(2) Independence

Carlson asserts that no possibility of conflict exists in this view because science and Christian theology are thought to share no common ground. Independence is represented by biology professor Jean Pond. She insists that it is appropriate to leave God out of science, but that science cannot declare God absent from reality.

Carlson explains that the main idea here is that Christianity and science work together as partners and so must influence each other. The contribution of each is valued and dialogue is sought. According to this view science does not threaten faith, but actually enhances it. Christian theology is considered an endeavor that can inform science. Those holding this view tend to accept around 15 billion years as the age of the universe, and evolution of life from very simple beginnings.

(3) Partnership or Integration

Carlson chose Howard J. Van Till to represent this view. He is a retired professor of physics and astronomy. Carlson sees this view as similar to the Independence view in that both value Christian theology and science as valid enterprises that work together to formulate a comprehensive picture of nature. Christianity is not threatened by science.

(4) Qualified Agreement

Stephen Meyer presents this position in Carlson's book. Carlson calls it Qualified Agreement because Meyer is in agreement with much of contemporary science.

> Here contemporary scientific data from cosmology and physics are used to build a new philosophical argument for the existence of a transcendent intelligent designer. In addition, the lack of an adequate (or complete) scientific understanding of how primitive life initially arose on our earth is also used to build an independent argument for intelligent design. In my opinion this is a type of conflict position, for Meyer sees science as a metaphysically naturalistic enterprise and as an alternative to a Christian worldview.[471]

These four views will be described briefly in order to explain the various approaches to the issues that are being advocated. Later I will introduce another view.

1. Creationism

Creationists in general accept the theological position that emphasizes the *trustworthiness* and *authority* of the Bible. Wayne Frair and Gary D. Patterson title their paper "Creationism: An Inerrant Bible and Effective Science." Wayne Frair was head of the biology department at The King's College, and is now retired. Gary D. Patterson is professor of chemical physics and polymer science at Carnegie Mellon University.

They take the same *theological* position that Whitcomb and Morris did. But they also say, "We strongly reject any scholarly program that ignores the clearly established data and conclusions of science or that rejects the authority of the Bible ... Both scientific and theological activities must be open to thorough examination and discussion."[472]

They are willing to interpret Scripture in a broader way than Whitcomb and Morris did. They believe that

> many controversies have resulted from the conflict between unsound interpretations of the Bible and unsupported or even incorrect conclusions presented in the name of science ... A genuine search for solutions requires a willingness to consider possibilities outside the narrow positions that sometimes emerge in such debates. There is much work to be done in both biblical exegesis and some fields of science, but it is our conviction that a proper understanding of the Bible will be consistent with all valid scientific observations of the world in which we live.[473]

They argue that "the initial act of God in creating the universe out of nothing is followed by the maintaining providence of God (Col 1:17) ... Any view that restricts His activity to one or a few miracles at the dawn of time is inconsistent with the biblical view of God's incessant involvement with his created world."[474]

They believe that how Christians understand the meaning of the Bible as it relates to the creation story has changed over time. They say that "the Bible cannot be interpreted apart from the valid tools of scholarly analysis. Evangelical believers must support the best analysis because we have pledged to obey the Bible; we actually care what it means!"[475]

The authors discuss the relationship of the Bible and cosmology, physics, chemistry, biology and history. They insist the big bang theory fits well with Bible teaching, but even if it should prove eventually to be false this will not affect their faith in the Genesis account of creation. They exclaim that their increased knowledge about the universe and the earth gives them greater appreciation of their Creator, and their faith has increased as a result.

They explain in some detail their conviction that developments in chemistry and in biology are only increasing their belief that scientific discoveries supplement and confirm the truths of Scripture. They are convinced that the belief that chemistry leads inevitably to life that is currently popular is really based on unsupported conjecture and nothing else.

They give a dramatic example to illustrate their belief that the cell must have been designed. "A common analogy that fully applies to the present subject is that it would be just as likely for a tornado ripping through a junkyard to produce a fully functioning 747 jet as it would be for random processes to produce all the interior structure of a cell!"[476]

They acknowledge how bravely nineteenth-century Darwinism is defended today, but are convinced the exquisite complexity seen in the world of life makes the argument for a designer more compelling than ever before.

They clearly reject the narrow interpretation of the creation account presented in *The Genesis Flood*, and are willing to consider that a much longer period was used in the creation process. But they also reject the broad interpretation of Van Till (which will be presented later). They insist that the Genesis account is historically accurate.

"The basic premise of our analysis," they say, "is that the Bible is an inerrant revelation from God."

> The more we learn about earth life forms—all plant life and living creatures—the more we appreciate the Creator of life. While the Christian belief in God as the Creator of life is established on clear exegesis of the Bible, the external evidence only strengthens this position. Christian life scientists should welcome all new observations of the structure and function of living organisms . . . It is our conviction that a compelling case can be made for the creation of life—the very complex structure and highly regulated functions of biological organisms have all the appearances of being created.[477]

They are convinced that with each new discovery in the life sciences their appreciation of the Creator of life will also increase.

Summary

Creationism is thought by many to mean the view popularized by Whitcomb and Morris in *The Genesis Flood*. This is the belief that God created the universe and all in it in six literal days of twenty-four hours each. Proponents of this belief generally admit that the Bible does not clearly say how long ago that was, but argue it could not have been more than ten thousand years or so ago.

But *creationism* is not limited to such a young earth viewpoint. Frair and Patterson represent a large group of Bible-believers who accept most of the conclusions of the scientific community, including the belief that the universe may be much older. They are similar to Whitcomb and Morris in their belief that the Bible must be interpreted literally and has theological authority. But they take the position that there is still much to be learned about the real facts regarding nature both in the Bible account of creation and in the physical world.

One of their conclusions is this: "We believe that a Christian does not need to abandon a biblical perspective in order to carry out effective science. Accurate exegesis and reliable interpretation of the Bible along with valid scientific conclusions are the goal of all scientists who are Christians."[478]

2. Independence

Carlson subtitled this "Mutual Humility in the Relationship between Science and Christian Theology." This view is presented by Jean Pond, a professor of biology at Whitworth College. Her doctorate is in microbiology from the School of Medicine at the University of South Dakota.

Pond suggests three approaches to avoiding conflict between science and Christian theology. One would be for the Christian to shun scientific and technical pursuits altogether. Another way to avoid conflict would be to seek to demonstrate that science and Christian theology are in essential agreement when both are correctly interpreted. A third way is to maintain an *independence* between them, because they are two very different pursuits. It is represented by Stephen Jay Gould, who argued that science and religion are not in conflict because their teachings concern very different domains.

Pond gives two practical differences between the practice of science and theology. In science new information is discovered daily, but not in theology. In science knowledge is provisional.

> Science and theology are different ways of acquiring different kinds of truths about the world. Science and theology differ in their areas of inquiry and in the methods they use. Stephen Gould has written that science and religion have a common goal: "Our shared struggle for wisdom in all its various guises." This is a sentiment with which I agree. But within that common goal the individual goals of science and theology differ. Science seeks an understanding of the physical or natural world, whereas theology seeks the fullest possible knowledge of God's actions in human history and God's purpose for our own lives.[479]

It should be noted that many Christians believe that theology covers a much wider spectrum than what Pond expresses here.

She gives two "primary requirements" for independence between science and theology to be maintained. First, she admits that science has strength and power when it operates within its own realm, but says it cannot deny the possible existence of other realms of truth. She gives biologist Richard Dawkins as a well-known example of prominent scientists who are secular materialists who do this, and mentions his often-quoted statement that "if God is non-existent, theology is the study of nothing. You cannot be ignorant of nothing."

She gives as a second requirement that believers should not make broad claims about religious implications of scientific theories, but we should have an understanding of Scripture that is dynamic and open to change.

She calls science a moving target. Scientific theories are always subject to change. It is unwise for Christians to base the proof of a scriptural interpretation on a scientific theory—because what is "scientific bedrock" today can be "quicksand" tomorrow. She gives the current big bang theory as an example, as some Christians say this proves, as Genesis 1 says, that God created the universe out of nothing. She argues that the most important scientific knowledge is realizing what we don't know, and says she is reluctant to believe that we will no longer get major surprises from our studying the physical universe.

Summary

Pond concludes by arguing that, with the independence of science and theology, science can make no claim to represent the sum of reality. But she adds,

> To ignore the evidence of science—to deny the fruitfulness and power of the big bang theory in physics or evolution in biology—is to make too little of the authority of science within its own domain. Historically this has not been a successful endeavor, and when the church has tried it, the church has been forced to retreat. The Christian's relationship with God does not grant him or her the authority to make pronouncements regarding the veracity of scientific theories.[480]

Pond represents many scientists that argue that science and theology speak about entirely different matters and should be kept completely separate. But this ignores the fact that the Bible *does* say much about the natural world. If what the Bible declares as factual could ever be shown conclusively to be false by confirmed scientific discoveries it would challenge the Bible's trustworthiness to declare "truth," including truth in matters that are beyond the reach of science.

3. Partnership: Science and Christian Theology as Partners in Theorizing

This view is represented by Howard J. Van Till, who until retirement was professor of physics and astronomy at Calvin College. He has a PhD in physics, and is a member of the American Astronomical Society and the American Scientific Affiliation.

Van Till says he accepts the *methods* of scientific naturalism but not its philosophy. He sees both Christian theology and science as valid enterprises. The proper role between them is where they work together as partners who influence each other. Both are valued and dialogue is encouraged. Science is seen not as something that threatens faith but as capable of enhancing it. Christian theology can inform science.

Van Till sees the universe and all in it to be the outcome of the effective will of the Creator that is revealed in the Bible. Because of this, he considers science and Christian theology as jointly contributing to the study of natural things. He believes that God created in such a way that nature is governed by natural laws that can be discovered by science, as contrasted by polytheistic deities who are thought to act capriciously. Environmental powers are not "deities," and it is the Bible that tells us this. He says,

> I have come to believe that the creation of which we are a part was brought into being, by God's effective will, from nothing about fifteen billion years ago and that it was gifted from the outset with all of the capabilities it would need to form the full array of both physical/ material structures and life forms that have ever existed.[481]

He goes on to explain that this creation from nothing already possessed at the beginning not only the *potential* for the development of all forms of life we can observe today, but also the *dynamic pathways* needed to guide this development. In other words, the processes of biological evolution were programmed into the original act of creation from nothing in such a way that God did not need to intervene later either for the development of mankind or any other higher forms of life. At the point of the "big bang" God set the processes into motion for all that has transpired since.

Van Till says he has made "a human judgment" that there is a wealth of scientific evidence to support the common ancestry of all extant life forms. By this he clearly means to acknowledge that this is a subjective judgment and not based on adequate scientific discovery.

Van Till seems confident that future scientific advancements will eliminate the "gaps" in nature that other Christians point to as necessitating God's intervention after his initial creation. He even includes the development of humans as needing no direct action by God. In other words, there was no intervention in Genesis 1 and Genesis 2 to create Adam and Eve apart from the animals previously created. He clearly means here that advanced life forms, including humans, are a development of a Darwinian-style evolution, though he sees it as programmed from the beginning by God.

Van Till believes God *can* act on His creation after it was first formed, but says that *we cannot know* if He does so or not. This reminds me of Michael Behe's comment:

> Some people (officially including the National Academy of Sciences) are willing to allow that the laws of nature may have been purposely fine-tuned for life by an intelligent agent, but they balk at considering further fine-tuning after the Big Bang because they fret it would require "interference" in the operation

PART 3—*What Do We Really Know about the World?* The Bible

of nature. So they permit a designer just one shot, at the beginning—after that, hands off.[482]

VanTill argues that this view of biblical creationism can partner effectively with science. Scientific discoveries cannot threaten Christianity in any way.

Summary

Van Till sees the picture that modern biology has given us in terms of the development of life forms, including humans, as a depiction of how God endowed the universe with potential at creation. So he sees science and theology as working together in partnership and not in any way in conflict with each other.

He says very clearly that his view—the "fully gifted creation perspective"—is that the initial creation by God set in motion natural processes that resulted in all the life forms that have ever existed without any further activity by God. Yet he modifies this definite position by adding that intellectual humility necessitates that we admit that it is possible that there were "gaps" between life forms that were bridged by direct intervention by God. Van Till admits that we will never know all the elements in "the creation's formational economy." All this edges Van Till into the theistic evolution camp.

4. Qualified Agreement: Modern Science and the Return of the "God Hypothesis"

> Contemporary philosophers of science and religion generally recognize that science and religion do represent two distinct types of human activity or endeavor. Most acknowledge that they require different activities of their practitioners, have different goals, and ultimately have different objects of interest, study, or worship.[483]

Stephen Meyer says his article in Carlson's book will reassert the classical view of agreement between theism and science. It will also argue that the scientific evidence does provide support, but not proof, for the theistic worldview of biblical Christianity. He calls his model of the relationship between science and Christianity "qualified agreement" or "mutual epistemic support." It maintains that the scientific evidence and biblical teaching support each other when correctly interpreted.

We are familiar with Meyer from chapter 12 on intelligent design. He uses the term "qualified agreement" here because he is in agreement with much of contemporary science. He goes on to say that advocates of "qualified agreement" agree that much scientific research and theory does relate to issues that are metaphysically and scientifically neutral. But they insist that some scientific theories have larger metaphysical implications. He insists that the best and most truthful (reliable) theories do not contradict a theistic worldview. He says, "In my view (and apparently the view of Friar and Patterson) scientists should seek the best explanation of the evidence and follow it wherever it might lead. Another view of science insists that scientists must limit themselves to strictly naturalistic or materialistic explanations no matter what the evidence might be. This view has, since Darwin's time, played an influential role in limiting scientific theorizing, not only in biology but in other fields as well."[484]

He points out that Sagan's famous comment that "the cosmos is all that is, or ever was or ever will be" expresses the assumption of naturalism well. Naturalism denies the existence of any source capable of actually explaining the existence of the universe. The big bang, along with general relativity, implies a single beginning for matter, space, time and energy. Therefore an entity capable of bringing about this singular beginning must exist and must transcend these four dimensions. This is exactly the kind of entity depicted in the Bible as being God the Creator. This is Meyer's argument.

Meyer insists that we should select the best explanation no matter where it leads. This approach seems very typical of how design theorists approach scientific investigation. I attended a meeting of the Socrates Club at Oregon State University the evening of May 22, 2006. The debate that evening was entitled "Evolution versus Intelligent Design: Scientific Assumptions in a Free Society." The debaters were professor of philosophy Michael Ruse and professor of biophysics Cornelius Hunter of Biola University. Ruse is a well-known proponent of Darwinian style, large-scale evolution.

Hunter, in this debate, took an identical position as Meyer. He argued that scientists should take an epistemological approach and follow the evidence wherever it leads. He believes that the evidence clearly leads to the existence of an "intelligent designer." He also stated that the theory of general evolution starts out with the assumption that all living things evolved from a simple beginning and evolutionists today interpret the scientific evidence from this assumption.

Meyer divides views on the relationship of science and religion into only two broad groups. One assumes the religious and metaphysical neutrality of scientific knowledge. He includes Jean Pond's independence and Howard Van Till's partnership in the first group. The other group is what he calls "the classical view" of Kepler and Newton, that the findings of science actually support important tenets of the theistic and Christian worldview.

Meyer points out that from around AD 1250 through 1750, scientists commonly supported the agreement between "the book of nature" and the "book of Scripture," and so he calls this the "classical view." He asserts that scientific discoveries in cosmology and biology support the theory of intelligent design even if they may not "prove" the existence of a Creator God.

Meyer says that the demise of the intelligent design argument was influenced *philosophically* by arguments by David Hume (1711–1776), and then *biologically* by Charles Darwin. It was further influenced *cosmologically* by the ascendance of belief in a universe that had no beginning, and supported by philosopher Immanuel Kant

(1774–1804). He argues that this brought about a "conflict" model for the relationship between science and religion.

Then he speaks of the return of the God hypothesis.

> During the twentieth century a quiet but remarkable shift has occurred in science. Evidence from cosmology, physics and biology now tells a very different story than did the science of the late nineteenth century. Evidence from cosmology now supports a finite universe, not an infinite one, while evidence from physics and biology has reopened the question of design.[485]

Einstein's theory of general relativity of space, time, matter and energy stimulated astrophysicists like Stephen Hawking to explore its implications to the age of the universe. Meyer says that "Hawking's result suggested that general relativity implies that the universe sprang into existence a finite time ago from literally nothing, at least nothing physical."[486] Hugh Ross also pointed out this conclusion of Hawking's.

Meyer says that even very slight alterations in the values of many independent factors would render life impossible. These include the expansion rate of the universe and the precise strength of gravitational and electromagnetic attraction. These "anthropic coincidences" and the "fine-tuning of the universe" have suggested to many scientists and philosophers the existence of design by a preexistent intelligence.

Meyer covers the same areas of science as we have discussed in chapters 6 and 7. He refers to *The Creator and the Cosmos* by Hugh Ross, which I discussed in chapter 6. He also refers to research by Michael Denton, whom we featured in chapter 7 on biology. He says that Denton, in *Nature's Destiny*, has documented many other conditions in chemistry, geology, and biology necessary for human life. He also refers to Michael Behe's research on the cell.

> Physics, astronomy, cosmology and chemistry have each revealed that life depends on a very precise set of design parameters, which, as it happens, has been built into our universe. The fine-tuning evidence has led to a persuasive reformation of the design argument, though not a formal deductive proof of God's existence.[487]

Meyer acknowledges that deism can also explain the "cosmological singularity and the anthropic fine-tuning." But he insists that clear evidence exists in the realm of biology to show that God has acted on His creation since the big bang. He says that fossil evidence indicates the origin of life on earth from some 3.5 to 3.8 billion years ago, and that this is clearly well after the origin of the universe. (He also argues that the view that life existed at the time of the big bang is very implausible.) He says the argument put forth by Francis Crick and Fred Hoyle that life was seeded on earth by a space alien or other extra-terrestrial agent is difficult to accept. For one reason, there is no scientific explanation for the existence of such beings.

As to biology, Meyer comments as follows:

> Developments in molecular biology have raised the question of the ultimate origin of the specific sequencing—the information content—in both DNA and proteins. These developments have also created severe difficulties for all strictly naturalistic theories of origin of the first cellular life. Since the late 1920s naturalistically minded scientists have sought to explain the origin of the very first life as the result of a completely undirected process of "chemical evolution."[488]

But he says scientists have been unable to do so. In concluding this he is in agreement with the conclusions of Booher and others as discussed in chapter 13.

Summary

Meyer's central argument is that, even though evidence from the natural sciences clearly provides strong support for Christian theism, science itself cannot provide certain proof. He says it simply is not in the realm of science to do so.

Meyer argues that scientists should follow the evidence wherever it leads, and there is a large body of evidence to demonstrate that the universe and life were designed. The fine-tuning of the universe and evidence discovered by modern biology abundantly demonstrate this.

> Since the fine-tuning and initial conditions date from the very origin of the universe itself, this evidence suggests the need for an intelligent as well as a transcendent cause for the origin of the universe. Since God as conceived by Judeo-Christian (and other) theists possesses precisely these attributes, his creative action can adequately explain the origin of the cosmological singularity and the anthropic fine-tuning. Since naturalism denies a transcendent and preexistent intelligent cause, it follows that theism provides a better explanation than naturalism for these two evidences taken jointly. Since pantheism, with its belief in an immanent and impersonal god, also denies the existence of a transcendent and preexistent intelligence, it too lacks causal adequacy as an explanation for these evidences. Thus theism stands as the best explanation of the three major worldviews—theism, pantheism, and naturalism—for the origin of the big bang singularity and anthropic fine-tuning taken jointly.[489]

Since naturalism cannot explain the existence of life anywhere, it cannot explain the existence of life on earth by suggesting life began outside our planet. He asserts that naturalistic theories have failed at the very point of explaining the origin of the *information* necessary for life's origin.

He insists that

> this does not, of course, *prove* God's existence, since superior explanatory power does not constitute deductive certainty. It does suggest, however, that

PART 3—*What Do We Really Know about the World?* The Bible

the natural sciences now provide strong *epistemological support* for the existence of God as affirmed by both a theistic and Christian worldview.[490]

Meyer's position as he explained it in Carlson's book does not seem at all to me to be a "conflict position" as Carlson labeled it. Meyer himself explains his view is seeking the best explanation for the known facts of modern science.

5. Another View

I want to refer to one more view here, though it was not included in Carlson's book. Bruce Waltke is a professor of Old Testament at Regent College. His article "The Literary Genre of Genesis, Chapter One" appeared in the December 1991 edition of *Crux*.

Waltke also argues for keeping science and Scripture separate. He looked at the creation account in Genesis 1 from the aspect of literature rather than as natural science.

> Moses aimed to produce through a true understanding of God a right perception of the universe and humans, including their relationship to God and one another, and to proclaim that truth in the face of false religious notions dominant throughout the world of his day.[491]

In presenting his case, Waltke points out several differences between the Genesis account of creation and science. Speaking of Genesis, he says that the *subject* is God, not the forces of nature. Their concerns differ because the language and purpose of Genesis are non-scientific.

Waltke argues that Genesis 1 is not scientific literature, so it cannot give a satisfying scientific account of origins. This is true, but it does not address the issue of whether or not the Genesis account is accurate as history. He says figurative language is used in Genesis and in other Bible passages that speak of the universe as God's creation. As a result many expressions do not easily equate with the precision of modern scientific language. This is also true.

He concludes from this that,

> since the biblical narrative is non-scientific, we draw the double-conclusion that it cannot be a satisfying scientific account of the origins of things and that it can be supplemented by scientific theories. The Bible and a scientific theory of origins clash only when the latter is set forth as the complete explanation of origins and the former is interpreted as a scientific treatise.[492]

Waltke comes to the same conclusion as Pond that science and religion are separate domains, but he does so as an Old Testament scholar.

Chapter Summary and Critiques

1. Frair and Patterson depict creationism very differently than the view popularized by Whitcomb and Morris in *The Genesis Flood*. They accept most of the conclusions of the scientific community, including the belief that the universe is much older than young earth creationists believe. They are similar to Whitcomb and Morris in their belief that the Bible must be interpreted literally and has theological authority. But they take the position that we still have much to learn about the real facts of nature, and hope for the day when enough facts are known that a full reconciliation of science and the Bible can be made.

Their view of creationism is shared by a very large group of Christians, even though the narrower view of the young earth creationists is often associated with the term "creationism."

They say the big bang theory is consistent with the story of creation in Genesis, but even if it is later proved wrong this will not shake their faith in the Bible. Developments in chemistry and biology, in their minds, support the Bible account of creation. In their view, design in nature supports creationism rather than intelligent design. The scientific evidence shows that God is continuing to be active in His creation.

Frair and Patterson's chapter was critiqued by the other contributors to the book.

Pond disagrees with Frair and Patterson's belief that Scripture is inerrant. She considers it inspired but not infallible. She says human reason is needed to understand how the Bible speaks to us today. This is a fundamental difference in Christian belief, and puts Pond poles apart from them in theology. She does not accept the Genesis account of the direct creation of the human race by God, and so prefers the idea of a long history of evolution from primitive life forms to mankind as we know it today.

Van Till also disagrees with Frair and Patterson on their belief in what he calls "biblical inerrancy and episodic creationism."

2. Pond believes that the best way to avoid conflict between science and theology is to maintain an *independence* between science and theology. She quotes Gould as writing that science and religion are not in conflict, because their teachings occupy distinctly different domains. If independence is maintained, science can make no claim to represent the sum of reality. Whether or not this universe has always existed or is the result of a big bang is not important to her. The various evolutionary theories held today to explain the existence of life on earth or the development of the first cell are not convincing to her. That evolution has occurred seems clear to most biologists, but no convincing explanation has been found of how the first cell originated. However, she maintains that science has the potential to develop much farther. She says we need to be humble in light of so much that is unknown to mankind.

Patterson and Frair in turn criticize Pond's thinking. They agree with her that the domain of scientific inquiry is limited. They say they share Pond's belief that the Bible should be studied in order to understand its message. But they believe that the acknowledgement that all creatures have a common *Creator* is what is important. They say the truths of the Bible are not dependent on the details of biology.

Meyer is much more direct in expressing his disagreements with Pond. First, he says that the Bible makes many factual claims about historical events, and also about the nature and origin of the natural world. These "domains" cannot be neatly separated, as Pond advocates. Second, scientists do not limit their theorizing to the material composition of the natural world and how and why it operates as it does.

Van Till and Pond appear in general agreement on their approach to science. Since Pond believes science and Scripture are in "different domains," she does not speak to the theory of how God created.

3. Van Till says the creation-evolution debate is destined for failure, and proposed a "robust formulational economy principle." He argues that it is "nonepisodic creationism" that is necessary to defend Christian doctrine in the light of scientific advancements. This means that God originally created the universe in a single act that included the processes of development so that He had no need to intervene subsequently in order to continue this process. Van Till seems confident that scientific advances will eliminate the "gaps" in nature that other Christians point to as necessitating God's intervention after His initial creation. He even includes the development of humans as needing no direct action by God.

In the background of Van Till's argument for "non-episodic creation" seems to be a concept of God very different from the one I understand is revealed in the Bible. Van Till does not reject all possibilities that God *can* act on earth subsequent to the initial act of creation, but maintains that God created initially in such a way that he need not.

Frair and Patterson emphasize their areas of agreement with Van Till. They say they agree that it is fitting to discuss the relationship of the Christian faith and science. They also agree that authentic knowledge is not a competitor to faith, but both science and theology have unique contributions to make. But they also point out that that the Bible is a profoundly historical book.

Pond says she is in full agreement with Van Till on a belief in the theory of biological evolution. She feels her main disagreement is on what level a partnership between science and theology should occur.

Meyer presents a detailed argument to demonstrate that the scientific evidence does not support Van Till's self-organizational properties of DNA. He also challenges Van Till's belief that we cannot detect divine action in the universe after the initial act of creation. He says that Van Till overlooks scientific evidences of God's involvement at certain points in the development of life.

4. Meyer is a prominent proponent of intelligent design, but calls his position in this book "Qualified Agreement." He believes that the natural sciences now provide strong epistemological support for the existence of God as affirmed by both a theistic and Christian worldview. His general conclusion appears to be that, if it cannot be proved scientifically that God created, at least the facts are consistent with the account

Other Views on Interpreting the Bible

of creation in Genesis. He says that the creation account is the best of all known explanations of the universe as it exists, and there appears to be no other explanation for such an account in the ancient literature of the Bible unless it was inspired.

Meyer insists that clear evidence exists in the realm of biology to show that God has acted on His creation since the big bang. Fossil evidence indicates the origin of life on earth well after the origin of the universe. He believes that science shows it is very implausible for life to exist at the time of the big bang. It is difficult to accept that life was seeded on earth by a space alien or other extra-terrestrial agent like some imagine. Scientists who advocate Naturalism have failed to explain the origin of the information necessary for life's origin.

All this sounds like old earth creationism rather than intelligent design. It appears that the difference in Meyer's thinking is that we should look at nature and realize that it was designed. Then we are to "follow the evidence" to reach logical conclusions that are outside of natural science. When we do, the best explanation for what we observe in the natural world is the Bible explanation of God as Creator and Sustainer.

Patterson and Frair gave a very detailed presentation of their disagreements with Meyer, even though they are the closest of the contributors to him in their theological approach to Scripture. They agree on an expectation, or hope, that science as it develops will move closer to what the Bible says about creation and that areas of apparent conflict will be eliminated. Their chief disagreement is that they believe we should start from Scripture and consider all the evidence that the universe and all in it are designed. This is in contrast to Meyer's approach of starting with the scientific evidence and inferring what it reveals about the necessity of an intelligent designer.

They also point out the inability of observations of the biological world to discover design that can explain biological imperfections and the moral depravity in the human race.

Pond takes an opposite approach from Patterson and Frair by asking how it would affect Meyer's arguments if science progressed to where the evidences of design that Meyer cited were proved to be in error—like if cosmologists came to believe that this universe actually had no beginning. She says that Meyer is arguing from the position that design is the best known explanation for apparent design but is not the only one. She thinks naturalistic explanations may improve to the point that the existence of a designer is no longer the best one.

Van Till predictably disagrees with Meyer's argument that science demonstrates both the need for a designer (God) as the initial Creator but also for God's intervention during the development stage. He agrees with Meyer that this universe is "fine-tuned" for the development of a habitable earth and life on it. But he disagrees with Meyer's assumption that any intervention was necessary after the initial creation because he believes God could have established at the very beginning a process whereby all could develop without subsequent action on God's part. Van Till, as we have pointed out, is not recognizing the description of God's subsequent activity in the Bible as reliable or authentic.

Perhaps it is fitting to end this chapter by emphasizing what Frair and Patterson believe is the most fundamental weakness of the naturalistic and evolutionary view.

> Although it has been speculated that RNA molecules were the ubiquitous primeval catalysts (for life on earth), no actual mechanism has been identified by which RNA molecules could have formed on the earth. Given the difficulty modern chemists have in reproducing the molecules of life, it requires remarkable faith to believe that they arose spontaneously under the hostile conditions that are believed to have existed on earth during the suggested time period![493]

22

Are Science and Faith Compatible?

Introduction

A book by C. John Collins written in 2003 caught my attention because he wrote it to cover many of the issues I treat in this book and, as he put it, "I am at heart a grammarian of Hebrew and Greek." His book is called *Science and Faith: Friends or Foes?*

Collins is professor of Old Testament at Covenant Theological Seminary in St. Louis. He earned his bachelor of science and master of science degrees at the Massachusetts Institute of Technology, a master of divinity at Faith Evangelical Lutheran Seminary, and a PhD at the University of Liverpool in Hebrew linguistics. This suggests he has the background in natural science and biblical studies to research and write on both fields.

Although I was drawn to read this book because I hoped it would increase my understanding of how key Hebrew words and phrases are used in the Genesis account of creation, I discovered many other interesting things in its twenty-one chapters and several appendices.

One thing quickly got my attention. Collins quotes from many of the same scientists and philosophers of science in his book that I do in mine. He refers to Michael Behe or his *Darwin's Black Box* fourteen times and to Hugh Ross four times. He mentions Michael Denton's *Evolution: A Theory in Crisis*. He also mentions Bernard Ramm's *The Christian View of Science and Scripture*. He refers to Henry Morris and Davis Young, and five other authors I quote in this book.

A clear emphasis of Collins's book is that "the Christian faith" (the Christian belief system) is rational, and faith in it is in no way opposed to reason. He is convinced that if Christianity is true, the world and life view presented in the Bible ought to equip us to live *better* on this earth than any other view can.

PART 3—*What Do We Really Know about the World?* The Bible

He says that "a world-and-life view is your basic stance toward the world, and we can express it by questions like these:

Where does this world come from?

Is the world good or bad? (How can we define good and bad?)

What does it mean to be human?

How should people live?

Should all people live by the same standards?

What should we do with our failure to live by these standards?

What is a reliable guide for answering these questions?

What place does God have in it all?[494]

I will begin by reporting some of Collins's observations regarding the meaning of faith and regarding science. Then I will discuss some of his comments on the creation account in Genesis, chapters 1 through 3. We will also consider how old the earth seems to be.

1. The Meaning of Faith

Collins says that faith and reason are not opposites (or alternatives) as some tend to think. We use reason in believing. He explains that faith is trust toward God because you are persuaded that He is trustworthy, and faith is *rational* because we become persuaded of God's trustworthiness. We do so because He gives us things to believe and reasons for trusting Him.

> One conclusion is that faith and reason are not at odds with each other. Faith is in fact rational behavior: given Who God is, and the reasons He's given for trusting Him, it's unreasonable *not* to trust Him. It's true that faith goes beyond what I can verify; but that's true of every kind of relational faith.[495]

He came to the conclusion that doubt is not always due to a lack of faith. If by "doubt" we mean wavering or divided loyalty, then he says it is dangerous to faith. However, if our loyalty to God is steadfast but we doubt our understanding of God's ways, then he points out that this often exists side-by-side with faith.

> Our faith supplies us with four strong motives for loving science.
>
> The first is, *to praise the Creator for His creativity* . . .
>
> Second, the sciences allow Christians *to enjoy God's goodness as we satisfy our curiosity*. Curiosity is part of the image of God . . .

Third, the sciences allow us *to serve mankind*.

Fourth, the sciences allow us *to answer unbelief*.[496]

2. Scientific Issues

He says a science is a discipline in which one studies features of the world around us, and tries to describe his or her observations systematically and critically.

Collins discusses *methodological naturalism* and quotes a description of science by the National Science Teachers Association (NSTA) that includes the following:

> (Science) assumes the universe operates according to regularities and that through systematic investigation we can understand these regularities ... Because science is limited to explaining the natural world by means of natural processes, it cannot use supernatural causation in its explanations. Similarly, science is precluded from making statements about supernatural forces, because these are outside its provenance.[497]

This argument seems reasonable to me until it is used to promote *philosophical naturalism*, but Collins says these statements have a very debatable premise. This is that in explaining everything by methodological naturalism, it is required that all explanations be in terms of natural causes only. He says this premise is questionable because it doesn't make a distinction between the facts of nature today and historical events.

He says that, since NSTA applies this approach to historical events, such as the origin of the universe, it includes the hidden assumption of philosophical naturalism—that natural causes are all there are. He calls this a "*theological* claim"—something the statement itself claims is "*outside* its provenance" or area of expertise or study. So Collins concludes that the NSTA contradicts itself. Perhaps more to the point is to recognize that many scientists willingly accept naturalism as a method but reject it as a philosophical assumption.

We have noted in earlier chapters that some scientists who accept both the biblical account of creation and valid science also accept naturalism as a *method* in scientific research. But Collins is concerned that many who accept "methodological naturalism" go beyond this. He observes that "the effort to promote *methodological* naturalism—appealing only to natural processes in your explanation—slides over into *philosophical* naturalism—the belief that natural processes are all there is."[498]

Collins states that our culture is "obsessed" with what you can measure, believing that this by itself guarantees objectivity. Then he points out the absurdity of ignoring the strength of a person's will when giving a "scientific" description of him.

Collins points out that many things that science accepts as facts are really based on inference. He reminds us that the inference is only as good as the chain of reasoning that produced it. For example, cosmologists think there is much "dark matter" in the universe, because they infer that this is the best way to explain certain gravitational effects that they see no visible explanation for. But he says we should be careful not to treat inferences on the same level as observed phenomena. "If we want to be really careful, we should say 'matter behaves *as if* it were made of protons and electrons and other stuff, and I don't see any reason to doubt that it really is'; 'molecules reflect X-rays *as if* they had such-and-such a shape'; and so on."[499]

He says many sciences will overlap with the content of our Christian faith: for example, when they deal with the origin of the universe (cosmology), or with the origin of man (anthropology), or with human nature (psychology). He believes the closer we get to what it means to be human, the more opportunities we have for overlap, and, as it turns out, the more one's personal commitments come into play in scientific theories.

Collins discusses the age of the earth, considering both cosmology and geology.

Cosmology

> *Cosmology* is the science that studies the origins of the universe, together with its history and physical structure as a whole. It's related to *astrophysics*, which focuses on the physics of stars and space, both in their inner workings and their interactions with each other.[500]

When scientists discovered in the twentieth century that the galaxies were actually moving away from each other they naturally concluded that the universe is *expanding*. If it really is expanding, it must have started from a small compact mass. Then space itself must expand as the material universe does.

Collins points out that the anthropic principle refers to the physical properties of the universe being so finely balanced that they can support life here on earth, and evidently this isn't true anywhere else in the universe. He refers to the anthropic principle as a "huge benefit that comes from our faith from modern cosmology." This of course refers to the "fine tuning of the universe" that Hugh Ross so aptly described and we presented in chapter 6. Collins refers to Ross's list of conditions needed to support life on earth.

Collins asks if the big bang theory points to the absolute beginning of the universe.

> Does the doctrine of creation from nothing imply that the universe had a beginning in time? The answer seems pretty clear to me: certainly it does. When Genesis 1:1 says "in the beginning" it refers to the beginning of the material universe. Further, when Revelation 4:11 says "for you created all things, and by your will they existed and were created," it likewise implies that there was a moment in which it all began.[501]

Collins's second question here is whether or not we need to have scientific proof that the universe had a beginning. He says we do not, because we already have God's word for it.

Geology

It was Charles Lyell (1797–1875) who first promoted uniformitarianism, and he had a profound influence on Darwin's thinking. Collins explains that uniformitarianism is the belief that the past earth history can be explained fully by the processes that we observe in operation now, and that it is not necessary to explain them by obsolete processes or by supernatural events.

> The substantive principle of uniformity says that the geological processes have run at the same rates, and at the same intensities, throughout the history of the earth. The methodological kind of uniformitarianism says that these rates and intensities may have varied over the course of the earth's history—which means that dramatic upheavals (catastrophes) may well have happened. This is the principle of uniformity that is most widely spread among today's geologists, and it makes no comment whatever on the possibility of miracles or of the biblical flood.[502]

"Methodological uniformitarianism" is the application of "methodological naturalism" to geology.

Collins makes the interesting comment that "as Christian (old-earth) geologist Davis Young put it, 'modern Flood catastrophists are really proceeding on the same principle as do modern uniformitarian geologists.'"[503] This suggests that what divides them is not the *approach* to the available geological data but only the interpretation.

Collins also goes into the issue of whether or not radiometric dating methods are reliable. He mentions a common argument of young earth creationists like Steven Austin who argued that radiometric methods yielded bad results when applied to some rocks in the Grand Canyon. Brent Dalrymple of the US Geological Survey, a leading expert in geological dating, challenged this. Dalrymple found flaws in Austin's analysis. Collins stated that he had no reason not to believe the standard theories of geology and that this includes their estimates of the age of the earth. In this way Collins rejects the arguments of the young earth scientists that the methods used to date the strata of rocks used to compute the age of the earth are completely unreliable.

Collins's conclusion, based both on cosmology and historical geology, is that the universe and the earth appear to be billions of years old. Here he is in complete agreement with Hugh Ross on the age of the universe and the earth, and with Davis Young on the age of the earth.

PART 3—*What Do We Really Know about the World? The Bible*

Special Issues in Science

Collins describes two key issues in biology that are opposed to the biblical account of creation in many minds. First is biological evolution. Collins reminds us that there are theistic advocates of the theory of evolution like Howard Van Till. "A theistic advocate of this theory would have to say that the natural events are God's action by way of 'ordinary providence'—that is, that God designed a universe so well that he could simply keep it in being and it would go on to generate life, and eventually us."[504]

Collins faults Van Till for saying that God made the world with all the capabilities built into it that would ever be needed so that he never needed to intervene in the history of earth or in the development of life. Also for Van Till insisting that if God had to intervene after the original creation it would show that he is an "incompetent craftsman." He says that Van Till is correct in stating that "modern biological theorizing" of neo-Darwinism "presumes that you can get from molecules to mankind by a purely natural process." And that it is acceptable to presume this without feeling they are required to prove it. Then Collins remarks, "But Van Till is dead wrong that the Christian doctrine of creation supports this presumption."[505] Collins refers to the fact that the creation account is "historical" (without necessarily being scientific) like we have discussed earlier.

Some young earth creationists argue that the Bible says that God created every species separately. But Collins argues that "kinds" in Genesis 1 does not mean "species." He says, "First, the biblical text simply says that those first plants bore seed according to their kinds, and that the first animals were created according to their kinds. It does not say that these are the only 'kinds' there ever were or ever could be. Second, it is notoriously difficult (notorious at least among Hebrew scholars, anyhow) to define just what the word 'kind' means: in any case it doesn't mean the same as 'species' (that's too narrow and too technical a meaning), it's more like 'category.'"[506]

Collins argues that the word "create" denotes a supernatural action, and the term "image of God" refers to capacities that no other animal possesses. Therefore, he says, the Bible shows that God was especially active both in the way various life forms began, and in the origin of mankind.

The second issue Collins raises is evolution and progress. He points out that even though Darwin envisioned that natural selection would cause organisms to "tend to progress toward perfection," the reality is far different. He quotes J. B. S. Haldane, one of the originators of the neo-Darwinian movement, as follows: "We are therefore inclined to regard progress as the rule in evolution. Actually it is the exception, and for every case of it there are ten of degeneration."[507] We reported earlier that Gary Parker discovered that most random mutations are actually harmful to the organism. This realization certainly weakens the support for Darwinian-style evolution.

3. The Biblical Arguments

> I have argued that we cannot get from the creation days any biblical position on how old the earth and the universe are supposed to be. All we can say for sure is that the beginning of the first day (Gen.1.3) may be some unknown amount of time after the absolute beginning of the universe (Gen.1:1), and that the creation "week" for earth (Gen.1:3–2:3) had to be longer than an ordinary week in order for Genesis 2:5 to make any sense.[508]

Collins believes that the three strongest biblical arguments for a young earth are that the days of Genesis are considered ordinary days; the genealogies of Genesis 5 and 11; and Jesus' statement in Matthew 19:4 and the parallel passage in Mark 10:6 that are thought to put Adam and Eve at the beginning of creation.

Collins calls the argument from Jesus statement in Mark 10 and Matthew 19 unsound. (Matt 19:4 says, "And He answered and said to them, 'Have you not read that He who made them from the beginning "made them male and female.""") The English used in Matthew 19 for "from the beginning" seems quite definite, but the original Greek is not. Collins refers to many places in the New Testament where the phrase "from the beginning" is used and shows that they refer to various times as suggested by the various contexts. This shows that "from the beginning" in Matthew 19:4 refers to the beginning of the human race and not the beginning of creation.

Regarding the genealogies, Collins demonstrates from Scripture that it is clear not all the men in a given genealogy are listed. He does this using examples like "Joran fathered Uzziah" in Matthew 1:8, and 2 Kings showing that Uzziah was actually Joran's great-great grandson. (But these gaps in the genealogies were acknowledged to some degree by Henry Morris also, as we saw earlier.) He also notes that the Bible says the Israelites were in Egypt 430 years, and Kohath is listed in Exodus 6:14–27 as the grandfather of Moses, and Moses was only eighty at the end of those 430 years. (Kohath was one of Levi's three sons, and he moved to Egypt with Jacob 350 years before Moses was born.) "Father" in the genealogies clearly can mean "ancestor" and "son" can mean "descendant." Collins concludes here that he cannot figure out how many gaps in the genealogies there are or how large they are.

Collins summarizes by saying the Bible gives six thousand years as the *lower* limit to possibilities of the age of the earth, but there is no way of calculating a possible *upper* limit.

The Biblical Doctrine of Creation

Collins goes into some detail to explain the doctrine of creation from Genesis 1 and 2, paying careful consideration to the usage of Hebrew in the Bible. In doing so, he addresses several important issues.

He doesn't believe that Genesis 1:1 is a summary and that the rest of the chapter is describing it, as some argue. He makes three points here.

> The first day begins in verse 3 (with "and God said"), and verses 1–2 are *background*—they describe the setting of day one. The most usual function of the kind of background statement you have in verse 1 is to give an action that took place some unspecified time before the narrative actually gets under way (as in Gen 16:1; 21:1; 24:1).
>
> All this means that the origin of the whole show gets taken care of in one verse, and the author moves on to focus on something else.
>
> Second, the word "heavens" and "earth" in verse 1 refer to "everything"; but after verse 2 they get narrower in their meaning: "heavens" narrows to "sky." This tells us that the author has narrowed his focus from the whole universe down to planet earth.
>
> Third, the high point of the narrative is the sixth day (which gets the longest description), and especially verse 27, the making of mankind . . .
>
> So the focus on the making of earth's different environments and inhabitants reaches its peak on the making of humans who are to rule over the whole earth. The earth is a good place for these people to live, love, work, and worship God.[509]

In separating Genesis 1:1–2 from the verses that follow Collins makes it clear that it is a mistake to consider these verses as part of the first of the six days of creation. This directly rebuts the young earth argument that the initial creation was part of the first day.

Then Collins asks the question,

> Now then: did the author mean us to take Genesis 1:1–2:3 as history? The answer is certainly yes, for two reasons. The first is its place in the book of Genesis, a book that is concerned with historical matters: it starts the whole thing off and explains why things are the way they are . . . The second reason is, that's the way people in the same culture read it.[510]

Collins does not consider the Genesis account of creation to be "scientific."

> So what do I mean by saying that the account is not "scientific"? Well we notice for one thing that it paints with broad strokes: except for man, no single species of plant or animal receives a proper name; we find no details about *how* the earth brought forth vegetation, or how the animals appeared in their different environments . . . The account describes things with suggestive terms, such as the "greater light" and the "lesser light" (strange names for the sun and moon, for which there were ordinary words in Hebrew).[511]

But if we say that Genesis account isn't quite "scientific," we haven't said enough when we're talking about the Bible and science. The way Genesis describes

Are Science and Faith Compatible?

God's work of creation lays a foundation for science and philosophy—for all sound thinking about the world. This is because it tells us that a good and wise God made the world for us to enjoy; and the things in the world have natures that are knowable (for example, the plants and animals reproduce "after their kind"). Our senses and intelligence allow us to say things that are true.[512]

Collins thinks Moses chose language that best fits his purpose. He says Moses chose words that were broad strokes, majestic in simplicity, allusive, but strict in pattern. He says this is well-suited to bring out a sense of wonder at the creativity and power of God.

The emphasis is on the fact it is *God* Who acts.

God says it, and then it happens—it doesn't say how long it took for it to happen, but that doesn't matter: what matters is that God's wishes get carried out. God expresses His power in the way He calls the universe into existence and then shapes the earth as a place for His human creatures to live.[513]

Genesis 1:1–2:3 says that God made all things; (a) from nothing; (b) by the word of His power; (c) in the space of six days; (d) all very good. This shows it is God's world from first to last.

He says that in this way the creation account is actual history, but not scientific in its descriptions. The first focus is on the *God* who created, and the second focus is on making a world suitable for *man*.

What Kind of Days Were They?

Next Collins asks what kind of days were the six days of creation? Were they "literal" days? Here Collins makes the point that a "literal" meaning of a text is to bring out literally what the author intended. He says the important thing here is to try to determine what a reader in that culture and time would have understood the text to mean.

Then he says that verse 2 explains what the conditions were on earth when the "creation week" got underway. It says that the earth was "without form and void," or "desolate and empty."

Collins argues from the Hebrew of the text of Genesis 2:5–7 that there must have been a long period of time since plants were first created before man was. (This is explained more in the appendix.)

Collins explains in detail why it is clear from Genesis 1 and 2 that at least some of the six days of creation were not "ordinary days," but were much longer periods of time. He agrees with many others that the things Moses listed on happening on the sixth day simply could not have been accomplished in one twenty-four-hour day. Then he points out that the Hebrew of Genesis 2:23 is best read, "And Adam said: 'This is *at last* bone of my bones and flesh of my flesh; she shall be called Woman, because she was taken out of Man.'" This suggests a very long period of time between Adam's creation and that of Eve.

Collins criticizes the argument that when the Hebrew word *yom* (Hebrew for "day") is qualified by "evening and morning," like it is six times in Genesis 1, that it can only mean a literal day. Collins replies that "day" is not modified by "evening and morning." He says there is no "rule" in Hebrew that would make this refer to a literal day in any case.

Collins argues that we should consider the six "days" of creation as periods of indefinite length, even though he disagrees with the "day-age" theory advocated by Davis Young and many others.

Conclusion

John Collins is well-acquainted with the Hebrew of the Old Testament, and his knowledge of Hebrew usage shines through in important places. For this reason his perspective on the issues discussed in part 3 can add to our understanding.

Collins says that faith and reason are not to be considered "alternatives," and that we do use reason in believing, whether it is in scientific or religious matters. Christians should love science, but they should never forget that there is much we do not yet understand about the natural world and about God's ways. Much of science is based on inference and not on actual direct observation.

He criticizes organizations like the National Science Teachers Association because they say that science cannot speak about supernatural forces, but they also promote philosophical naturalism which includes the hidden assumption that natural causes are all there are. So much of the argument for the validity of scientific theory based on modern Darwinist thought is based on the faith conviction that "nature is all there is." Collins is one of many who encourage people to look for human biases in the presentation of truth claims.

He believes the Genesis 1:1–2:3 creation account is historical but not "scientific"—because it presents the account in broad strokes and not in the details needed for scientific description. Yet the description lays a foundation for science and philosophy, in that it attests that nature can be understood and humans are equipped with the intelligence to understand it. He also rejects the idea that "kinds" in the creation account means "species," and adds that it is notoriously difficult to define what the word "kinds" means.

Collins is convinced that the six "days" of creation are literal days in the sense that they are not allegorical. They must be periods much longer than our twenty-four-hour days because of the description in Hebrew of what happened on certain days. He says that the Hebrew makes it clear that plants must have existed and gone through several cycles of rain, plant growth, and dry seasons before mankind was created. He shows from Genesis 2:5–7 that there must have been a long period of time between the time plants were first created and Adam's creation. He also quotes the Hebrew of Genesis 2:23 to show that a long period of time must have transpired between the creation of the man and that of the woman. He points out that so much happened on the sixth day that it must have been a very long day.

I think he would wholeheartedly endorse Rimmer's comment that the Hebrew word "day" is one of the mysteries of Scripture, and that no scholar is so erudite that he can be dogmatic about its meaning.

Collins argues against the gap theory because it is not natural to read the Hebrew of Genesis 1:2 to mean "the earth *became* without form and void." He accepts the idea that the days of creation are periods of indefinite length but does not accept the "day-age" way to explain this.

This explanation allows him to consider that our earth and the universe may possibly be as old as the astrophysicists believe. He believes with many others that the genealogies in the Bible are incomplete. His conclusion is to consider six thousand years the *lower limit* for the possible age of the earth and says there is no way of calculating the *upper limit*. Yet he does believe it is old.

Collins's thinking is consistent with the big bang theory. He says that the Bible clearly teaches that the universe had a beginning, and that this beginning was determined by an act of God's will.

Collins's contribution to this book is at least twofold: First, his analysis of the meaning of the creation story can be considered important because of his training in the usage of the Hebrew language. And second, he accepts much of what science has discovered while cautioning that we must be careful of human biases in the truth claims we accept—whether these are in the field of science, in philosophy, or regarding religion.

He concludes one chapter by referring back to the issue of faith versus reason.

> Sometimes people want to contrast faith and reason or faith and science . . . but genuine Christian faith makes us want to be true to reason, and to reason well in science. It's not my faith that denies reason and that reasons poorly from the world: it's neo-Darwinism.[514]

23

An Evaluation of Evolutionism and Creationism

We introduced Harold Booher in chapter 13 when we reviewed various fields of science and the development of the science supporting current theories. Booher also discussed the various creationist views in his book *Origins, Icons, and Illusions*. The book is subtitled *Exploring the Science and Psychology of Creation and Evolution* and covers views on creation as well as on the science supporting evolution and intelligent design. We noted that he sought to be objective, as an expert in systems analysis.

The Credibility of Scientific Creationism

> Without the belief that credible witnesses observed supernatural events recorded in history, there would be none of the monotheistic religions of today. Without a belief in a literal Exodus with attendant miracles, Judaism would not have come into existence. Without the credible report of eyewitnesses to the death and resurrection of Christ, Christianity would not be.[515]

Here Booher acknowledges a fact about both Judaism and Christianity. This is that they are rooted in a belief that reliable witnesses reported supernatural events that happened on earth at specific historical times. He does so without revealing whether or not he is a believer in any religion.

> For the believer in intelligent design and in the existence of a designer capable of both intervening into history and communicating to humans certain characteristics of His design, the book of Genesis is a valuable document. At the very least it provides clues for the philosopher and the scientific investigator of origins.[516]

Here Booher recognizes the validity of studying the book of Genesis for information on the origin of the universe and life. Again he does so without declaring faith in the supernatural origin of the Bible.

> Although neither religion nor science would wish to be held to specific interpretations, there is a rough approximation between what science believes today with what is told in Genesis. There was a beginning; there was progress from material to plants to animals to human; each type of living thing is distinguishable even in the fossil record and reproduces "after its kind"; the universe is in a decay process toward greater disorganization, and geology largely reflects catastrophe through flooding and volcanic eruptions.[517]

Although Booher does not mention this here, these are some of the things that believers in divine inspiration point to as a demonstration that God revealed the truths of Scripture to men. There appears to be no other logical explanation why Moses would know these things if they were not revealed to him by a supernatural source. They were written down thousands of years before mankind developed the scientific methods and tools to investigate such natural relationships and facts.

Booher reports that recently there has been an increase in interest in what can be called scientific biblical creationism. This has not only been seen in the United States and in Britain, but also in Russia since the dissolution of the USSR. He says it is characterized by two things. First, it relies on ancient human documents to discover clues to past cosmological and biological events. Second, it is almost totally excluded from mainstream scientific publications, and government funds are not available for research on creationist theories. Creationists are not even knowingly assigned to research positions. He points out that most of the prominent scientists who are leading the new look at creationism have appropriate credentials and training in their fields, and have different research interests and findings just like evolutionists do.

Booher examined the more recent research of many of these creationists and found it to be of the highest academic quality.

> In general, creationist statements on what the observational data base actually is appears as sound as any the evolutionist scientist would make. Even the most rigid of their members do not deny that mutations occur and are heritable, that there is a process of natural selection, or that fossils show a general trend from simple to complex in the geological strata. Most also accept the same geological and biological data that supports the new biology and the catastrophe paradigms. Where they seem to differ from evolutionists of any kind, even theistic ones, is [in] how old the earth (and the universe) is and whether there was a world-wide Noachian flood.[518]

As we saw when we discussed the writings of scientists like Hugh Ross and Davis Young, many creationists who take the Bible as the verbally inspired Word of God accept an old age for the earth. Many also consider it is quite possible the flood in Noah's

PART 3—*What Do We Really Know about the World?* The Bible

day was limited to the areas where humans were at the time. They cannot in any sense of the world be called "evolutionists."

> The young earth biblical creationists . . . seem to have a model which can work within the constraints of the fundamental observable natural processes (including micro-evolution), which accepts the fossil record and existing species classification system, and is generally consistent with the new paradigms for geology and biology.[519]

While saying these things, it is clear that Booher is speaking from a viewpoint outside of the orientations that he evaluates.

Five Clues from Genesis

Booher refers to five clues from Genesis that suggest what a "generalized creation model" might consist of.

> The primary components of a generalized creationist model could comprise 1) a first cause which over a period of time brought into being the universe and all living things; 2) an introduction of the second law of thermodynamics sometime after the original creation which effects the entire universe including life on a path from low entropy to high entropy; 3) a special creation and division of life into separate "kinds"; 4) an extra special creation of human kind; and 5) a general ordering of the fossil record of kinds in the geological record from simple to complex, due to catastrophic events.[520]

He adds that this model postulates a supernatural intelligence as the initiator at each sub-stage, but that scientists can use naturalistic models to investigate the observable effects of each of the five supernatural intrusions.

Booher explained his "five clues from Genesis" as follows:

The first clue is when God created heaven, earth, light and energy, vegetation, the sun, the moon, the stars, and living creatures, and ended with the creation of man. After each major creation "God saw that it was good." He explains that this records the concept of ancient people that the universe was originally perfect. It supports the idea that the operations of nature were fully functional when they came into existence and did not need further modification in order for them to function. It does not rely on any cosmological theory of evolution.

The second is the fall of man, which brought disintegration. God cursed nature at that time because of man's sin, and Paul in Romans 8:21 speaks of the whole creation becoming subject to the "bondage of corruption" or "decay."

Booher suggests this can explain the laws of thermodynamics.

> From these two clues of the Bible which arguably complement the first two laws of thermodynamics, a simple creation model for degree of complexity over time can be contrasted with the evolution model. The creation model supposes ideal order and organization to be the condition from the beginning. With passage of time, there can be only a decrease in overall degree of complexity.[521]

Booher seems correct in saying that the Bible is consistent with what we observe in the natural world—even though Genesis was written long before science discovered the laws of thermodynamics. He contrasts this with the evolutionary model, that predicts an increase of complexity with time through natural processes. He concludes that evolution as a present-day theory does not match the facts as well as Genesis does. In other words, creationism seems to explain better what we actually observe in nature.

The third clue from Genesis is the idea of the separation of "kinds," and he says this is supported by modern biology and geology. But he adds that exactly defining what "kind" is presents a problem to creationists. He thinks that "family" might be close to its meaning.

Booher explains that if "family" is the meaning, this could mean foxes, wolves, dogs, coyote, jackals and hyenas could all have descended from one kind. But it would predict that there would be no transitional forms between families.

> Fishes would not be linked to reptiles, reptiles to birds, or apes to man either directly or through the aid of some common ancestor. This, the model predicts, would *not* be shown either in the present array or in the fossil record.[522]

Booher says that the fourth clue relates to psychology. He notes the fact that humans have a unique intelligence and free will is because of the human mind. The clue to this is that "something was implanted in people which was over and above that which separates the other 'kinds' from each other." God created man in His own image.

> Mortimer Adler says it is this distinction of man in the image of God which forms the basis for both the Jewish and Christian faiths of the unique relationship of man with God and ideas of the immortality of the soul. He argues there is considerable scientific support for any belief where a touch of "spirituality" is uniquely manifested by the human intellect (not intelligence). In particular psychological research on animal behavior shows the human mind to be distinctly different in kind from that which can be found in any animal.[523]

Booher calls the universal flood the fifth clue from Genesis.

> A central issue which distinguishes some creation and all evolutionary models from the biblical creationist is in interpretation of the cause of the geological strata. The literal Genesis creationists believe that the entire surface of the earth was once disturbed by a world-wide deluge—that the crust's formation

and the dispersion of fossils within the strata can largely be explained by a single catastrophic flood.[524]

Booher calls the intelligent design theory "which combines the literal Genesis record with field research and laboratory observation" an extreme intelligent design theory. But then he adds that it is intriguing that scientifically derived data supporting the Genesis hypothesis is greater now than it was in Darwin's day.

Booher is convinced that it would not be impossible for an omnipotent God to create a perfect world in six days, later add a universal disintegration process, and sometime later sweep the earth with a catastrophic flood.

In the above critical analysis Booher does not point out anything written in Genesis that appears to contradict the state of nature as we understand it today.

> Perhaps the strongest criticism against the research of modern biblical creationists has been their general willingness to accept too quickly nearly any claim if it could be used against evolution . . . The problem of not well screening claims seems to be disappearing with those creationists who maintain a high respect for the scientific method.[525]

Views on Theistic Origins

Booher reports that theistic scientists have postulated three general positions on how God has been involved in the origins of living creatures.

The most well known he calls special creation. This includes the belief in God's special activity in creating fully formed organisms from the beginning. (I would add that it often includes God's occasional intervention later in the processes of nature and limited biological evolution. This view is similar to Gordon Mill's position that Booher describes below, but its adherents would not agree on being called "theistic evolutionists.")

The second is called theistic evolution. He points out that there are only a few differences between this position and a completely naturalistic scientific perspective, and one is that they believe in a principle of an "in-built potentiality of all-that is." They believe that God originally created the universe with all the potential needed for all the developments that have transpired since. This is similar to Michael Denton's view. We considered this viewpoint when we introduced Van Till's thinking in chapter 21.

Booher considers the third position a new theory of theistic evolution suggested by Gordon Mills. He quotes Mills, who says that "in the history of the origin and development of living organisms, at various levels of organization, there has been a continuing provision of new genetic information by an intelligent cause."[526]

> Especially important for the creation/evolution debate is the fact that Mill's theory suggests a positive approach to scientific research that is not bound to

the naturalistic model. For example, he is not bound to assumptions of *initial simplicity*. An intelligent cause could have provided genetic information for whatever degree of complexity that was necessary.[527]

Booher considers the third position a new theory of theistic evolution suggested by Mills thus does not assume like Van Till and others do that the original organisms were simple and evolved in complexity. But he also allows for intervention by the designer at any point in the development of organisms.

> Mills' theory covers such possibilities as (a) rapid diversity of species (as might follow mass extinctions); (b) macroevolutionary events requiring a number of new genes and control factors; and (c) the dormant retention of genetic information until needed again hundreds, thousands or possibly millions of years later.[528]

I'm not sure why Booher calls this a "new theory of theistic evolution." As described above, creationists like Hugh Ross would surely agree with it. So would Davis Young and many other creationists whose views are described in this book. These scientists consider themselves creationists and not "theistic evolutionists."

The Problem of "Imperfection in Creation"

Booher highlights what he considers the most difficult problem with the creationist position.

> It is difficult to conceive of a tougher philosophical or theological question for the believer in design (by a supreme intelligence) than that of human or animal suffering. Cruel and awful things totally outside of human control befall the good as well as the wicked. Catastrophic floods, earthquakes, mud slides, and volcanic eruptions wipe out thousands of people including innocent babies and children without warning every year. Great portions of the continuance of life depend on animals eating each other. Ugly, horrid diseases maim, blind and slowly draw life from millions of people year after year after year.[529]

But this does fit into a design framework, if the designer does not concern himself with moral issues, according to Booher!

> Strictly speaking the system of "eat or be eaten" readily fits within the concept of design. Such a design seems indifferent to individuals but provides for variety and survival of species as a group. It is only when we require the designer to be a benevolent one that nonmoral appearing nature creates such a shock to our sense of "goodness" in design.[530]

Booher thinks it is *the problem of evil* that challenges the concept of a personal, moral God. He says, "The Problem of Evil creates one of the greatest unsolved mysteries for monotheistic religions."

> Briefly, the problem is "If God is omnipotent, omniscient, and all good, how is it possible for evil to exist in the Universe? If He is all good, He should desire there be no evil. If He is omniscient, He should have been aware of evil His creation would undergo. If He is omnipotent, He could have created a world which did not have evil."[531]

Booher does not think this issue has been solved philosophically. Indeed it does seem hard to understand why a moral God would create humans with free will that He knew would disobey Him and corrupt the human race morally. The Bible says that Christ was crucified in the plan of God "from the foundation of the world" and His crucifixion was for the purpose of allowing God to "be just and the justifier" of those who believe on His Son. This shows that God knew before He created Adam and Eve that they would fall into sin. Yet God created them with a free will that could choose to sin. His nature is holy, and to be true to His nature He could not just pardon sin and simply void the penalty for sin. But why God would even knowingly allow such a tragedy is difficult for the human mind to fathom.

Booher says that the famous Catholic philosopher Thomas Aquinas proposed as a solution that God permitted evil to enter the world to bring about a greater good. Part of this greater good, he says, is human free will.

Booher suggests there are three arguments given by theists to explain why an intelligent Creator Who loves His creatures would make a system with so many imperfections.

First is what he calls the "Bigger Picture." This says that the imperfections are few and small compared with all in the universe that has been designed to function so well. Scientists say that earthquakes and floods actually do have beneficial results. God allowed diseases to exist, but also provided humans with the capacity to advance medical technology to develop cures. Some argue that genetic "mistakes" and "deviations" are unavoidable consequences of processes that contribute somehow to the overall development of the basic designs. (Bernard Ramm argued that God allows natural disasters and sickness to exist to give mankind "tasks" to overcome, as we saw in chapter 17. To Ramm, these problems would have existed even if mankind had not fallen into sin.)

Another argument is "God Is Not Omnipotent." Some have concluded that disease, accidents, and even murder are—for some unknown reason—things that God cannot control.

Booher concludes that these first two arguments are from those who believe in design and also in macroevolution. To say that it takes earthquakes, floods and disease to make His system work, or that God needed billions of years of evolution that include evil in nature, in order to make a system to His liking, is simply unconvincing. Booher says he finds it difficult to believe any theologian would put God in what he calls "such a restricted box" unless he was convinced that this is the only way to reconcile the facts of evolution with his theology.

The third is the "Genesis Story."

> The Genesis story of creation, however, starts out the way our consciences tell us it should be. The earth and all its inhabitants were at peace. Animals did not eat each other or fight for individual or species survival and people did not eat or kill the animals. This God prescribed as "Good." The Genesis writer describes God's idea of good and ours are not that different. But as time went on flaws came into the good design.[532]

Booher mentions such flaws as man's fall into sin, and the flood, and refers to Paul's statement on present conditions: "For we know that the whole creation groans and with birth pangs in birth pains together until now" (Rom 8:22).

> Although there were flaws introduced into the good design, God has been working diligently, but patiently to correct them. Up until the past few thousand years, specific examples of His direct working in history are recorded.[533]

Booher feels this last argument is different from that of the "Bigger Picture" and that of "God Is Not Omnipotent" in several ways. It acknowledges that what we sense are imperfections are really that; they are not "moral" as he puts it. Also, these imperfections weren't part of the world that God wants.

He says there is no indication in the Genesis story of God being either not omnipotent or inactive. The imperfections are temporary. God is working patiently to diligently to correct the flaws. "Finally, in a not incomprehensibly distant future, an even better design is promised, one which not only includes animals and people at peace, but also people freely and joyfully loving one another as well as God, their Creator."[534]

There is at least one weakness in this argument. According to the Bible, all nature will be "redeemed" from the curse put upon it someday. But not all people will participate in this redemption. Only those who have been personally redeemed by God's grace and faith in Christ will. Many will exist forever away from God's presence and favor. We really do not know why God allowed sin to enter the human race, with the result that many are "lost" forever. The only answer I can conceive of is that somehow allowing mankind to choose or reject Him is consistent with the eternal nature of God himself.

> Regardless of the basic philosophy endorsed, creationists and evolutionists alike find that some combination of design and also chance is necessary to account for all of nature. There is perfection and harmony and also evil and catastrophe operating together within the same system.[535]

Is There a Middle Ground?

Only two primary theories have ever been presented to explain the diversity and origin of life. These are special creation and evolutionary creation.

Many civilizations have creation stories, some roughly similar to the Genesis account, and Booher believes that opposing views have coexisted as philosophical choices from the beginnings of civilization.

But Booher adds that many today are seeking some middle ground between a totally atheistic evolutionary viewpoint and a theistic view that rejects evolution altogether.

> For the first time in history, with the advent of theistic evolution from the theologians and teleological molecules from the micro-biologists, the black-white distinction between world views has grayed. There can now be seen a regression from each opposing belief toward the center, each attempting to have the best of both world views. Theologians of the late nineteenth century developed a new theology to reconcile their faith with what they thought was proven scientifically while biologists of the present have discovered [that] pure randomness [is] inadequate to explain the overwhelming complexity and synthesis of even the simplest life forms in nature.[536]

He says that theistic evolution sees God the inventor of evolution and that "teleological materialism" attributes God-like qualities to evolution. (By teleological materialism Booher means a materialistic philosophy that embraces the concept of purpose or goal-directedness in evolution.)

> It is possible to be intellectually honest, true to scientific fact, and yet believe in a God who designed original "kinds," some of which persist until this day. Similarly it is possible to see all of nature from a strictly materialistic point of view, believing evolution is the glue which holds it all together. Differing combinations of these basic beliefs are also possible—like God making evolution or material taking on intelligent-like qualities. What is not legitimate is the claim that any of these central beliefs should either be excluded or be given special consideration on the basis of science.[537]

In this way Booher insists that, in considering various philosophies of science, the belief in creationism should be recognized as legitimate along with the belief in evolutionism, and various mixtures of these basic orientations should be recognized as well.

> The examples and arguments in this book show neither of the two old opposing views of nature can rely on science to convince the "non-believer" that it is necessarily closer to the truth. One view must rely on revealed truth, as the ultimate basis for confidence, while the other has its faith founded on a grand illusion which defies demonstration. Both of these concepts are all

encompassing, fundamental to all thought. Both, with the present state of observational data, lie outside scientific verification. Any conceivable data yet to be discovered can be placed in either camp. Consequently this debate is unlikely to be resolved through further scientific discovery ... each new finding confirms *design* to the believer in a supreme intelligence. Conversely, to the believer in evolution, the same discoveries only confirm awe of natural selection.[538]

Then he asks, "Is evolution a religion?"

For a belief so fundamental and so often seen as competing with religion, it is not unreasonable to ask whether evolutionism is as much religion as creationism. In the central role that evolution plays in the minds of so many people it certainly can be likened unto a religious one.[539]

Booher reports that some evolutionary scientists have argued that physics will permit the resurrection to eternal life of everyone who has lived, is living, and will live. That thinking certainly puts evolution into an area that is commonly considered religion! But Booher would "rather leave it where it belongs—as a philosophy." This is because it does not rely on certain attributes of religion like belief in God or some supernatural entity.

In any case, there are physicists and other scientists who explain the possibility of an unseen world in scientific terms. I once heard physics professor Sugeno of Tokyo University explain that there can be unseen worlds around us right now. He explained that the table right in front of him looked solid, but that it was mainly empty space—the atoms that make up the table take up much less space than the empty space between them. Then he said that God might exist in a fifth or sixth or seventh dimension and be right here—and we would not be able to know this by our human senses. He gave an example: When Jesus walked through a closed door in John 20:26 he could simply have walked through the empty spaces between the atoms of the door, since he was in a higher dimension. Yet he could enter our dimension and be seen in it, and he could even eat our food.

Even some scientists who do not believe the Bible do believe that a physical resurrection of humans is scientifically possible. Booher mentions the omega point theory as one model to explain how evolution can explain creation and also predict life everlasting.

Although evolutionism is a philosophy and not a religion, it serves in many ways in place of religion in the minds of scientists like Carl Sagan and Ernst Mayr. Sagan even considers it is possible that there are other intelligent beings somewhere out in space who will come and save us from self-destruction. Booher explains how evolutionism, to scientists like Sagan and Gould, has its own cosmology, doctrine of salvation and doctrine of the future. He says there are believers in a purposeful and progressive evolution who anticipate a wonderful future for the world, one that includes the ultimate end of disease, poverty, overpopulation and wars. Evolutionism has its own system of ethics as well. It certainly does function as a substitute for religion in the world view and moral-ethical systems of many.

Booher is saying three main things here. First, there are two broad divisions of thought or world views. Second, that macroevolution and creationism are both all-embracing systems that encompass the natural, social and ethical worlds and give meaning to our existence. Third, that today there are many who seek a middle ground that embraces concepts from both worlds in order to find a more satisfactory explanation for natural and spiritual phenomena. Yet, they do so while maintaining their basic loyalty to one or the other of these two worldviews.

What Should We Believe about Origins?

> With no past scientific hypothesis being verified and no new one capable of being tested, the debate can continue indefinitely. Under the conditions where creationism relies on revelation from God and evolutionism requires evidence that cannot be verified under uniformitarian conditions, there appears a philosophical standoff. In fact, one is hard pressed to envision any evidence now being sought that, if found, would really settle the issue once and for all.[540]

Behind the controversy between evolution and creation there is a basic question: "what can we know about anything?" Where scientific experiments produce consistent results time after time we think we can know the nature of reality to some extent. But when it comes to origins, we cannot confirm answers today scientifically any better than the ancient Greeks and Hebrews could.

> Any person today who wishes to make a choice among alternatives of special creation and evolutionism will find him or herself choosing between two opposing philosophical theories of knowledge—we can conceive of no other, and compromise, while approachable for the socialization of science and education, is too dissonant for the individual's belief system.[541]

It seems Booher means that those who seek a "middle ground" in order to take into account what seems to be true in both the theories of evolutionism and creationism do so by adapting one or the other of these belief systems—rather than compromising their core belief.

Booher says that both theories are able to acknowledge three characteristics of how people think that are fundamental to directing behavior on the highest levels. These are rationality or intelligence, conscience or a knowledge of what we ought or ought not do, and faith in the sense of a strong central belief.

> All three characteristics have been argued as (a) having evolved, or (b) provided by God. When performing at their best, the three work in concert; intelligence

keeping faith rational and conscience keeping faith and intelligence sociable. But it is "faith" which gets to the heart of "believing" itself. There is something within us (either evolved or given by God) that compels us to have a central belief which provides the focus and organization for all other beliefs.[542]

Booher concludes that, if we think faith has *evolved*, then faith is a characteristic of the brain that has value for survival. But if we think faith comes from God our conclusion is different.

> In this faith, we must be as children, not moved to substitute our own creations, but open to learn from God about Himself, about His creations, about us. If we are to *know* God-given answers to our questions we must be receptive and attentive to His words and works using mental qualities given by Him.[543]

Since science cannot prove which theory is right about origins, he says we will be pulled toward one of these ways of thinking by our faith.

> What can we know about our past or our future? It depends on the path of faith we take—the path of evolutionism or the path of the revealed Word. On questions of origins, science can be a useful tool in the hands of either faith; but science cannot be the judge of which is closer to reality.[544]

But Booher seems to use "science" in a metaphysical meaning here. Shouldn't we be aware that both verifiable science and clearly-expressed revelation each have their own vital contribution to our knowledge? Then, if we admit the limitations of each, we may be ready to accept the unique contributions of each. Certainly seeking to know and experience *reality* should be our quest as we seek to "follow the truth wherever it leads."

Summary

Booher emphasizes in his book that there are only two dominant theories of origins that have wide support today. These are evolutionism (evolutionary creation) and creationism (special creation). The first is based on naturalism, and the second on belief in biblical revelation.

Booher does not profess in his book personal faith in either Judaism or Christianity. But he acknowledged that neither religion could have come into being and endured through the centuries without a belief in the literal words of the books of Moses for Judaism, and in credible reports of eye witnesses to Jesus' death and resurrection for Christianity. He is convinced of the validity of studying Genesis for reliable information on the origin of the universe and of life on earth. He suggests a "generalized creation model" as consisting of five things.

1. A first cause which, over time, created the universe and all living things.

2. Later an introduction of the second law of thermodynamics which affects the entire universe including life.

3. A special creation and a division of life into separate "kinds."

4. A special creation of humans. This explains the superior intellect of people.

5. Catastrophic events, which brought about the fossil record which shows kinds developing from simple to complex. He thinks a "single, catastrophic flood" can account for the fossils that are dispersed throughout the world.

He says most scientists who profess a belief in creationism and do research have been adequately trained and have appropriate credentials in their specialties. Their research methods and scientific observations are sound. But most are excluded from the mainstream scientific community.

He thinks young earth creation scientists have a scientific model which fits observable natural processes, accepts the fossil record, and is consistent with the scientific data that gave rise to the new paradigms for biology and geology. Booher thinks a plain reading of Genesis suggests literal creation days. It suggests a young earth and a universal flood. But most of the measures of geological time do not seem to support the young earth thesis. However, he has found nothing in the Genesis account that seems to be contradicted by what we observe today in nature.

The totally naturalistic viewpoint that acknowledges evidence of design in nature attributes it to natural selection. But there is no scientific evidence to support this assertion, and there is evidence that strongly challenges it. The design viewpoint attributes design to an intelligent source outside of nature. Booher believes this theory has a legitimate place in scientific theorizing. But Booher does not think it has adequately explained the evidences of chance or the imperfections in nature.

Booher does acknowledge the common argument that the designer had no obligation to make perfect designs. He says this may be true in theory when we don't know the nature of the designer, but feels it usually offends human sensibilities. He acknowledges that the biblical doctrine of the fall of man and the resulting curse does answer this problem. But this answer comes from revelation and not from science. The evidences for randomness and chance in nature may be because we do not understand the purpose for these random events—but does that really mean that an all-knowing God has no purpose at all in allowing them?

There are many today who believe in naturalism but recognize the failure of current evolutionary theory to explain many of the things we observe in nature and are seeking new evolutionary explanations. Others who hold creationist views seek to understand how new discoveries in cosmology and biology might shed light on the teachings of the Bible.

Booher explains how we humans seek to fit what we observe, experience and believe into internally consistent "cognitive maps" or worldviews. Thus Gould ponders how natural selection can produce design in nature. Einstein comes to believe in a transcendent God because what he observed in nature logically demands it. But he does so without embracing the doctrine of God taught by the Jewish and Christian Scriptures.

It is possible for both scientific and also what Booher calls "mystical realities" to exist. But naturalism rejects the supernatural. Humanism can provide a kind of meaning to our existence. But he says it cannot explain the evil in the natural world or in human nature. And it fails to provide real meaning because something in the human mind seeks life beyond the grave. A truly satisfying worldview must include a philosophy that explains nature as we experience it. It must also explain the quest for something beyond this present existence to bring satisfaction to the human mind.

Booher agrees that one can be intellectually honest and true to the facts of science, and believe in God and creation. One can also believe in a strictly materialistic philosophy of evolution, or in a combination of these two. He believes that creationism should be recognized as a legitimate philosophy of science along with evolution, and various mixtures of these basic orientations should be recognized as well.

Will the scientific community and mankind ever agree on one worldview and one theory of origins? Booher thinks this is doubtful. There are two opposing worldviews, they are mutually exclusive, and no real alternative seems possible to Booher. He says we can adopt an "and/also" inclusivism based on what we see in nature, but must do so within one or the other of these two opposing positions. If we attempt to include in our worldview beliefs that are inconsistent with our overall belief system, psychological tension will result—though this may not be in our conscious awareness.

How the universe and life originated cannot be verified by the scientific method. It is important to separate in our thinking what we know by verification through the scientific method from what we infer from this knowledge. It is also important, I believe, for creationists to separate what we know from the clear statements of Scripture from what are only our interpretations and applications of scripture. To quote Rumsfeld again, there "are things we don't know we don't know."

We can either accept the Bible account of creation or accept some theory of evolution. But Booher says that science cannot be the judge of which is closer to reality. Our worldview is what predisposes us to choose, and we can be true to what is known about nature, be intellectually honest, and choose either view. (But he adds that we should do so while being aware of the psychological gimmicks that the scientific establishment resorts to in order to project the illusion of evolutionary connectedness that is not supported by scientific observation and verification.)

In sum, Booher evaluated the scientific theories and theistic positions that guide scientists and philosophers of science today. The young earth creationist view seems to explain the geological columns better than current evolutionary theories. But the geological record and radiometric dating seem to favor an old earth in more ways than it does a young one.

He criticizes the current state of development of all the theories of origins. He criticizes the theory of intelligent design as a field of investigative science, though he thinks that it fills the requirements for a viable philosophy of science better than its alternatives. But he is hardest on Darwinians. He does not mean all Darwinians; not even the vast majority of those who are doing research—but those who promote naturalism as an all-embracing philosophy. He criticizes it as a "religion," and then backs off and says that it is not a religion. This is because he feels all religions include a belief in a god or gods or some sort of transcendent power. He claims it is being made a substitute for religion by its zealous advocates. He also claims that its advocates are using specific psychological tools to make their arguments appear more valid than they actually are.

I think Booher has done an excellent job both in reviewing the state of our scientific knowledge today and how well various theories of origins are supported by what we actually know about the world. But, to a Bible-believing Christian, what the Bible says about supernatural things and about life beyond the grave is far more important than what it teaches about natural things.

If knowing the "truth" about the world of nature is sufficient to enable us to live wisely in this life and there is nothing beyond this life—then what we believe does not have eternal consequences. But if the Bible is true then it does have much to say about living effectively now and preparing for the life beyond. Considering whether or not the Bible speaks "truth" about the natural world and origins is very important in that it is a test of whether it truly is a revelation from God. But I believe the real importance of this knowledge is to encourage us to know the Bible and experience what it teaches about spiritual things.

24
Response

We have considered quite a few ways to reconcile the verified realities of science with what the Bible says about the natural world. In this chapter I will seek to evaluate these with two criteria in mind. First, are the scientific facts presented unambiguous? Second, does the Bible clearly support the interpretations of the preceding chapters when the rules of bible interpretation are carefully followed?

"Following the truth wherever it leads" requires an openness to new information that challenges our long-held biases. We all have biases, and generally tend to be unaware of them and how they influence our openness to learn.

We are constantly bombarded with facts and bits of information. Rather than overload our memories I think we tend to find a pattern or conclusion when faced with information we deem of some importance to us. It is so much easier to remember what we concluded about something than it is to remember the jumble of facts and assumptions that led to the conclusion. I call this a tendency to seek "closure." It is a tendency to "complete the picture" when confronted with a puzzle that has missing pieces. It is very often an unconscious process. Since our conclusions are made so often from incomplete information and are usually remembered as "fact," this can lead to mistaken thinking.

In seeking "the truth wherever it leads," I think it is important to acknowledge freely that many things that are considered factual are based on incomplete knowledge. We need to remember, in Rumsfeld's words, there are "things we don't know we don't know." This kind of intellectual humility should help us to understand our world more accurately.

Science claims to discover *reality* in the natural world. The Bible claims to proclaim *reality* about both the natural and the spiritual worlds. If these claims are true, then where the Bible and science speak about the same things, there should essentially be agreement on the facts.

Is there such a thing as objective reality? Is there "truth" that is reality at all times and for all people in all places? How can we know what is real from what is not? We discussed this in part 1.

The chapters in part 3 have been presented to clarify what it seems we can embrace as true as verified on scientific grounds and also on what is clear from the Bible regarding nature when the laws of bible interpretation are carefully followed.

A. First We Will Review the Ideas Presented Earlier in Part 3

1. Philosophy of Science

Bernard Ramm taught both the philosophy of science and biblical interpretation on the university level. He explains several ways where the language of the Bible is different from that of science.

He points out that the first feature of the biblical view of nature is that "God is the Almighty Creator of heaven and earth." Therefore nature exists fundamentally for spiritual purposes and there is a purpose in nature because it is ordered by a Personal Intelligence.

He also points out that the same God Who created all things preserves it and gave us the moral law. Because God is the reality behind nature, morality, redemption, and reason, all these are established to accomplish His purposes.

First, regarding nature, Ramm says the Bible uses the everyday language in which people converse. Second, the language is *phenomenal* or "pertaining to appearances." It doesn't explain the "how" or even the "why," but limits itself to the action itself. Third, the Bible does not theorize as to the actual nature of things. It also is expressed in the culture of the times when it was written. It is *pre-scientific* and *non-scientific* because it can be understood by pre-scientific people, but it is not *anti-scientific*.

For this reason we cannot expect the Bible, written thousands of years before the development of modern science and the vocabulary of science to give clear answers to all the questions of modern science. Another important point to remember is, as Ramm emphasized, the Bible was written primarily to communicate that *God* is the Creator of all, and that God made humans like Himself in some distinct ways ("in the image of God"). God is still active in His creation and seeks a positive relationship with humans that is in keeping with His moral nature.

2. The Gap Theory

The gap theory includes three beliefs. These are a belief in a literal interpretation of Genesis; a belief in a very old but undefined age of the earth; and a belief that there was a time period and events between Genesis 1:1 and 1:2 that can explain the origin of the older fossils and geologic strata. Some who hold this view believe that during this time period the devil sinned and was cast down to earth and this caused chaos on earth and made it "waste and void."

Harry Rimmer says the word "day" is one of the mysteries of Scripture as it has various meanings in the Bible. For this reason we cannot know from Scripture how old the earth is or how long life has been on it.

The theory that Genesis 1:2 can refer to a very long period of time makes good sense. It says, "The earth was without form, and void; and darkness was on the face of the deep. And the Spirit of God was hovering over the face of the waters." The word "hovering" or "brooding" (like a bird over her eggs) suggests a period of time passed, and the condition of the earth as "without form and void" can also refer to an extended period. So I believe a "gap" of millions or even billions of years when the universe is expanding and changing is certainly well within the possible meanings of this verse. As to how long ago life might have been on the earth, the word "brooding" in verse 2 seems to suggest the Spirit was watching over the beginning and development of sea life long before the six days of creation.

The Hebrew word in Genesis 1:2 for "form" is *tohoo*, and means confusion, empty space, without form, nothing. The Hebrew of "void" is *boohoo*, and means emptiness, void. It also can mean "indistinguishable ruin." These Hebrew words are consistent with the theory. But the interpretation that "the earth *became* without form and void" is suspect. As John Collins explained, there is a more natural Hebrew word for "became" that was not used in verse 2.

3. Ramm's View

Bernard Ramm argues for progressive creationism. He says this view recognizes two types of creation: creation *ex nihilo* (from nothing) by nothing more than God speaking (by fiat); and a process over a considerable length of time that followed the laws of nature that God established in order to bring about the forms of nature he preordained. He explains that progressive creation teaches the transcendental activity of God. Ramm acknowledges that the large divisions in the animal and plant worlds were created separately. The fact that this appears to be true is demonstrated by the fossil record, as we have seen in part 2. It is also consistent with Behe's research on the complexity of the cell.

Ramm believes that Genesis 1:2 reports a lengthy "blank and void" state; and that the "six days of creation" refers to a very long process of development toward the condition of the earth that we observe today. He rejects the young earth view, and says we cannot know from Scripture the age of the earth.

Ramm rejects the idea that *all* death was a result of man's sin. He asks why carnivores would have such huge teeth and sharp claws if they did not eat other animals in order to live even before man sinned. He says, "The Bible ascribes death from sin to man alone."

I think this is another place where it may be wise to leave our minds open. One thing that seems to lend support to the view that sin brought physical death only into the human race is the wording of Genesis 3:22–23.

Genesis 3:22–23 says that God cast Adam and Eve out of the garden of Eden so they would not eat of the tree of life and live forever. This suggests that they had to eat this fruit to prevent their deaths. However, this is an issue that has not been considered in any of the books that I have read.

(Another interesting thing is that the book of Revelation, in verses 22:2, 22:14, and 2:7, mentions a "tree of life" in heaven. Since Revelation reports a vision that God gave John, the meaning here could well be symbolic. First Corinthians 15:42–54 uses terms like "imperishable" and "spiritual" for the transformed bodies believers will have in heaven, which suggests that no "tree of life" will be needed to sustain life in the future.)

Ramm believes that God created the earth in such a way that there were "imperfections" like diseases, and storms from the beginning—even before man sinned. When man was created he was given the task of learning how to cure diseases and protect himself from inclement weather. He also had to learn to protect himself from lions, tigers and parasites. Ramm saw this as the situation even before Adam and Eve sinned.

This view of Ramm's is unique among authors of books on science and the Scriptures that I have read. Whitcomb and Morris call Ramm's comment "rather dubious" that "God did not say that creation was perfect, but that it was good." But Ramm's view does not seem to violate the principles of biblical interpretation in any way. Its weakness seems to me to be in not dealing with the verses in the New Testament like Romans 8:18–22 and Revelation 21:1–3 that refer to the "curse" the world is under now. Genesis 3:14 suggests God put the animal kingdom under the curse caused by the sin of Adam and Eve as well as that God "cursed the ground" in Genesis 3:1–18. Yet there seems to be no indisputable declaration in the Bible that shows there was no entropy in the universe or cruelty or deterioration in the world before sin entered in Genesis 3.

4. Young Earth Creationism

In *The Genesis Flood* the authors based their arguments for a young earth on a few beliefs. They believe that the universe is around ten thousand years old, that God in six literal twenty-four-hour days created the earth, plant and animal life, and created mankind on the sixth day.

John Whitcomb and Henry Morris take the Bible as the revealed truth from God and that it is to be taken very literally in all its references to creation. The Bible and not science is considered infallible and without error.

They argue that plants, animals, and humans were created in their present "mature" condition, and there has been no advancement in their condition as described by the theory of evolution. They say that the original creation had "the appearance of age," in that fully mature organisms were created at the beginning. They believe no plant or animal died before Adam sinned, so all fossils are a result of death that occurred after the fall. A worldwide flood in Noah's day accounts for the geologic record that scientists point to when arguing the earth is billions of years old.

They point out that Genesis 3 and Romans 8 declare that the natural realm on earth was "cursed," and became in some way or ways different than it was before Adam and Eve sinned. But to argue, as Whitcomb and Morris do, that there was no decay, no entropy, no death at all in the world does not follow inevitably from this argument.

Young earth creationists often point to Genesis 2:4, which says, "This is the history of the heavens and the earth when they were created, in the *day* that the Lord God made the earth and the heavens." They argue that this says the universe was created in one day of twenty-four hours, and not over an extended period of time. I see two weaknesses in this argument. The first is that the Hebrew word here, *yom*, can be either singular or plural. The other is that the word *day* does not appear until Genesis 1:5, *after* the creation of the universe and when God changed the atmosphere around earth to make half of each day light and half of it dark. It reads, "God called the light Day, and the darkness He called Night. So the evening and the morning were the first day." This seems to greatly weaken the argument that our planet is no more than a few tens of thousands of years old.

Whitcomb and Morris argued that "we say, on the basis of overwhelming biblical evidence, that every fossil man that has ever been discovered, or ever will be discovered, is a *descendant* of the *supernaturally created* Adam and Eve."[545]

I believe that this argument is persuasive, if we limit it to fossils of actual humans. If the human race that exists today did not have as its beginning the direct creation by God of the first man and woman, then the Bible that is the foundation of Christianity is not the inerrant Word of God. If either of the two basic foundations of Christianity are proven conclusively to be false—the veracity of the Bible as we have it today or the physical resurrection of Jesus Christ—then "the entire edifice of Christian theology" is undermined like they claim. (As to the historical veracity of Christ's resurrection, let us remember the disciples' testimony of finding His tomb empty and Paul's words in 1 Corinthians 15:6 of over five hundred people seeing the risen Christ with their own eyes and that many were still alive when Paul wrote that so they could confirm the fact.)

But, as some scholars insist, if some of these fossils are of "pre-humans" or hominids, then the discussion changes directly—though this is a possibility that Whitcomb and Morris reject. If we do not assume the days of creation were twenty-four-hour days, might not beings somewhat similar to man have existed long before Adam was created by God? If so, they must have been different from humans today in a very significant way, though there is no conclusive way that science can demonstrate this. The issue then becomes, "what makes humans uniquely human?" It seems science cannot answer this; it is in the domain of philosophy or religion. Chapter 25 looks at this issue.

5. Geology and the Old Earth

Davis Young insists that the six days of creation are *historical*, that they are "indeterminate stretches of real, historical time." During these long "days" animals and plants evolved. There is also a strong possibility that some events listed on a particular day could have overlapped another day. However, mankind was directly created by God as one man and one woman. He bases his concept of long days of creation and biological evolution on the findings of modern science, while retaining a belief that the Genesis account is historical and literal. He says, "The words and real teachings of Scripture are never in ultimate conflict with the real facts of science or history." But he insists that scientific discovery can aid us in understanding the true meaning of Scripture—as it was not written in scientific language.

In this way Young interprets the creation account in Genesis very differently from the way Whitcomb and Morris and other young earth creationists do. Yet he also follows the widely accepted rules of bible interpretation as he does so.

6. Cosmology and the Old Earth

Hugh Ross believes the big bang theory is a valid basis for understanding the universe and its creation. Astrophysicists figure this occurred about 13.73 billion years ago, and he believes that the earth is 4.6 billion years old. He says that as the earth aged the rotation of the earth slowed. This affected life on earth and there have been several times when life became extinct and only fossils remain. He sees no conflict between these ideas and Scripture.

He thinks the Spirit may have created life in the oceans before the events of the six creation days begin, when He "brooded" over the face of the waters (Gen 1:2). He is convinced that the six days are long periods of time.

Ross says that God established the laws of nature in such a way that only minor biological changes have occurred since God created each kind of life. He argues that the Bible clearly denies that any modern species descended from simpler forms of life. Human beings are distinct from all other animals, including the bipedal primates that preceded them, in that humans alone possess body, soul, and spirit. God made man, and only man, in His image.

Ross explains that when God created "the heavens and the earth" this refers to the entire physical universe. It refers to all of the matter and energy and all else in the universe.

Ross's interpretation of Scripture follows his knowledge of astrophysics, but does not violate the laws of biblical interpretation. He says that "no author writing more than 3,400 years ago, as Moses did, could have so accurately described and sequenced these events, plus the initial conditions, without divine assistance."[546] He goes on to argue that this gives us reason to believe that God also guided the other writers of the books of the Bible to communicate truth without error.

7. Other Views

(a) Wayne Frair and Gary D. Patterson argue that the realities of the natural world demonstrate the truthfulness of the Bible. They believe that advances in science will increasingly support a biblical view of nature. They say the big bang theory fits well with Bible teaching, but even if it should prove eventually to be false this will not affect their faith in the Genesis account of creation.

(b) Jean Pond insists on the independence of science and religion. But this assumes that religion does not speak about the realities of the natural world. She is right in that science cannot measure the supernatural and is not qualified to speak about it. But as Bernard Ramm has explained, the Bible speaks clearly about nature, and presents God as the creator and sustainer of the natural realm.

(c) Howard Van Till sees science and theology as working together in partnership and not in any way in conflict with each other. He seems confident that scientific advances will eliminate the "gaps" in nature that other Christians point to as necessitating God's intervention after his initial creation. He even includes the development of humans as needing no direct action by God.

In the background of Van Till's argument for "non-episodic creation" seems to be a concept of God's involvement in nature that is very different from the one I understand is revealed in the Bible. Van Till does not reject all possibilities that God *can* act on earth subsequent to the initial act of creation, but maintains that God created initially in such a way that he need not.

However, the Old Testament reports how God intervened often in events on earth even well after the creation period. Many psalms like Psalm 78 and Psalm 106 list how God intervened during the exodus from Egypt and the time in the wilderness—but in the literary style of Hebrew poetry.

Did God act directly to create Adam and Eve subsequent to creating the heavens and the earth? Genesis clearly says He did. Does God intervene in the affairs of mankind on earth? The Bible clearly says that in many places in history He did. His intervention is seen in the flood of Noah's day, in the dividing of the waters when Moses and His people crossed the sea in Exodus 4:21–31, when Joshua crossed the Jordan River (Josh 3:14–17), and in the "long day" when Joshua defeated the forces surrounding Gibeon (Josh 10:7–14). It is also seen when Jesus died on the cross and when He was resurrected.

Yes, there are visions recorded in the Bible where God is said to have revealed Himself to individuals on earth. But there are also numerous accounts of where God spoke to individuals directly and told them to do specific things. If what is said in the Bible can be trusted, then it is clear that God has intervened many times in human affairs, and also changed the course of nature to accomplish specific tasks. Whether God's appearances are called "the Angel of God" or a similar description, they are said to be the intervention of God on earth. Either these things are to be taken literally—as history in time and space—or they are not. I believe a plenary verbal understanding of Scripture demands that they be taken literally. This is totally consistent with the message of the whole Bible, and also with the way the Gospels depict Jesus as quoting from and referring to the Old Testament. If the accounts in the New Testament of the historical Jesus are factual, then how the Founder of Christianity interpreted the Old Testament is vital to a true understanding of Scripture.

(d) Stephen Meyer insists scientists should seek the best explanation of the evidence and follow it wherever it might lead. He is convinced that the best and most truthful (reliable) theories do not contradict a theistic worldview. He argues that physics, astronomy, cosmology and chemistry have each in their own way revealed that life depends on a set of very precise design parameters that are built into the universe.

This supports the design argument, even though he admits it is not a scientific proof of God's existence and activity. It simply is not in the realm of science to provide such proof. He says that naturalistic theories have failed to explain the origin of the information necessary for the origin of life.

He believes that the natural sciences now provide strong epistemological support for the existence of God as affirmed by both a theistic and Christian worldview. His general conclusion appears to be that, if it cannot be proved scientifically that God created, at least the scientific facts are consistent with the account of creation in Genesis. In fact, the creation account is the best of all known explanations of the universe as it exists, and there appears to be no other explanation for such an account in the ancient literature of the Bible unless it was inspired by God.

(e) Bruce Waltke says the testimony of science and the Bible about creation complement each other, but they cannot and should not be harmonized. This is because the Bible is literature, and not a scientific book.

(f) John Collins, a professor of Old Testament, says we cannot know from the Bible how old the universe and the earth are. Genesis 1:3 reports the first day, which was an unknown amount of time after the absolute beginning of the universe reported in Genesis 1:1. He also says that the creation week (Gen 1:3—2:3) had to be much longer than an ordinary week in order for Genesis 2:5 to make any sense. He also explains from the Hebrew words that "father" in the genealogies clearly can mean "ancestor" and "son" can mean "descendant." He concludes that there is no way to determine how big the gaps in the genealogies are and therefore there is no way of knowing the age of the human race.

(g) Harold Booher acknowledges that Judaism could not have come into being and endured through the centuries without a belief in the literal words of the books of Moses. And that Christianity could not have endured without credible reports of eye witnesses to Jesus' death and resurrection. He is also convinced of the validity of studying Genesis for reliable information on the origin of the universe and of life on earth. He said he has found nothing in the Genesis account that seems to be contradicted by what we observe today in nature.

He points out that creationism must rely on revealed truth, while evolutionism must rely on a grand illusion which defies demonstration. Both lie outside scientific verification.

He criticizes the current state of development of *all* the theories of origins, though he thinks that intelligent design fills the requirements for a viable philosophy of science better than its alternatives. He criticizes those who promote naturalism as an all-embracing philosophy, saying it is being made a *substitute* for religion by its zealous advocates.

8. Is the Creation Account in Genesis Historical?

The Jews throughout Old Testament history relied on the Bible as a historical document. Jesus also had a literal understanding of Scripture. Examples are when Jesus spoke of Jonah being "three days and three nights in the belly of the great fish" (Matt 12:40) and that Scripture will be fulfilled down to the smallest detail (Matt 5:18). The Apostle Paul was clearly referring to Genesis 2:7 as a historical record when he wrote the following in 1 Corinthians 15:45: "And so it is written, 'The first man Adam became a living being.'"

The Bible contains poetry, visions, and parables that have their special rules of interpretation, and contain symbols and figures of speech that cannot always be interpreted literally. But the Bible also contains twenty-two books that claim to be historical. These are the first seventeen books of the Old Testament and the first five of the New Testament. The creation account in Genesis is part of this solid tradition to recognize the books of Moses and the other Old Testament historical books as accurate history. I believe the evidence points clearly to Edward Young's conclusion that "Genesis is sober history."

As to the hotly argued issue of how long life has been on earth, it seems that neither science nor the Bible gives a clear answer. The oft-mentioned problems with dating the fossils (not knowing the original conditions, whether there was a constant process rate, and possible lack of isolation) precluded clear scientific assessment. The Bible does not clearly reveal this either, because of the very broad use of the word "day" in the Old Testament, and the gaps in the genealogies.

So yes, Genesis can be trusted to be a historical account revealed by the Creator as an "eye witness," but it is good to remember that these real events were not expressed in the language of modern science.

B. In Seeking to "Follow the Truth, Wherever It Leads" While trying not to let our natural biases color our vision, how will we respond to the material presented in this book?

1. First, Are the Scientific Facts Presented Unambiguous?

One thing that surprised me when doing the research is that science has evidently progressed to where I think we can confidently conclude that some things *did not* happen! Consider Michael Behe's research on the complexity of the cell that shows evolution by random mutation and natural selection is severely limited. The evolutionary scientists who challenged his conclusions failed to counter them on scientific grounds. This was not unexpected. But the frank acknowledgement by many leading evolutionary scientists that the fossil record shows that complex organisms "appeared suddenly" at certain important points in biological history was. The evidence in the twenty-first century is so clear that leading evolutionary paleontologists like Stephen Jay Gould rejected Darwin's theory of gradual, incremental change and sought to imagine a form of evolution that could proceed "in leaps and bounds." This was something I hadn't anticipated. The reason I didn't anticipate the testimony of the fossils that slow, continuous development from simple to complex life didn't occur was the fact that so many evolutionists point to the fossil record as "proof" of macroevolution.

How far do random mutations and natural selection go in developing new biological forms? It seems unambiguous that organisms can sometimes change through mutation and selection to where they can cope better with their environments and prosper in the "survival of the fittest." It is clear that new varieties can develop both by human design and natural selection, and evidently new species can also. But there is no evidence that anything much beyond that can.

But many are loath to use the word "evolution" for these changes, because this implies improvement. Several scientists have stated that most mutations are actually harmful to the organisms, and those that are beneficial are rare and produce biological change to a very limited degree.

The big bang theory is clearly the dominant theory in cosmology today. There seems overwhelming evidence from astrophysics that the universe had a beginning—when all of energy, matter, space and even time itself suddenly came into being. But, as Harold Booher pointed out, the scientific data that shows the evidence cosmologists rely on for this theory is actually only obtained from the prisms of light obtained by our telescopes. Hugh Ross and evidently most astrophysics infer that the universe is about 13.73 billion years old, but Booher points out that the scientific evidence of this does not measure up to the tradition standards of scientific proof. He says only three key observations support the big bang theory. These are Hubble's red shift, the abundance of helium, and the cosmic background radiation. He says that none of these observations requires a big bang interpretation. He concludes from his research that there is no convincing way to estimate the age of the universe.

The testimony of current astrophysicists like Hugh Ross as to the age of the universe and earth is not at all ambiguous. But since the verifiable factual support for the current big bang theory does not as yet meet the standards of hard scientific proof, I think it is safer to acknowledge that we do not know with any assurance the real age of the universe or that of the earth.

Neither does the Bible tell us these things. The wording of Genesis 1:2 clearly allows for the possibility of millions or billions of years between the creation of the universe and the processes that brought about living things on earth. The broad use of the Hebrew word used for the six days of creation also allows for very long days, but does not preclude the possibility that the last few days may have been just twenty-four hours each. We simply do not know with any confidence from science or the Bible the process and speed of the development of life.

At least two other assumptions that are taken as factual by modern science do not have any support in verifiable science. One is the theory that somehow life began on earth by natural means. It is reported that every experiment made so far to demonstrate how primitive life could develop on earth from non-life has failed completely. The other assumption is that intelligence in the cell could be produced by a non-intelligent source. Stephen Meyer reminds us that science has faced insurmountable problems in trying to demonstrate how this could possibly happen. He says that, without exception, in human experience only intelligence has been able to produce intelligence. Human nature and human intelligence also cannot be explained at all by naturalistic means.

Yet the Bible explains quite simply and rationally how the intelligence in the cell and in humanity came into being. It was created by the God Who knows all things and can do all things.

If we cannot know with confidence the age of the universe or how long there has been life on earth, are there things we can feel justified in assuming we know for certain? Yes!

I believe it is clear that we can know for certain that this universe had a beginning. It is clear that life did not develop from non-life, and that the intelligence in the cell had an intelligent source. We can be sure that human nature is not a product of random mutation and natural selection from any simple forms of life.

I believe we can accept the conclusion as fully rational that the Bible, written thousands of years before the advent of modern science, speaks truth and can be relied on in its pronouncements about nature. We can be confident that the Bible speaks truth about matters that are beyond the realm of science also. Then, knowing this, can we go on to learn from the Bible things of a spiritual and moral nature that will impact our lives now and in the future?

As we "follow the truth wherever it leads," are we fully justified in going to the Bible to find answers to the important questions of life? I refer to such questions as: What in life is really important? Is this life all there is to my existence? Does it really matter how I live my life?

2. Second, Does the Bible Clearly Support These Interpretations When the Rules of Bible Interpretation Are Carefully Followed?

The Bible needs to be interpreted, using the common rules of language, but all the while remembering it sometimes describes things beyond human experience and is thus hard to explain by human language. Exegesis, or explaining the meaning if a passage in the Bible, follows clear laws and is by no means a subjective or arbitrary activity. It follows the laws of biblical "hermeneutics." This is the science of interpreting the Bible.

The rules of proper biblical interpretation can be briefly summarized as follows:

1. Careful consideration must be given to context.

2. The meaning of words is often determined by the time period and cultural context.

3. Consideration must be given to the form of the biblical text; whether it is literal or figurative, poetry or prose.

4. Often the Bible interprets itself. Especially when the passage is difficult to understand it is important to consider other passages that deal with the same theme. Doing this can often clarify the meaning of a passage.

Of these four rules, the first, third, and fourth clearly apply to the Genesis record of creation. The creation story in Genesis is written in historical form, and the rules for understanding history are fairly straightforward. Historical records are to be understood literally and in context.

I think the fourth rule listed above is worthy of special mention. Here the story of creation has strong support as historical and to be taken literally. There are many hundreds of references in the Old Testament and the New Testament to show how a literal, historical understanding of the creation story in Genesis shaped the worldview of the various authors of the Bible.

I believe most of the authors we have studied in part 3 have followed these rules and can be trusted to give insight into what the Bible reveals about natural things.

What Do We Know from the Bible?

What things are clear about creation from the Bible? Recognizing that equally dedicated Bible-believers disagree on what the Bible means on some things, can we make a minimum list of what we can confidently say we are sure about regarding creation? If so, what will be on this list? I suggest at least five things.

(1) God created the universe out of nothing. He used no pre-existing material because there was none.

(2) God created fish, birds, and animals separately as fish, birds, and animals.

Genesis 1:24 says, "Then God said, 'Let the earth bring forth the living creature according to its kind: cattle and creeping thing and beast of the earth, each according to its kind'; and it was so" (see also Gen 1:11; 1:21).

Although Genesis 1 says that plants and animals were to reproduce "each according to its kind," we have no clear understanding of the biological classification of "kind."

(3) God created humans as one adult male and one adult female. They were essentially at the same stage of development as modern humans.

(4) God created man "in the image of God." This means, at a minimum, that humans are like God in having an intellect and a will, and are social beings who possess personality and moral natures. They also have the capacity to know God and to worship Him.

(5) Later, after Adam and Eve sinned, God "cursed the ground," so that the world we see today is in important ways different from the garden of Eden where they first lived. (This can help answer the criticism that a Creator would not design imperfect beings.)

Francis Schaeffer reminds us that the word for "create out of nothing" is used very sparingly in Genesis.

> The word *created* (Heb. *Bara*) is used only a limited number of times in Scripture. This is specially true of the specific form used at three crucial points. The first of these is the point at which God created out of nothing (Gen. 1:1); the second point at which God created conscious life (Gen. 1:21), and the third the point at which God created men (Gen. 1:27).[547]

The five times this word is used in Genesis 1 are all speaking of these three events. This suggests that the original creation of the universe, the creation of sea life and birds, and the creation of humans are especially important for understanding God's revelation about origins. The emphasis is clearly on the creation of mankind. In verse 27, referring to the creation of the first male and female, "create" is used three times. (In the first five books of the Bible there are five verses related to God creating mankind.)

(6) I believe it is important to add one more point here. God continues to supervise His creation and intervenes directly at times to enable it to fully carry out His will. We referred earlier to a few of the many places in the Old Testament where God intervened on earth to deliver and protect humans.

If we accept what is written in the book of Genesis as the inspired revelation from God we can confidently assert that the above six things are historically accurate. Accepting these six points does not resolve all the issues about how to harmonize science and Scripture, but I believe they establish a foundation for seeking such a resolution as further truth in nature is discovered.

25

When the Pursuit of Truth Hits a Speed Bump

So far the pursuit of "truth" as it relates to the natural world has proceeded smoothly. When we limit our investigation to what is clear in science and in the Bible I believe we have made great strides. But now, in the tenth year of studying and writing in this area, I am facing a "speed bump" on the heretofore smooth path to our destination.

We first looked at the universe through the eyes of astrophysicist Hugh Ross and others, and could appreciate the way modern scientific discoveries have supported the clear statements of Scripture that the universe had a beginning and that our solar system was uniquely "fine-tuned" for advanced life on earth. This demonstrates that a Creator God exists Who is pre-existent and transcendent.

Then we looked at the world of living things through the eyes of several scientists, and were greatly impressed with the research results and explanations of biochemist Michael Behe. It seems clear that living organisms are irreducibly complex and could not have developed incrementally from simple life forms by random mutations and natural selection like Charles Darwin proposed.

After that we looked at geology and the fossil record, and then considered how the human race might have originated. All that we studied can fit well with the teachings of the Bible.

Our path seemed relatively smooth and unidirectional when we sought to separate what seems clear from science and the Bible from what does not and were content to relegate a few conclusions to the category of "we don't really know." The Bible does not tell us how old the universe is, or even when life originated, but it does tell us clearly that God created it all.

Then, after almost finishing the book, I asked several knowledgeable people to review it. When I asked a biochemist to review my chapters on biology and intelligent design he questioned if intelligent design is really a *scientific* theory, and introduced me to a book by Francis Collins called *The Language of God*, published in 2006.

Francis Collins is an interesting person. He calls himself "a rigorously trained scientist" and also "a geneticist, a biologist, and a believer in God." He is one of the world's leading scientists. Collins received a PhD in physical chemistry from Yale University, and then attended medical school and obtained an MD. He led the International Human Genome Project that sequenced all of the human DNA and determined aspects of its function. He was awarded the presidential Medal of Freedom in 2007, the nation's highest civil award, and the National Medal of Science in 2009. He is a member of the National Academy of Sciences, and is currently director of the National Institutes of Health.

He says he was an atheist until a neighbor who was a minister suggested he read *Mere Christianity* by C. S. Lewis. When he read Lewis's description of how the "moral law" was a universal feature of human existence he began to deeply consider the uniqueness of humans from other animals.

Collins says that for years he was content with his atheism, but then says that two things led him to forsake his atheism and embrace theism. The first was what he and Lewis call the "moral law," which he felt challenged this belief. He observed that "it is the awareness of right and wrong, along with the development of language, awareness of self, and the ability to imagine the future, to which scientists generally refer when trying to enumerate the special qualities of *Homo sapiens*."[548]

The second was the anthropic principle, which recognizes that the universe is uniquely tuned for human life on earth. Collins refers to Arno Penzias as "the Nobel Prize-winning scientist who co-discovered the cosmic microwave background that provided strong support for the Big Bang in the first place." Penzias wrote, "The best data we have are exactly what I would have predicted had I nothing to go on but the five Books of Moses, the Psalms, [and] the Bible as a whole."[549] Collins concluded from this that only one universe exists, it is precisely tuned to make intelligent life possible, and that this reflects the action of the One Who created it.

He returns again and again to the "moral law."

> Consider a major example of the force we feel from Moral Law—the altruistic impulse, the voice of conscience calling us to help others even if nothing is received in return. Not all of the requirements of the Moral Law reduce to altruism, of course; for instance, the pang of conscience one feels after a minor distortion of the facts on a tax return can hardly be ascribed to a sense of having damaged another identifiable human being.[550]

He describes altruism as "the truly selfless giving of oneself to others with absolutely no secondary motives." Then he calls this love by the New Testament Greek word *agape*—and distinguishes it from affection, friendship and romantic love.

> Agape, or selfless altruism, presents a major challenge for the evolutionist . . . It cannot be accounted for by the drive of individual selfish genes to perpetuate themselves. Quite the contrary, it may lead humans to make sacrifices that lead to great personal suffering, injury, or death, without any evidence of benefit.[551]

He describes this love as one that seeks no recompense, and claims it is an affront to materialism and naturalism. Then he adds that it is the sweetest joy that anyone can experience.

He quotes the famous philosopher Immanuel Kant as saying that the two things that constantly and increasing filled him with admiration and awe are "the starry heavens without and the Moral Law within." Collins concluded from these two facts of nature that a creator God exists, and that the Bible must be true.

The impact of these two things on Collins caught my attention, because this is exactly what the book of Romans says about how God reveals himself to mankind. Romans 1:20 says the existence of the universe attests to his "invisible attributes"—His eternal power and divine nature. Romans 2:15 shows that human beings have a "moral law" written in their hearts. The testimony of the philosopher Kant and the scientist Collins are clear acknowledgement of the truth of God's word here, even though both were probably unaware of these Bible verses when they first believed in God.

The moral nature of humanity is a testimony to the uniqueness of the human race, that we are "created in the image of God" as Genesis 1 and 2 says. Collins doesn't deal with the issue of man's evil nature—though it too is a testimony to the truth of Genesis 1 to 3. When writing this the radio news was filled with reports of a medical student who took four guns into a movie theater and killed about a dozen people who were not known to him and wounded many more. I don't think evolutionism can explain this either. But the Bible does.

Collins concludes, "Clearly, the scientific worldview is *not* entirely sufficient to answer all of the interesting questions about the origin of the universe, and there is nothing inherently in conflict between the idea of a creator God and what science has revealed. In fact, the God hypothesis solves some deeply troubling questions about what came before the Big Bang, and why the universe seems to be so exquisitely tuned for us to be here."[552]

Collins said he started his journey of intellectual exploration to confirm his atheism. After considering what he calls the "moral law" and other issues he began to "admit the plausibility of the God hypothesis." He concluded that faith in God now seemed more rational than unbelief.

He writes that "this book aims to dispel that notion [that a rigorous scientist could not also be a serious believer in a transcendent God], by arguing that belief in God can be an entirely rational choice, and that the principles of faith are, in fact, complementary with the principles of science."[553]

He says, "It also became clear to me that science, despite its unquestioned powers to unravel the mysteries of the natural world, would get me no further in resolving the question of God. If God exists, then He must be outside the natural world, and therefore the tools of science are not the right ones to learn about Him."[554]

Francis Collins led the Human Genome Project, the team of scientists that uncovered the human genome sequence. He said of this, "For me, as a believer, the uncovering of the human genome sequence held additional significance. This book was written in the DNA language by which God spoke life into being."[555]

But in the book *The Language of God*, Collins also challenges some of the conclusions I have made in previous chapters about the irreducibly complex nature of the cell, and about the fossil record. He also presents a somewhat different conclusion on the origin of the human race than I made earlier. For this reason I think it is a must read as we seek to "follow the truth wherever it leads."

This chapter will cover Collins's rejection of intelligent design and his challenge to Behe's *irreducible complexity* arguments. It will also present his view that the fossil record does not support a "God of the gaps," his belief that Darwinian evolution is seen in the human genome, and his attempt to reconcile his views with the first few chapters of Genesis.

It will also cover his spirited defense of Darwinism and theistic evolution, which he covers under the name Bio-Logos. Collins is the founder of Bio-Logos, a group of natural scientists who are also Bible-believers.

Collins Scientific Objections to Intelligent Design

Collins says,

> Intelligent Design fails in a fundamental way to qualify as a scientific theory. All scientific theories represent a framework for making sense of a body of experimental observations. But the primary utility of a theory is not just to look back but to look forward. A viable scientific theory predicts other findings and suggests approaches for further experimental verification. ID falls profoundly short in this regard."[556]

This is a different way to formulate the issue than the one that many advocates of intelligent design use when they say it isn't inferior to evolution just because it can't be falsified, and argue that Darwinism has been falsified.

Collins is right by claiming that advocates of intelligent design primarily look back to support the theory. This was pointed out in chapter 12 when we discussed Stephen Meyer's claim that intelligent design meets all the tests necessary for it to be considered a truly scientific theory.

Collins is also right in the sense that Darwinism has given scientists a framework for interpreting the results of their scientific experiments. But ID advocates are also correct in saying there is no scientific proof to date that evolution by random mutations and natural selection has caused or even is capable of causing the biological complexity that exists today. In this regard Collins says ID lacks because it provides no mechanism to show how the postulated supernatural interventions give rise to complexity.

Collins Challenges Behe's Conclusion

Collins gives a zealous defense of Darwinism and macroevolution. As I reported in chapter 8, Michael Behe concluded that the many biological irreducibly complex systems like he described in his two books could not possibly have been formed by numerous, successive, slight modifications. Behe argues from this that Darwinian style macroevolution did not happen. He says small mutations don't pile up and become large mutations without being designed. Behe emphasized that "if there is not a smooth, gradually rising, easily found evolutionary pathway leading to a more complex biological system within a reasonable time, Darwinism won't work."[557] In this way he argued that the cell is so irreducibly complex that no conceivable pathway exists in nature to accomplish large-scale evolution.

Collins argues that the human blood-clotting system, that Behe cited as irreducibly complex and thus could not have developed incrementally over time, could actually have done so. But Collins offers no evidence to suggest it did. He also tries to undercut the argument that the bacterial flagellum could have developed incrementally by natural selection, but he admits that scientists are still far from filling in the whole picture. In this way Collins rejects Behe's arguments that the complexity of biological systems shows that Darwinian evolution cannot explain how they came into existence.

Speaking of the human blood-clotting system, Collins says,

> The system appears to have begun with a very simple mechanism that would work satisfactorily for a low-pressure, low-flow hemodynamic system, and to have evolved over a long period of time into the complicated system necessary for humans and other mammals that have a high-pressure cardiovascular system, where leaks must be quickly stopped.[558]

But it is important to note here that Collins is not trying to refute Behe on the basis of scientific experimentation, like others have tried to do and so far failed. Collins is arguing on theory of what he believes could have happened. Collins and Behe give their main arguments from different ways to study nature.

However, Collins's main argument for Darwinian evolution and against intelligent design comes from his studies in the human genome, which is his particular area of expertise.

Collins concludes that intelligent design fails as a science because it provides "neither an opportunity for experimental validation nor a robust foundation for its primary claim of irreducible complexity."[559] He also argues that the distinction between macroevolution and microevolution seems an arbitrary one.

But Collins does not attempt to answer the arguments against naturalism and for intelligent design by Stephen Meyer that I report in my chapter 12. He does not refer to Stephen Meyer's position that both Darwinism and intelligent design are similar in that neither are testable today by experimental research.

So here we have two highly respected and competent scientists who are experts in how DNA functions and are taking opposite sides on an issue that lies at the very heart of the argument against even the possibility of Darwinian-style macroevolution! Both are believers in the God of the Bible, and both acknowledge that God has been active in the world since the initial act of creation.

In spite of the emphasis in the book that Collins disagrees with Behe on how the biological complexity in the world came into being, the titles of their books is a striking illustration of what they do agree on. The title of Collins book *The Language of God* reminds me of Behe's title *Signature in the Cell*. Both authors are referring to the "language" of intelligence.

The two men are actually in agreement here! Both authors are saying the same thing in the titles of these two books. Both are declaring that only a supernatural being (God or a powerful Designer) created the intelligence in the cell that brought about the development of complex biological life on earth.

Collins Argues that the Cambrian Fossils Do Not Demonstrate the Intervention of a Supernatural Force

In chapter 9 we reported Strobel's discussion with Jonathan Wells about the Cambrian explosion. They concluded it was extremely unlikely that fossils will ever be found that predate Cambrian that show the same complexity as those found dating to the Cambrian era. They also concluded that the fossil evidence doesn't support evolution above the species level.

We may remember that Strobel changed his mind about Darwin's Tree of Life after Wells explained to him the significance of the Cambrian explosion. He said this totally contradicts Darwinism, which predicted the slow, gradual development of organisms over time. Leading evolutionary paleontologists like Stephen Jay Gould have acknowledged this problem and sought for alternative explanations of the fossil record. But they have so far not proposed any theory that has found wide acceptance.

Collins argues that the gaps in the fossil record don't prove anything. He says that "while attempts have been made by certain theists to argue that the Cambrian explosion is evidence of the intervention of some supernatural force, a careful examination of the facts does not seem to warrant this. This is another 'God of the gaps' argument, and once again believers would be unwise to hang their faith upon such a hypothesis."[560]

But Collins does not present any "facts" to support this, other than by saying, "The vast majority of organisms that have ever lived on Earth have left absolutely no trace of their existence, since fossils arise only in highly unusual circumstances," and, "The so-called Cambrian explosion might, for example, reflect a change in conditions that allowed fossilization of a large number of species that had actually been in existence for millions of years."[561]

Collins does have a point. But here I would rather take the side of an internationally known paleontologist like Gould than a geneticist like Collins. Collins acknowledges that "Gould himself questioned how evolution could account for the remarkable diversity of body plans that appeared in such a short span of time."[562]

How the Human Genome Research Project Convinced Collins that We Are Products of Both Evolution and God's Working

Are We Related to Other Primates and Mammals?

Collins was impressed with how closely the gene sequences of primates like chimpanzees and even mammals like rats resembled that of the human race. He says this is true of both active and inactive genes and protein. Knowing this, Collins asks why the genomes of mammals and humans are so similar if they do not have a common ancestor? He says, "The study of genomes leads inexorably to the conclusion that we humans share a common ancestor with other living things."[563]

He makes a point of explaining that humans have twenty-three pairs of chromosomes while the chimpanzee has twenty-four. This is because two chromosomes in humans are fused together. Then he says the place of this fusion "is truly remarkable" because it is found "right where evolution would have predicted . . . The fusion that occurred as we evolved from the apes has left its DNA imprint here. It is very difficult to understand this observation without postulating a common ancestor."[564]

He asks why God would put such a nonfunctional gene in this precise location if he created humans by a supernatural act. "For those like myself working in genetics, it is almost impossible to imagine correlating the vast amounts of data coming forth from the studies of genomes without the foundations of Darwin's theory."[565]

PART 3—*What Do We Really Know about the World?* The Bible

The Origin of the Human Race

> Population geneticists, whose discipline involves the use of mathematical tools to reconstruct the history of populations of animals, plants, or bacteria, look at these facts about the human genome and conclude that they point to all members of our species having descended from a common set of founders, approximately 10,000 in number, who lived about 100,000 to 150,000 years ago. This information fits well with the fossil record, which in turn places the location of those founding ancestors most likely in East Africa.[566]

By estimating there were about ten thousand in a "common set of founders" of the human race, he seems to be referring to humans who lived at the same time that Adam and Eve did. He says, "The biblical texts themselves seem to suggest that there were other humans present at the same time that Adam and Eve were expelled from the Garden of Eden. Otherwise, where did Cain's wife . . . come from?"[567]

This is the assumption of Collins and many other geneticists, and clearly it is an assumption. The matter of human origins was explored in chapter 11, long before I heard about Francis Collins and his book. It is clear from the research I did and report in chapter 11 that the scientific evidence is not at all as conclusive as is often assumed on many of the issues related to human origins.

How does Collins reconcile this theory with the creation story in Genesis? First, by suggesting that Genesis 2 may be a "poetic and powerful allegory of God's plan for the entrance of the spiritual nature (the soul) and the Moral Law into humanity." He adds, "Many sacred texts do indeed carry the clear marks of eyewitness history, and as believers we must hold fast to these truths. Others, such as the stories of Job and Jonah, and of Adam and Eve, frankly do not carry that same historical ring."[568]

Collins presents two main challenges to the conclusions presented so far in this book. One is that the human body today is the result of Darwinian evolution. The second is that there were many humans in existence when Genesis 2 reports the "creation of Adam and Eve, or at least at the point where they fell into sin. This second challenge is easier to respond to—because it does have a clear answer in the Bible.

My Response

Collins is proposing a theory that Adam and Eve may have been two of *many humans* who lived back then, and this is related to an important rule of biblical interpretation that I introduced in chapter 24. This rule is that when a passage is difficult to understand it is important to consider other passages that deal with the same theme. Doing this often clarifies the meaning of a passage.

It is true that consideration must be given to the form of the biblical text; whether it is literal or figurative, prose or poetry. The language of the creation story in Genesis

When the Pursuit of Truth Hits a Speed Bump

does seem to have a poetical flare, but does this preclude it from being "sober history"? Most of the book of Genesis is clearly presented as factual history. The issue at hand is whether or not this is also true of Genesis 1 and 2.

First, what does the New Testament have to say about whether or not Adam and Eve were the ancestors of all humanity? First Corinthians 15 is one text that clearly answers this. Paul says in 1 Corinthians 15:45, "And so it is written, 'The first man Adam became a living being.' The last Adam became a life-giving spirit."

This is part of a long presentation that Paul gives with two major themes. The first is that Adam was created by God as the first human, the beginning of the whole human race. The second is that Jesus Christ, by His divine birth, sinless life, death on the cross in our place for our sin, and resurrection from the dead, brought in a new beginning for the human race.

Paul also refers to Adam being created from dust, like Genesis 2:7 tells us. "The first man was of the earth, made of dust; the second Man is the Lord from heaven" (see 1 Cor 15:45–49).

But the strongest argument against Collins's position here is the clear teaching of the Bible that all of humanity came under bondage to sin when Adam and Eve sinned.

The Apostle Paul, who was ordained by God to explain the "systematic theology" of Christianity, explained why a new beginning for the human race was necessary. He wrote, "Therefore, as through one man's offense judgment came to all men, resulting in condemnation, even so through one Man's righteous act the free gift came to all men, resulting in justification of life. For as by one man's disobedience many were made sinners, so also by one Man's obedience many will be made righteous" (Rom 5:18–19). Note also Romans 5:12, "Therefore, just as through one man sin entered the world, and death through sin, and thus death spread to all men, because all sinned."

Paul earlier in 1 Corinthians refers to another central truth in the Bible. "For since by man came death, by Man also came the resurrection of the dead. For as in Adam all die, even so in Christ all shall be made alive" (1 Cor 15:21–22).

This clearly refers to the temptation and fall of Eve and then Adam in Genesis 3. Although Eve sinned first, the responsibility for morally corrupting the human race is placed squarely on Adam. The moment Adam sinned he died spiritually—his holy and righteous and life-receiving spiritual relationship with God was severed. As a result, all human beings born after that were born with sinful human natures. As a result, we have all sinned (Rom 3:10, 23). We sin in our actions (1 John 1:10) because we have a sin nature (1 John 1:8). This is immediately seen in Genesis 4 where Cain killed Abel.

This is what made the virgin birth, sinless life, and death on the cross of Jesus for the sins of Adam and Eve and of all humanity necessary for a right relationship to God and for eternal life to be possible to "all who believe." Actually, Jesus dying on the cross as the "seed of the woman" was prophesied in Genesis 3:15 and planned by God the Father even before that (1 Pet 1:18–21; Acts 2:23).

PART 3—*What Do We Really Know about the World?* The Bible

Collins bases his argument that other humans must have existed of the same generation as Adam and Eve on something that does not stand up to sound biblical interpretation.

He gives the argument that other humans must have existed then because Cain and Seth had wives, and the Bible does not say that Adam had any daughters at that time. He thinks it strengthens has argument to point out that incest is prohibited later in the Old Testament.

However, women are not listed in the genealogies except in specific cases, as when the narrative mentions them in some context. (Only males were heirs to property and family name. Women married and entered the families of their husbands. Even the fact that Jesus had sisters is only mentioned on one occasion [Matt 13:56; Mark 6:3], and their names are not even mentioned although the names of Jesus' brothers were.) We must also keep in mind that Genesis 5:4 informs us that "after he begot Seth, the days of Adam were eight hundred years; and he had (other) sons and daughters." This does not clearly say that Adam had daughters while Cain was still living at home, but there is no reason to doubt that his wife was a daughter of Adam and Eve.

It seems clear the Jewish custom and the common tendency in the Old Testament was to not mention the names of daughters unless they were connected to a specific event in the biblical narrative. So it is common and very natural to infer that Adam and Eve did have daughters even while Cain was still living with them, and that Cain and Seth married their sisters. There is nothing written in the Bible to preclude that Cain may have been married when he killed Abel.

The argument legitimately cannot be made that there were humans on the earth who were not descendants of Adam but they all died in the flood of Noah's day. This is because it was for the wickedness of mankind that God sent the flood (Gen 6:5–7), and it defies logic to argue that the humans who were not affected by the fall into sin of Adam and Eve would be universally wicked by the time of Noah. Romans 5:12–13 says, "By *one man* sin entered the world," and that one man is clearly Adam.

Therefore, I am convinced that the belief that Adam and Eve were two of many *humans* on earth when God did His creative act in Genesis 1:26-27 is not scriptural. This is because if there were other humans on earth when Adam was created the whole human race would not have come under the bondage of sin and the condemnation of God when Eve and Adam sinned.

The second scriptural argument against Collins's position may not be so conclusive. He argues that Adam's body was a product of biological evolution. My investigation so far into the wording of the Bible and the scientific literature has led to the confident conclusion that the facts do not support this argument. But don't we need to be open to the science Collins presents that he says supports his view, as we "follow the truth wherever it leads"?

There are four references in the Bible of man being made of dust or dirt or earth (Gen 2:7; 1 Cor 15:47; 15:48; 15:49).

Was Genesis 2:7 simply emphasizing the human body is made of natural material like that of all other living things? Genesis 2:7 says, "And the Lord God formed man of the dust of the ground, and breathed into his nostrils the breath of life; and man became a living being."

How figurative is this? The creation story in Genesis is clearly in poetic language to some degree—but does it include "poetic license"—expressions which are not to be taken literally?

If Genesis had said, "God prepared a body for Adam," then it would be similar to a reference to Jesus' human body. (Jesus' body seemingly developed in the womb by natural biological processes. Evidently He had his mother's DNA. But He did not have Joseph's DNA—Mary was a true virgin when Jesus was conceived "by the Holy Spirit." Heb 10:5 says, "Therefore, when He [Jesus] came into the world, He said: 'sacrifice and offering You did not desire, but a body You have prepared for Me.'")

If this had been said of Adam, then to explain it as God creating hominids and then at some point creating a soul and spirit in one of them to create the human race would be easier to understand. This is because the Hebrew word "create" here (*bara*) means to create something that hasn't existed before, but does not always mean "from nothing" (*ex nihilo*).

Genesis 2:7 does not say God "prepared a body for Adam." It says He formed man from the dust of the ground. Collins believes the human genome is so strikingly similar to that of other primates that it proves Adam's DNA came from the same ancestors as that of chimpanzees and other primates. But is this sure proof?

Again, how was Eve created? Did God literally take a rib from Adam and fashion a woman from it, as the wording of Genesis 2:21–23 says? I have always wondered how literal these passages are that speak of how Adam and Eve's bodies were created, but a literal interpretation does fit the context.

The Apostle Paul, in 1 Corinthians 11:7–12, makes the argument that Eve came from Adam. "For man is not [or 'does not originate'] from woman, but woman from man."

On the other hand, if Adam and Eve's bodies were created separately from other living beings, why would God arrange the biological evidence to show what Collins and other believers in human evolution say is "proof of human evolution"? Collins is convinced that the similarities between the human genome and the mouse genome are so great that they point to "a common mammalian ancestor." He writes, "Unless one is willing to take the position that God has placed these [ancient repetitive elements] in these precise positions to confuse and mislead us, the conclusion of a common ancestor for humans and mice is virtually inescapable. This kind of recent genome data thus presents an overwhelming challenge to those who hold to the idea that all species were created ex nihilo."[569]

Strobel asked Jonathan Wells about this problem. He asked, "What about recent genetic studies that show humans and apes share ninety-eight or ninety-nine percent of their genes? . . . Isn't that evidence that we share a common ancestor?" Wells replied that apes and humans share 100 percent of the DNA that controls the formation of their bodies and yet their bodies are so different and this is as compatible with common design as it is with common ancestry.[570]

It may be that the "kinds" mentioned in Genesis 1 include one of all primates and land mammals—as Collins says the genome sequence of humans and rats are remarkably similar. In that case, hominids might share much of their genes with other primates and even rats. If such were the case, creation by separate "kinds" like the Bible says could fit into Collins's argument.

On the other side of the argument, there is no evidence of evolution in the human race in the record of the fossils considered by paleontologists as "human" as is explained in chapter 11. But if God acted to create Adam from the body of a hominid, then the discussion changes. There may have been any length of time when the bodies of hominids "evolved" before the creation of Adam and Eve. The Bible is of course silent on this, because it does not mention hominids at all. They may have evolved from animals that are mentioned earlier in the creation record in Genesis.

But it should also be kept in mind that Collins simply believes that the *bodies* of humans are a result of evolution, not the human soul and spirit. He says, "The comparison of chimp and human sequences, interesting as it is, does not tell us what it means to be human. In my view, DNA sequence alone, even if accompanied with a vast trove of data on biological function, will never explain certain special human attributes, such as the knowledge of the Moral Law and the universal search for God."[571]

No matter how many hominids may have existed then the Bible reports that God "breathed" into Adam only, and later created Eve directly somehow.

I don't know which way the reader will go on this one! Collins has a powerful argument here, but God's ways are higher than our ways and His thoughts are often beyond our comprehension. Why would God design the human genome so similar to that of other mammals if humans have no common ancestors with other animals? Conversely, why does Genesis say that God formed Adam's body from the dust of the ground if the word "create" here does not include his physical body?

I think it is fitting here to quote another member of Bio-Logos, the biochemist Sy Garte, who like Collins was drawn by the realities of nature from atheism to faith in God and the Bible.

> We are able through science to find magnificent and overwhelming evidence of God's intervention and on-going engagement in our world, from its creation to our everyday lives, in every aspect of reality, including in our ongoing discoveries of the secrets of the natural world. We now know that the universe was not always here. It had a beginning. It was created. This is Gospel, but it is also science.

> Although we find many pointers to divinity, God so designed the world that his hand in its creation can never be proven beyond doubt. If that were not true, then free will and the beauty of faith would disappear. Faith is a gift to be accepted by an open heart and an open mind. The knowledge of God's grace cannot be forced on anyone by the discovery of any irrefutable fact that proves his existence. But the converse is also true. No scientific endeavor will ever prove the absence of God, and so we are free to believe.[572]

Yes, you can believe in macroevolution and still be a Christian, and lead a life that pleases God. Theistic evolutionists who believe the gospel and have experienced God's saving grace are heirs to all the wonderful promises God has made to these he saves—like everlasting life.

Collins says, "I do not believe that the God Who created all the universe, and who communes with His people through prayer and spiritual insight, would expect us to deny the obvious truths of the natural world that science has revealed to us, in order to prove our love for Him." He continues, "In that context, I find theistic evolution, or BioLogos, to be by far the most scientifically consistent and spiritually satisfying of the alternatives. This position will not go out of style or be disproved by future scientific discoveries."[573]

From what Collins writes, it is clear that he has faith in his personal relationship with God—that God is working in the world and in his personal life—today. This is much more than the faith some theistic evolutionists have—that God exists but does not involve Himself in nature or in humanity on a day-to-day basis. Clearly Collins has personally experienced the power of the Living God in his life!

So the issue here is not one of Christian orthodoxy, but of correctly applying the laws of biblical interpretation. It would be a serious error to refuse to embrace as a "brother in the faith" those who take the opposite side from oneself on this issue. This reminds me of Psalm 119:63, which says, "I am a companion of all those who fear You [God], And of those who keep Your precepts."

Conclusion

Collins believes that Darwinism is adequate to account for our human bodies, but cannot account for the soul and spirit that makes us so different from all other animals. He calls the intelligence in DNA "the language of God" because he is convinced it clearly had a divine source.

He rejects Behe's argument that the irreducibly complex nature of biological organisms cannot be explained by random mutations and natural selection like Darwin proposed. But Collins rejects it by arguing from theory and does not present experimental results like other scientists have attempted and failed to do. Collins also argues that the sudden appearance of fossils of complex organisms in the Cambrian period does not support the intelligent design argument because the conditions necessary for fossilization are so rare. These are arguments of two well-known and competent scientists that probably never will be settled by solid scientific research. But I for one believe that the evidence so far supports Behe.

When it comes to biblical interpretation, Collins argues that the creation account in Genesis is poetical and may be allegorical rather than literal history. He believes with complete confidence that Genesis presents a powerful and poetic narrative that recounts the historical account of God's creative actions.

In this regard, his assumption that the bodies of Adam and Eve developed over a long period of evolution comes primarily from his discovery of how close the human genome resembles that of mammals, including mice. He asks, if God actually created the species separately, why would He have done so with their genome sequences so similar? From his perspective as a person who became a scientist well before becoming a Bible-believer, it is understandable that he is seeking to fit the creation story in Genesis into a strong worldview based on modern science, which is based on naturalism and macroevolution.

Collins acknowledges that evolution alone cannot account for the existence of the human race. The moral law within each individual and other things that distinguish people from all other animals on earth make it crystal clear to him that human nature was directly created by the God of the Bible.

Collins says he is a "theistic evolutionist," but says this is a "terrible name." So he coined the term "BioLogos" to refer to his belief. The term *bios* is the Greek word for "life," and *logos* is the Greek word for "word." ("Word" in John 1:1 refers to God.) He says "BioLogos" expresses the belief that God is the source of all life and that life expresses the will of God.

What if Collins is right that the bodies of humans are a product of evolution but the souls and spirits are a result of God's creative action long ago? Would this prove the Bible is in error? No! It would show that the account of creation in Genesis 1 and 2 are more poetical and less literal that I believe them to be. It would also suggest that some of the references elsewhere in the Bible to the creation account in Genesis 1 and 2 include poetic language. But it certainly is not an indication that most of the Bible is anything but accurate history and trustworthy in its teachings. Collins does recognize that much of the Bible is authentic history.

While acknowledging the alternative explanation may possibly be true, I am more comfortable with believing that God created Adam's body directly in Genesis 1:27 and 2:7 without Adam having any biological ancestors. Collins is clearly more comfortable with believing that Adam's body was a result of evolution over many generations through natural biological processes.

Who is right in this? Let the reader choose! In any case, which of these we believe need not affect the quality of our spiritual lives in any way. What is crystal clear from the Bible is that what makes humanity human—our soul and spirit that were created "in the image" of God—was directly created by God and makes us different from all other biological life.

Our "speed bump" should not knock us off course. At best, it should slow us down. Collins himself calls the issue here a "dilemma."

It seems to me the information presented in *The Language of God* is important enough for us to revisit some of the conclusions presented earlier in this book, but is not sufficient to reject any of them. Collins does not present any solid evidence that human evolution has actually occurred. His challenge to Behe's careful research is that the complex biological organisms that exist today *might possibly have* developed incrementally. His reason for concluding this is the discovery of similarities between the human genome and DNA with that of the genomes of other mammals. He does not present any evidence from either genetic research or the fossil record to show this actually occurred. This seems to be an illustration of how we start out seeing the world from our own perspective—our worldview—and while trying to be open to truth continually seek to fit new knowledge into that perspective.

Collins's Dilemma

Perhaps it is best to end this chapter focusing on a dilemma. As Collins expresses it,

> The real dilemma for the believer comes down to whether Genesis 2 is describing a special act of miraculous creation that applied to a historic couple, making them biologically different from all other creatures that had walked the earth, or whether this is a poetic and powerful allegory of God's plan for the entrance of the spiritual nature (the soul) and the Moral Law into humanity.
>
> Since a supernatural God can carry out supernatural acts, both options are intellectually tenable. However, better minds than mine have been unable to arrive at a precise understanding of this story over more than three millennia, and so we should be wary of staking out any position too strongly.[574]

PART 4

How Significant Is This for Our Daily Lives?

Preface to Part 4

How important has it been for us to learn from science? To learn of the existence of electricity, that led to the invention of the light bulb and many other things? To discover aerodynamics, that led to the development of the airplane? To learn of the discoveries of science that led to new technologies and medical treatments? Openness to learn from scientific discoveries has changed our daily lives and revolutionized the way we live. It has greatly increased our effectiveness and that of our society.

How important is it for us to learn from God's revelation in the Bible? Does openness to learn from biblical truth improve our daily lives in the same way—or in an equivalent way? Theologian Henry Thiessen argues that we can accept the claims that the Bible makes, that it speaks truth, if what it says about other subjects, such as history and science, are found to be true. Why is this important? The Bible was written millennia before modern science led to the discovery that the universe had a beginning and the cell is irreducibly complex. This is a dramatic and objective testimony to the fact that the Bible speaks the truth.

If the Bible can be trusted to reveal truth (reality) in things that science cannot test or measure, this in turn should encourage the serious seeker of truth to study it, pay close attention to what it says about the meaning of life and other issues, and seek to experience the power of spiritual truth.

There are other realities that demonstrate the authenticity of the Bible as the Word of God and demonstrate its truthfulness of course. But I think for most of us who were raised and educated in modern society this demonstration that scientific investigation supports what the Bible says about the natural world is strong and helpful testimony to us today.

To live life to the fullest don't we need to be aware of both *natural* and *spiritual* realities?

26

Where Does Truth Lead Us?

Isn't the most important thing in a study of science the discerning between clearly verified scientific information and firmly held theory that is not clearly supported? Some "truth about nature" seems firm, but much promoted as truth does not.

In a similar way, isn't the most important thing in a study of the Bible and science the determining of what the Bible says clearly and unambiguously about nature and origins from what is simply human interpretation? Such a clarification should put us on firmer grounds to try to judge where modern science and the Bible agree and disagree in respect to where they speak about the same issues.

What can we be sure is true and will not be challenged by new knowledge? Can we really be sure that what we accept as reality today will stand the test of time?

If, as the Bible claims, the eternal God inspired its writers and guarded them from all factual error, then what the Bible says clearly can be relied on as unchanging reality (truth).

But many wonder if this "truth" that we are seeking is only an illusion.

It has been a reoccurring theme of this book that some important assumed "knowledge" about our world does not stand the test of scientific investigation. There are things we don't know that we don't know. This begs the question, "What do we really know for sure?"

In contrast, has God given us a clear message from the Bible—one that applies equally to all of us? Is there something like "ultimate truth" that is reality for all time and all people everywhere? The Bible says that the answer to this is a resounding *yes*!

Will a confidence that there is "truth" that is truth for all time be helpful to us? The philosopher Francis Schaeffer answered it this way, "We need absolutes if our existence is to have any *meaning*." He then asked, "How can we be sure that what we think we know of the world outside ourselves really corresponds to what is there?"[575] I would add to this the question: How can we be sure what we rely on as true today will stand the test of time?

We tend to look at all of life from our own worldview, whether we are conscious of it or not. Our worldview or philosophy of life helps us answer the big questions in life. It gives us a sense of identity (What or who am I? Where did I come from?). It helps us understand how to live (Is there any real meaning to my life? Does it really matter how I live my life? Is this life all there is to my existence, or is this life only a preparation for a more permanent existence?). Our worldview is what we look to for the answers to these important questions. It informs our *values* and our values guide our *actions*.

There are two main worldviews today: naturalism as expressed in evolutionism, and in contrast to this an awareness of spiritual realities as expressed in creationism and the Bible.

We tend to turn our worldview into an "absolute." Philosophers of science emphasize the tentativeness of scientific knowledge and theory, as we saw in chapter 2. Christians tend to turn our faith in the Bible into an absolute also—but is there objective justification for this? I believe we are justified in doing this only as long as we carefully separate the clear teachings of the Bible from our assumptions and biases.

How Do We Know Which Worldview Is Closer to Reality?

The worldview of naturalism claims its legitimacy from modern science. Science gets its authority from observing facts about the natural world, measuring and systemizing them, proposing and testing theories to explain them, and then it seeks to discover and verify general laws of nature. It includes information gained from direct observation of nature and by controlled experiments. Experimental results can be supported or contradicted by other experiments. In this way scientific theories can be tested.

But such verification of natural events in the distant past is impossible. Biological experiments today can only determine if certain biological transformations are possible. The fossil record can support evolutionary theory when it shows unbroken development from simple organisms to more complex ones. In this sense, attempts to reconstruct past events are not without meaning. But shouldn't we avoid claiming a theory has been proved in relation to past events based on fragmentary and inconsistent geologic records, even when biological possibilities are known?

Science is validated by how accurately it seems to "speak truth" about the universe and our natural world. Scientific investigation, then, has limits in its attempt to reconstruct events in the distant past and in its attempt to study phenomena that cannot be directly observed and measured.

The worldview of creationism comes from the belief in the revealed truth in the Bible. If "religion" is a seeking of spiritual meaning as part of our worldview and our human experience and reasoning, then how can we know if it expresses something real? What is it based on—where does it get its authority?

Here is where I believe the Bible stands out from all other sources of religion. If what the Bible teaches is reality and is authentic truth, then the situation changes dramatically from a man-centered search for truth. This is because the Bible claims to be a self-revelation from the God Who created all things, knows all things, and is in ultimate control of both the natural and spiritual worlds. Then shouldn't the first thing be to focus on how we can be sure that what the Bible presents is authentic and reveals "truth" truthfully?

27

The Bible as Truth

How Can We Know If What the Bible Reveals Is Objective, Enduring Truth?

1. One demonstration of the accuracy of the Bible is emphasized earlier in this book. This is that the Bible's clear statements about the origin of the universe and of life on earth are supported by modern scientific investigation. If the universe had a beginning it had a *cause*, and only the act of a pre-existent Creator seems to be an adequate cause. Science has not demonstrated how life could have developed spontaneously on earth—or anywhere else in the universe. The Bible's explanations of the beginning of the universe, of life on earth, of the development of complex life, and of human nature are consistent with what we know of potential causes of these things through modern science.

We will remember that philosopher of science Stephen Meyer says he systematically analyzed the various scientific theories formulated to explain the biological and geological evidence of events on earth in the distant past. He studied all the current theories of origins and concluded that the intelligent design theory provided the best explanation of all the scientific theories. He concludes that intelligent design passes the test of being a *general* scientific theory and this is the theory that the modern view of creationism comes under.

He emphasizes that it cannot be proved scientifically that God created the universe and all that exists within it, but the scientific facts are consistent with the account of creation in Genesis.

Does it take *faith* to believe that God simply spoke and the universe was created? Yes! Hebrews 11:3 says, "By faith we understand that the worlds were framed by the word of God, so that the things which are seen were not made of things which are visible."

Hebrews 11 lists many historical events that were preceded by God foretelling an event that led people to prepare for what later happened. One example is Noah obeying God by spending years to build a large ship to prepare for a flood that actually occurred.

John Collins emphasizes that faith is a rational response when you are persuaded that God is trustworthy. He also points out that faith in God is relational, and so goes beyond what we can verify.

2. Another is the internal consistency of the Bible. It was written over a fifteen hundred-year period by over thirty authors. Yet it is consistent throughout in its revelation of God, the nature of mankind, and the nature of spiritual things.

3. Third is the evidence of its authenticity in reporting historical events. Harold Booher emphasizes this when he wrote, "Without a belief in a literal Exodus with attendant miracles, Judaism would not have come into existence. Without the credible report of eyewitnesses to the death and resurrection of Christ, Christianity would not be."[576]

4. Another is the Bible's testimony about itself.

In chapter 3 we mentioned the theologian Henry Thiessen's insightful assertion that "if we can prove the genuineness of the books of the Bible and the truthfulness of the things they report on other subjects, then we are justified in also accepting their testimony in their own behalf."[577]

We have had this in mind throughout this book when exploring the current state of scientific knowledge today and investigating whether or not these support the clear statements on nature and origins in the Bible.

The Bible claims it is speaking "truth" that can be relied on when the rules of biblical interpretation are carefully followed. Psalm 119:60 says, "The entirety of Your word is truth." Jesus said of God the Father in John 17:17, "Your word is truth." Psalm 117:2 adds, "the truth of the Lord endures forever."

The word "truth" is used over two hundred times in the Bible. God is called the "God of truth" four times in the Old Testament (Deut 32:4; Ps 31:5; Isa 65:16, twice). God's Word—the Bible—calls itself "truth" five times (Ps 119:142, 151, 160; Isa 25:1; John 17:17). The Holy Spirit Who the Bible says inspired the Bible writers to write without error is called "the Spirit of truth" four times in the Bible (John 14:17; 15:26; 16:13; 1 John 5:6). Jesus is also called the "truth" (John 14:6).

When the Bible speaks of "truth" it speaks of objective reality. God's truth is *enduring*. "The truth of the Lord endures forever" (Ps 117:2). It speaks of spiritual reality to all humans everywhere on earth.

5. Fifth is the personal experience of those whose lives have been transformed by believing the gospel and seeking to live in obedience to its teachings. Lee Strobel's wife is an example of this transformation. He said her values transformed, and she became more loving, caring and authentic.

But here there is need for much caution. If the Bible is "truth" in the sense that it accurately expresses the realities of the natural world and the spiritual world, then there is an objective standard for spiritual truth. Human experience at best can only *illustrate* and *demonstrate* spiritual truth, and subjectively confirm and interpret it to the one experiencing it or observing its effect on others.

Do we experience spiritual truth in the same way we experience the effects of natural laws such as the law of gravity? Yes, if the spiritual law applies to all people at all times. But many of the statements about spiritual truth in the Bible come with clear conditions. For example, Romans 10:9 says, "If you confess with your mouth the Lord Jesus and believe in your heart that God has raised Him from the dead, you will be saved." This verse gives two clear conditions related to salvation. There are many statements in the Bible that promise definite results that are dependent on clearly stated conditions.

28

Life's Questions

As we come to the end of the book, I hope I have established that there is clear objective evidence to conclude that the Bible has truthful answers about how to live more realistically and effectively day by day. What are some of the more important "truths" that the Bible teaches clearly?

1. Who Is God and How Is He Described? What Does the Bible Say?

God is so awesome, and in so many ways beyond our human understanding, that I feel inadequate to describe Him! However, I'll list here some of His characteristics that are abundantly clear from the Bible.

The first verse of the Bible introduces the Creator Who is *Elohim*, or "the Supreme God." The Bible is fittingly called "God's self-revelation." Its primary purpose is to reveal to mankind God's nature and acts, and His relationship with humans.

In His essential nature, God is "spirit" (John 4:24) and thus invisible and has no body (1 Tim 1:17). He is "self-existent" (Exod 3:14); and "eternal" (Ps 90:2). He has personhood, which includes personality and the desire and ability to form personal relationships.

The Bible says that God, in His spiritual presence, is everywhere. He is both transcendent (distinct and separate from His creation) and imminent (and thus present wherever we are) (Ps 139:7–12; Acts 17:28).

God is the Creator. God is immensely powerful. In fact, He is "omnipotent." The Bible says, "Nothing is impossible with God" (excepting only what is contrary to His moral nature). The Bible says there is no limit to His power. Matthew 19:26 says, "With God all things are possible" (see also Gen 18:14; Jer 32:17, 27; Luke 18:27).

God is aware of everything that happens and has happened and will happen in the universe. He is "omniscient." He knows all things in the present, the past and the future. Psalm 147:5 says, "Great is our Lord, and abundant in strength; His understanding is infinite."

He knows what all of us are doing, like Proverbs 15:3 says: "The eyes of the Lord are in every place, keeping watch on the evil and the good." (Figures of speech like "the eyes of the Lord" and "the arm of the Lord" are very common in the poetical books. They are "anthropomorphisms"—the attributing to God of human characteristics in order to make some characteristic or act of God understandable to us.) God knows at all times what each of us is thinking.

God is "holy." This means He is exalted above and separate from all creation, and it also means He is separate from all moral evil and imperfection. He is righteous in His nature and actions, and hates injustice and all that is sinful.

God is benevolent, loving, gracious and merciful. God's grace is shown in His willingness to forgive and save all who come to Him in genuine repentance and faith. His mercy is seen in His reaching down to help people in their misery, suffering and distress.

The God of the Bible is a God Who wants a personal relationship with humans and created them "in His own image" to make this possible. God's love of mankind is seen throughout Scripture, and is expressed in the most famous verse in the Bible, which is John 3:16: "For God so loved the world that He gave His only Son, that whoever believes in Him should not perish but have everlasting life."

God is truth. He is the basis of all reality, and His words in the Bible perfectly conform to reality in both the spiritual and natural realms. He is called "a God of truth and without injustice; righteous and upright is He" (Deut 32:4). "His truth endures to all generations" (Ps 100:5). The Bible also calls itself truth from beginning to end, as was discussed in chapter 27 (Ps 119:60; 119:151; John 17:17).

God is the Judge of all. The Bible reveals that a day is coming when all of us will come before God as our righteous Judge, and His judgment will determine our eternal fate (Rom 2:12–16; 1 Cor 4:5; 2 Tim 4:1; Rev 20:11–15).

2. God and Nature

The primary message of Genesis chapters 1 and 2 is that *God did it*. *God* created the universe and all in it. He did it directly, without intervening causes. Initial creation of the universe may have been instantaneous, even though a long process evidently ensued to bring about the condition of the universe and world to become as we know it today. He was in control of the process; and His control continues to this day. This does not mean that changes do not occur without His direct intervention. We may assume that changes in living things and in the cosmos occur constantly as a result of the natural laws He has established. He is in control of the results.

3. What Is Man?

The Bible is very clear about human nature. We are created in the "image of God." Of course this is not a physical image, because God has no body. It is a mental, social, and moral image. God has *personality*. He has an intellect, emotions, and a will. God has a social nature—He created humans and wants a positive relationship with us. We have an intellect, emotions, and a will. We are social beings, and seek positive relationships with others. We were created with the capacity to know God and to worship Him. God has a moral nature. He is holy and righteous in all His thoughts and deeds. We also have a moral nature, even though unlike Him we can be and are unrighteous.

All humans have a body, a spirit, and a soul. Our souls and spirits are eternal. The Bible teaches that man has both a "soul" and a "spirit," and this makes humans different in kind from all other physical beings on this earth. We have a "spirit" that is capable of believing in God and communing with Him. It is clear from Scripture that the spirits of believers will continue to exist beyond the grave. Physical death seems to be when the spirit (*pneuma*) of a person departs from the body. (See Luke 8:55). "Spirit" here means the rational soul, and the word is used of angels as well as of humans.

Francis Collins, when interviewed on Public National Broadcasting on February 20, 2013, discussed how neuroscientists are beginning to make a map of the neurons in the human brain like he and his teams did of the human genome. They are trying to discover the functions of the neurons to create thought. Collins called the human brain "the most complicated organism in the universe" and suggested this research was an advance of that which sequenced the human genome and was a new frontier in science. A syndicated newspaper recently reported that there are more genes in the human brain than there are stars in the universe.

But science has never come up with a convincing answer to how the human brain could possibly have developed without it being created by God. Collins called the language in the cell that guides the function and development of organisms "the language of God." In this sense it can be said that science studies the language of God in nature the same way theologians study the language of God in the Bible.

We have a moral nature. Before humans sinned this seemingly included righteousness. All men have a conscience. The purpose of conscience is to bear witness when we do something contrary to our sense of what is morally right. Romans 2:15, speaking of the conduct of people who do not yet know God, says they "show the work of the law written in their hearts, their conscience also bearing witness, and their thoughts alternately accusing or else defending them." This reveals two things about the moral nature of all mankind. First, God has placed in our hearts some knowledge of His moral law. Second, the function of conscience is to reveal to us when we violate our sense of what is moral or righteous. Conscience shows whether or not our acts and heart attitudes conform to the standard of morality we believe is required of us.

It serves as an alarm clock that sounds an alarm when we violate our sense of what is morally right or wrong.

I believe one of the most remarkable manifestations in human beings of the nature of God is His capacity to love. We have the capacity for unselfish love,

- even when offended;
- even when grieving over the actions of the one loved;
- even when it means personal loss and sacrifice of things one values.

I am speaking here of *agape*, not of love out of dependency or fear; and not love as infatuation.

There is a sensitivity of loving, a vulnerability of loving someone that seems to work against the "survival of the fittest" mentality. I think this is a truth that naturalism cannot explain. An example of the human tendency toward compassion for other humans is when the bombs exploded near the finish line of the April 15, 2013, Boston Marathon. It was reported that many people ran *toward* the blast scene to help the wounded—an act which cannot be explained by the human survival instinct.

4. How Does God Reveal Himself to All Mankind?

According to the Bible He provides a witness to all mankind in two ways. Romans 1:20 says, "For since the creation of the world His invisible attributes are clearly seen, being understood by the things that are made, even His eternal power and Godhead, so that they are without excuse." This says that all mankind has some true knowledge of the Creator because of creation. The existence of this world testifies that a powerful Creator exists. They may ignore or reject this witness, but it is there for all to see.

The second way was discussed above under mankind's moral nature. According to Romans 2:15 we instinctively know that some things are morally right and other things are morally wrong. This is a kind of testimony to the moral nature of our Creator.

God also reveals Himself in the Bible, including the reports in the four Gospels that relate the nature of Jesus and how He lived a human life on earth.

5. What Is the Relationship between God and Man?

The God Who created man is a moral, righteous God and demands righteousness in humans. The Bible says that Adam and Eve were created directly by God. They were given a direct command by God not to do something. Tempted by God's enemy they disobeyed. This is the clear message of Genesis 3. Later the Bible clearly says that when they disobeyed they became *sinners*—their moral nature changed from being righteous to being sinners, and all humans born since were born with a "bent toward sinning"—a sin nature.

God created mankind to have fellowship with Himself, so He created man "in His own image." It is very clear that God created humanity as an object of His love, and seeks a personal, positive relationship with each individual.

When God commanded Adam not to eat of a certain tree, He declared the penalty of disobedience: God told him he could freely eat of all but one tree in the garden of Eden where Adam was. He told Adam if he ate of this one tree, "in the day that you eat of it you shall surely die" (Gen 2:17). We know from other verses in the Bible that when Adam and Eve disobeyed God in the garden they died spiritually. This means the wholesome, spiritual life-giving relationship with their Creator-God was severed.

God expects and demands a response from individuals that is opposite from the one Eve, and then Adam, made in Genesis 3. When Eve ate the forbidden fruit she did so because she did not believe God's clear word that "you shall surely die." God's enemy led her to believe God gave that command to keep her from something that was to her benefit. He said, "You will not surely die. For God knows that in the day you eat of it your eyes will be opened, and you will be like God, knowing good and evil." She doubted God's word and disobeyed Him. The response that God seeks of us is that *we believe God's Word and act upon it*. God leaves it up to us to want to restore the healthy spiritual relationship with Himself—yet the Bible says He takes the initiative in seeking this relationship.

God has been actively working ever since to restore that righteous relationship that was broken by the sin of Adam and Eve. The Bible likens us to lost sheep (Isa 53:6). God is a God Who seeks the "lost." Jesus put this very beautifully when He said, speaking of Himself, "For the Son of Man has come to seek and to save that which was lost" (Luke 19:10).

How can we be restored more fully to the "image of God" that He desired when He created humanity to express His nature as a moral and social being? The Bible, from beginning to end, shows that God loves us and takes the initiative to reach out to us and desires us to respond to Him. The spiritual result of responding is, as Ephesians 2:1 says of believers, "And you He made alive, who were dead in trespasses and sins."

But then Ephesians 2 goes on to say, "For by grace you have been saved through faith, and that not of yourselves; it is the gift of God, not of works, lest anyone should boast. For we are His workmanship, created in Christ Jesus for good works, which God prepared beforehand that we should walk in them" (vv. 8–10). When we respond to God and God's Word by repentance and faith, God does everything else and saves us from the eternal consequences of our sin by His grace. The Bible says we are to do good works because we have a new relationship to God. This is very different from believing that our good works will restore a positive relationship to God.

Who acts first to restore this relationship? God always does. Revelation 3:20, speaking of the resurrected Christ, says, "Behold, I stand at the door and knock. If anyone hears My voice and opens the door, I will come in to him and dine with him, and he with Me." Christ wants to enter every human heart by His Spirit for fellowship and to enable us to live as we should. But we must respond; we must "open the door of our hearts" and invite Him in. Dining together, in Jewish culture, was the way Jews had fellowship with each other in an intimate way.

Even our turning to God is motivated by "self-gain" when we include God and the teachings of the Bible in our understanding of what is reality. We seek forgiveness of our sins from God to be delivered from a sense of guilt. Or we seek God's help in order to cope with life more effectively. We read in the Bible that the souls of those who are not "saved" go to Hell when they die, and we fear such a terrible fate. Our felt need which brings us to God may be different for different people, but in every case we turn to God because we believe it is for our good.

And the Bible tells us this will always be for our good! God promises that if we live for Him and for what is right we will not lose as a result. We may not gain in possessions, position, or prestige that motivate people (1 John 2:16–17), but God will meet our human needs and give us a richer and more fulfilling life on earth and for eternity. People tend to seek meaning and satisfaction and express this in activities or possessions or recognition. But there is an "emptiness of the heart" that can only be truly satisfied by knowing God and finding spiritual health. We are "created in the image of God."

6. What Is the Purpose of Life on Earth?

I think what the Bible teaches about this can be likened to school and the school curriculum. This is never to be an end in itself but a transition time and a process to prepare children for adult life. For example, 1 John 2:17 says, "The world is passing away, and the lust of it; but he who does the will of God abides forever."

A good verse to explain the Bible emphasis briefly is Jesus' words in John 6:27, "Do not labor for the food which perishes, but for the food which endures to everlasting life, which the Son of Man will give you, because God the Father has set His seal on Him." Mark 8:36 puts it this way: "For what will it profit a man if he gains the whole world, and loses his own soul?" John 12:25 says, "He who loves his life will lose it, and he who hates his life in this world will keep it for eternal life." The Bible teaches us to labor in order to have food, clothing, shelter and the other necessities of life, but while doing so to value eternal and spiritual realities higher than physical things (see 2 Thess 3:7–12; 1 Thess 4:9–11; 2 Tim 2:6).

The Bible clearly says that life on earth is short in the light of the afterlife, that it is beset by weakness, temptation, and physical deterioration as humans grow into old age. The Apostle Paul speaks of this in graphic terms in 2 Corinthians 4:16—5:7. He describes the process of aging in the light of the afterlife in phrases like: "our outward man is perishing, yet the inward man is being renewed day by day," and "for our light affliction, which is but for a moment, is working for us a far more exceeding and eternal weight of glory." The emphasis is clearly on the transitory nature of this life and the importance of preparing for life after death. Paul ends this passage with the words: "for we walk by faith, not by sight."

7. How Should I Live My Daily Life on Earth to Do "What's Best for Me"?

The Bible answer to this is very clear. Since what we believe in this life and how we live now affects our situation in the after life and eternity, we should value the spiritual more than the physical. Because it is our spiritual nature—our soul and spirit—that will continue in the afterlife, we are exhorted to put spiritual values first. Jesus said, "But seek first the kingdom of God and His righteousness, and all these things shall be added to you" (Matt 6:33).

The Bible also says, "Do not love the world or the things in the world. If anyone loves the world, the love of the Father is not in him. For all that is in the world—the lust of the flesh, the lust of the eyes, and the pride of life—is not of the Father but is of the world. And the world is passing away, and the lust of it; but he who does the will of God abides forever (1 John 2:15–17).

8. How Can I Live—or What Can I Do—to Best Help Those I Love?

I believe one characteristic of Christian growth is unselfish love, the love that seeks the highest good of the object of one's love. When Lee Strobel's wife became a Christian he soon noticed a difference in her character and in her relationship to him. This led him to remark, "Her values underwent a transformation," and "she became a more loving and caring and authentic person." The Bible describes growth as a Christian by using such terms as "love, joy, peace, longsuffering, kindness, goodness, faithfulness, gentleness, self-control" (Gal 5:22–23). Ephesians 5:9 adds, "For the fruit of the Spirit is in all goodness, righteousness, and truth."

Compassion for the suffering and for the homeless is another characteristic of growth as a Christian. This is because as a person grows spiritually his character begins to gradually reflect the values and character of God. God has compassion for the poor and the suffering. "The Lord is gracious and full of compassion, slow to anger and great in mercy" (Ps 145:8; also Pss 111:4; 86:15; Lam 3:22, 32). God also asks His people to show compassion on others (Zech 7:9; Matt 18:13; 1 Pet 3:8).

PART 4—How Significant Is This for Our Daily Lives?

9. How Can I Please God? What Does the Relationship Look Like That God Wants with Me?

One day a lawyer—an expert in Old Testament law—came to Jesus and asked, "'Teacher, which is the great commandment in the law?' Jesus said to him, 'You shall love the Lord your God with all your heart, with all your soul, and with all your mind. This is the first and great commandment. And the second is like it: You shall love your neighbor as yourself. On these two commandments hang all the Law and the Prophets'" (Matt 22:35–40).

What place does God have in my daily life? Is it true that God seeks my relationship to Him to *trump* all other relationships I have? This is clearly what God is asking of me as my Creator!

Although our understanding of the natural world changes, the realities of nature do not. In the same way the truths in the Bible about spiritual things do not change even though our understanding of them does change with experience. In this sense we can expect our understanding of both the natural and spiritual worlds to increase as we continue on the upward path of increasing our knowledge of truth.

29

Is Truth a Destination or a Process?
Experiencing Truth

Mankind has observed the existence of the moon from earth for millennia. We knew the truth that the moon is out there. With the invention of modern telescopes our understanding of the moon has deepened and expanded. But when the astronauts landed on the moon and actually walked on it, their experiences gave them a new understanding of the truth that the moon is real and what it is like. This is a common experience when experiencing natural truth personally.

How much faith in science did the astronauts have who set out in a rocket to the moon, expecting to land on it and come back to earth alive? In this regard, it can be said that even in the natural world sometimes faith precedes experience.

I believe the same thing can be said of reality in the spiritual world. Faith in God's promises is needed to experience them. Then our personal experience of spiritual reality deepens our understanding and conviction that what the Bible says is true and that we can wisely build our lives on its teachings. As C. S. Lewis so aptly expresses it, "I believe in Christianity as I believe the sun has risen, not just because I see it, but because by it I see everything else."

It has been said that religious knowledge is more demanding than scientific knowledge—because it calls for a response; a commitment to do something. In this it is similar to the commitment that the astronauts made who first landed on the moon.

John Collins argued that faith in God is a *rational* faith. This is because we become persuaded of His trustworthiness, for He gives us things to believe and reasons for trusting Him. Then we take God's word seriously and listen to it and respond in active faith.

The Christian life can be likened to an upward path that is entered by one's personal choice through a gate. It can lead to many new and meaningful experiences, but it must be entered first. "Enter by the narrow gate; for wide is the gate and broad is the way that leads to destruction, and there are many who go in by it" (Matt 7:13).

A *commitment* is needed by the individual that leads to a path of learning by experience and finding increased confirmation about what is truth (reality) in the spiritual world. The path is by its nature a process of growth and inner change. It is entered by faith and continues as a walk of faith, with experiences along the way that confirm and deepen our understanding of truth and spiritual realities.

The fact that modern science has not come up with a persuasive explanation of how the universe could have suddenly come into being should certainly suggest to an open mind that it is feasible that "in the beginning God created the heaven and the earth"—but it does not prove that God did it.

To believe God exists and to motivate us to seek Him a measure of faith is needed. "But without faith it is impossible to please Him, for he who comes to God must believe that He is, and that He is a rewarder of those who diligently seek Him" (Heb 11:6).

So yes, it can be said that experiencing truth is a process of growth both in knowledge and in experience. According to the Bible it has a destination where sin and suffering and pain and disappointment are no more—but we do not arrive at the destination in this life.

The Apostle Peter admonishes us to "grow in the grace and knowledge of our Lord and Savior Jesus Christ. To Him be the glory both now and forever. Amen" (2 Pet 3:18).

30

Living Life to the Fullest
Can Clarifying One's Worldview Give New Meaning to Life?

Living life to the fullest includes finding meaning and purpose to our daily existence. What is meaningful and purposeful is surely different for different individuals.

We reported at the beginning of the book how impressed Lee Strobel was when he read that Bertrand Russell decided that science has revealed the world as being "purposeless" and "void of meaning." Strobel also concluded with William Provine that there's really no evidence that God exists, or evidence of life after death; an absolute standard for right and wrong; ultimate meaning of life; or that people really have free will.

Lee Strobel, facing the prospect that life has no inherent purpose or meaning, and that there is no life beyond the grave, set about creating a meaningful life for himself by seeking to excel as an investigative reporter for the *Chicago Tribune*. Years later he found new meaning when his beliefs changed and he came to understand that Christianity and the spiritual world are not "myths." His worldview changed so profoundly that as a result he spent several years interviewing scientists whose recent research greatly affected the intelligent design versus evolution debate and then he wrote about it in his best-selling book series.

Were you and I *created* or are we just a product of mindless biological processes and natural selection? If created, are we created in "the image of God"—a personal God Who brought us into being by biological processes to be sure, but also for a purpose He wants us to fulfill in life?

Earlier we mentioned the experiences of several authors whose worldviews changed greatly when they openly and honestly explored the claims that the Bible is

PART 4—How Significant Is This for Our Daily Lives?

true and experienced new meaning and purpose in life when they accepted and acted on its teachings.

Examples of a Dramatic Change in Life's Direction

Perhaps the best historical example of a person who became convinced of the veracity of the gospel and made a dramatic change in the direction of his life is the Apostle Paul. He was of impeccable lineage, and this was very important to his countrymen in that day. He was highly intelligent and had an excellent education in both the Hebrew and Greek cultures. Paul was exposed to two worldviews, one religious, based on the OT, and one secular, grounded in Greek learning and the Roman social order. He was zealous for the cause he believed in, and progressed upward in the Jewish hierarchy faster than his peers (Gal 1:10–14). It seems striking to me how similar Strobel's testimony is to Paul's here. They were young, ambitious, and eager to make their mark in the world.

Paul, once he was convinced that Jesus was not a dead heretic but a resurrected Savior, said he determined to "know Him [Jesus] and the power of His resurrection, and the fellowship of His sufferings, being conformed to His death, if, by any means, I may attain to the resurrection from the dead" (Phil 3:10–11). Then Paul admonishes his readers, "Therefore let us, as many as are mature, have this mind" (3:15).

Paul became totally convinced that Jesus' words were true, and that there is life after death that is unending. As a result of this belief he spent the rest of his life on earth traveling around the Roman Empire proclaiming the gospel and writing letters explaining how believers are to live in the light of spiritual realities.

Did Paul think he was "living life to the fullest"? He testified to this effect when he said in Philippians 1: 21, "For to me, to live is Christ, and to die is gain."

Moses was well-educated and highly intelligent like Paul, and adopted into the household of Pharaoh, the ruler of all Egypt. Evidently when he was very young and staying temporarily in the home of his Jewish birth parents, Moses learned how God had redeemed the Jews and established a special relationship with them. Moses also was exposed to two different worldviews. When he was about forty, and a member of the household that ruled over Egypt, he killed an Egyptian taskmaster who was beating and abusing a fellow Jew. As a result he had to flee Egypt for his life.

Hebrews 11:23–26 explains what motivated Moses to risk the privileges and safety of his high status in Egypt to do so. These verses show he clearly compared the transient benefits of being in the royal family with the eternal benefits of following God. He must have had a clear understanding of the coming Savior and of eternity in heaven, for verse 26 says he esteemed "the reproach of Christ greater riches than the treasures in Egypt; for he looked to the reward."

Most people who profess to be Christians do not have a dramatic and sudden experience of conversion like Paul did. But the Bible says that if our faith is genuine, like Paul's was, our eternal destiny—the ultimate reward of saving faith—will be the same as his. It seems Strobel's conversion was gradual as he proceeded to investigate the truth claims of Christianity. I get the impression from Hugh Ross's account of his conversion as reported in *Creation and Time* that it was gradual during the couple of years he spent studying the Bible. Sudden or gradual, if a person has really experienced the new birth, as Jesus used the term in John 3, there results a dramatic change in one's understanding of oneself and of the purpose of life. This leads to a transformation of one's values, priorities, and behavior.

Clarifying One's World View Can Give New Meaning to Life

I was not raised in a Christian home, and did not attend church or have a Bible. But when still in high school my oldest brother invited me to an evangelistic service. While there I felt an overwhelming sense of God's presence. I have always believed there was a God "out there somewhere" but this belief had little effect on my daily life.

But at that meeting *I surrendered to the God I felt was speaking into my heart*, and went forward after the service and accepted Jesus Christ into my heart as my Savior from sin. Then, for the first time in my life I felt spiritually clean!

Since then God has been very personal to me, not "Someone out there somewhere." I have learned to read the Bible daily to find out how God wants me to live. I attend church regularly to worship God and grow in my faith, and to fellowship with and encourage other believers.

I am convinced that God has been with me all these years, guiding and providing and answering very specific prayers. As an example, I was summoned into the Army for the Korean War, and our whole division was sent to Japan and then later to the front lines in Korea. Then I transferred back to Japan while still in the military, and while there married my wife—a missionary from Oregon, where I was also from. While in Japan I applied to Columbia International University in South Carolina, and asked that my acceptance letter be sent to Fort Lewis, where I was scheduled to be discharged from the Army. I asked God for a sign, telling him that if I didn't get the letter of admission at Fort Lewis I would give up any plans to go to Bible college.

When I arrived at Fort Lewis I expected to be there a couple of weeks or more, but was there only two days. The day after I arrived I went to the base post office and asked if I had any mail. I was told that they didn't handle the mail of those who were just being processed for discharge, but always returned the mail to the sender. Then the man at the post office asked me my name and rank and then said, "This just now came in for you"—and handed me my letter of admission to the school! Did this just happen? Or was God answering my specific prayer in a very marvelous way?

I had done a really stupid thing, being naïve, but I believe God in his mercy and grace answered my prayer anyway.

This changed my whole life from that time on. I studied to be a bible teacher and missionary, and we returned as a young family to Japan—where we served for over forty years. While there I was a pastor, a school teacher and a college professor. I am convinced that God in His love and grace has guided every step of the way, and enriched my life and given real meaning to it.

Our life has been what I describe as walking with God—a very personal God Who has guided us and answered hundreds of very specific prayers. I believe God has been true to the many promises He has made in the Bible, and that His Word can be completely trusted.

Circumstances beyond my control may have shaped my life, but I believe they have been orchestrated by a personal God Who knows all things and loves me enough to want what's best for me.

Gradually as I studied the Bible and sought to understand it and apply its teachings to my daily life I began to understand myself and my world in a new way—a way that helped me understand the significance of my own experiences and those of others. I believe I developed a coherent worldview that helped me understand reality and better cope with the problems of life. My new worldview became a reference point or framework to help me make wise decisions.

What Does Our Worldview Do For Us?

Our worldview speaks to every aspect of life and helps us to understand what is true and what is false in both the natural and spiritual worlds. It gives us insight in what is important in life, and what is worth living for. It helps us to understand ourselves, others, and our personal human relationships.

Our worldview helps us make sense of our world and our experiences in it. It helps us make sense of history, science, culture and society, and personal experience. It informs our *values* and our values motivate and guide much of our *behavior*. It affects our reasoning, our emotions, and our responses to life's situations. It helps us understand who we are, and can be a basis for finding real meaning in life.

A world and life view must answer important personal questions, such as What am I? Is there any real meaning to my life? Is this life all there is to my existence? Does it really matter how I live my life? What should we do when we fail to live by the standards we set for ourselves? Is this life all there is, or is this life only a preparation for a more permanent existence? Science cannot answer these personal questions.

1. It Helps Us Understand What Is True and What Is False

Francis Schaeffer emphasizes that we need absolutes if our existence is to have any meaning. He asks how we can be sure that what we think we know of the world outside ourselves corresponds to what is actually there? If modern science supports what the Bible from ancient times has declared about the realities of nature, then can we believe that the Bible also speaks truth about the meaning of life on earth and the existence of an afterlife?

This brings to mind Henry Thiessen's point that we are justified in accepting the claims the Bible makes of itself that it is inspired by God and speaks truth if we can prove its genuineness and the truthfulness of the things they report on other subjects. This includes what has been verified as true in history and in the natural world.

If the teachings of the Bible are in agreement with the known and verified facts of nature then this demonstrates they are truly a revelation from God. These are not just the facts that were known to ancient wise men when the Bible was written, but also to realities that have only been discovered recently. An example of this agreement is that recent discoveries in the cosmos have clearly shown that the universe is not eternal, but came into existence and has been expanding ever since.

Both geneticist Francis Collins and Hebrew scholar John Collins emphasize that faith in God and the Bible are rational just as faith in operational science is rational, because they are based on both personal experience and evidence in the world we observe. John Collins adds that every kind of rational faith goes beyond what can be verified, and Francis Collins argues that the principles of faith in God and the Bible are *complementary* with the principles of science. Francis Collins also states his belief in the limitations of science when he says, "It also became clear to me that science, despite its unquestioned powers to unravel the mysteries of the natural world, would get me no further in resolving the question of God. If God exists, then He must be outside the natural world, and therefore the tools of science are not the right ones to learn about Him."[578]

It says in the Gospel of John that a willingness to obey God is a precondition for knowing spiritual truth. Jesus said, "If anyone is willing to do His will, he shall know of the teaching, whether it is of God or whether I speak from Myself [on My own authority]." (John 7:17 NASB).

2. It Helps Us Understand Both the Natural and Spiritual Worlds

Modern science discovered that the universe that exists today had a beginning. Space, energy, matter and physical time came into existence at some point. The Bible says that God created it, and science has provided no alternative plausible explanation. God exists outside of space, energy, matter and time. His existence is *eternal*—He has no beginning and no end.

The Bible does not reveal much about the *how* and *when* of the origins and processes that caused the universe and life on earth to come into being. It focuses rather on that it was *God* Who is the Creator and that the *spiritual* proceeded the physical and will endure beyond it in time.

PART 4—How Significant Is This for Our Daily Lives?

The Bible says God created the human race "in His image." Unlike God a human is a finite being, but he was created sinless. The first man and woman fell into sin, and this morally corrupted their nature. All of their descendants were and are born with a sin nature, and as a result sin has affected their thoughts, words and deeds. The Bible says that "all have sinned and fallen short of the glory of God."

God created humans with body, soul and spirit. Our bodies will die, but our souls and spirits will exist forever. This life is in a real sense a preparation for the unending existence which follows it. Our conduct and decisions in the here and now greatly affect our situation in the next life. In this sense life on earth is like a school curriculum. It is not an end in itself, but a preparation for the life that follows graduation.

This physical world and the universe as we know it will end. Cosmologists today expect the universe to keep expanding, more and more rapidly, until it ends somehow. Peter, in 2 Peter 3, prophesied that "the heavens will pass away with a great noise, and the elements will melt with fervent heat; both the earth and the works that are in it will be burned up." Then he admonishes, "Therefore, since all these things will be dissolved, what manner of persons ought you to be in holy conduct and godliness, looking for and hastening the coming of the day of God, because of which the heavens will be dissolved, being on fire, and the elements will melt with fervent heat?"

Then Peter adds, "Nevertheless we, according to His promise, look for new heavens and a new earth in which righteousness dwells." Then there will be no pain, suffering, tears or death. Revelation 21:4 adds, "And God will wipe away every tear from their eyes; there shall be no more death, nor sorrow, nor crying. There shall be no more pain, for the former things have passed away."

3. It Helps Us Understand Ourselves

What am I, really? How can I know my true nature? If the Bible is truly the inspired, inerrant Word of my Creator to me as a human, then I can come to know myself best by studying His Word, and by being open to the truth about myself from all sources. These sources include the Bible, and also what God Himself shows me about myself and about the significance of my experiences.

What am I? A finite creature, with a mortal body and an eternal soul and spirit, who is part of one human race that was created by the infinite God. I, as a human, was created in the image of almighty God—"in His image" because He wants a personal relationship with me. This is my true identity as His creation.

Is there any real meaning to my life? I found real meaning to life when I met God years ago, and this led to an ever-growing sense of self-identity. I believe this can become the experience of each reader of this book.

Is this life all there is to my existence? A biblical worldview is based on God's revelation that this world is transitory, and that we have an eternal soul that will endure forever.

Does it really matter how I live my life? Hebrews 9:27 says, "It is appointed for men to die once, but after this the judgment." The Bible clearly says that this life is only a preparation for a more permanent existence. What we do and experience in this life has eternal consequences.

An openness to the truth about ourselves brings us to confront the truth that we all fail to live up to the standards we set for ourselves, much less to meet the standards that God sets for us in his Word. "If we say that we have no sin, we deceive ourselves, and the truth is not in us" (1 John 1:8).

What should we do when we fail to live by the standards of the Bible? In reply to this 1 John 1:9 says, "If we confess our sins, He is faithful and just to forgive us our sins and to cleanse us from all unrighteousness." God in His love and mercy will reach out to each and any of us when we repent, confess our sins to Him, and seek to please Him in our thoughts, words and deeds.

The Human Quandry

Living true to what I really am—a human being—
Made in the image of God, but also part of fallen humanity.
Striving to be what I was created to be—but with so many distractions!
Failing, then returning;
Growing, but never fully arriving.

Created in the image of God—my Creator—but so unlike Him!

Our final destination:

> "I press toward the goal for the prize of the upward call of God in Christ Jesus" (Phil 3:14).

> "Behold what manner of love the Father has bestowed on us, that we should be called children of God! Therefore the world does not know us, because it did not know Him.

> "Beloved, now we are children of God; and it has not yet been revealed what we shall be, but we know that when He is revealed, we shall be like Him, for we shall see Him as He is.

> "And everyone who has this hope in Him purifies himself, just as He is pure" (1 John 3:1–3).

PART 4—How Significant Is This for Our Daily Lives?

4. It Helps Us Understand Others and Our Personal Human Relations

The Bible says that God created mankind in His own image. This means that man has a *spiritual* nature like God has, and in this sense is different from all else in the physical creation. We have personality—we have an intellect, will and emotions. We have awareness of time, and some concept of the distant past and the distant future. We have a capacity to relate to God.

God is *relational*. Once in Japan I heard the philosopher Francis Schaeffer remark that love existed before the human race came into existence, because there has always been love among the three members of the Trinity. We humans are born into relationships and seek positive, wholesome relationships with others. So we are like God in our social nature.

The *family* seems to have a special place in the heart of God. This begins with God declaring Adam and Eve "one flesh" in Genesis 2:4

Even the term *father* is a reflection of God's relationship to us. Paul wrote, "I bow my knees to the Father of our Lord Jesus Christ, from whom the whole family in heaven and earth is named" (Eph 3:14–15).

God called His relationship to King David and to his son Solomon that of Father and son. Jesus referred to God as "your Father in heaven" when addressing His followers. Paul refers to "God, the Father" (1 Cor 8:6; Eph 5:20), and "one God and Father of all, who is above all" (Eph 4:6). In the Old Testament book of Hosea, God says of His people, "You are sons of the living God!" John exclaims about New Testament believers, "Behold what manner of love the Father has bestowed on us, that we should be called children of God!" (1 John 3:1). In this sense, we are the most blessed of all His creation!

God treats us as His loving children, Whom He also disciplines for our own good and His glory. Hebrews 12:5–10, in the New American Standard Bible, says, "He disciplines us for our good, that we may share His holiness" and "it afterwards yields the peaceful fruit of righteousness." We are also to discipline and train our children for their good (Prov 22:6; 29:15).

Scripture puts high priority on providing for the needs of one's family. First Timothy 5:8 says, "But if anyone does not provide for his own, and especially for those of his household, he has denied the faith and is worse than an unbeliever." But providing for one's family does not mean seeking wealth and material possessions for their own sake. As an illustration that human relations are more important than an abundance if possessions in the experience of most humans, I am reminded of the Greek legend of Midas. I am referring to the Greek legend of the king who received from a Greek god the power to turn everything into gold that he touched. But even his food turned into gold, and when his daughter touched him she turned into gold. So he begged the god to take the power back.

In Luke 12:16–31, Jesus tells the parable of a rich farmer who centered his life on increasing his wealth. In response to the rich man's attitude of wanting to increase his wealth even more and live well and rest on his laurels, God replied, "Fool! This night your soul will be required of you; then whose will those things be which you have provided?" Then Jesus adds, "So is he who lays up treasure for himself, and is not rich toward God." Then Jesus applies this parable to His disciples by saying, "And do not seek what you should eat or what you should drink, nor have an anxious mind. For all these things the nations of the world seek after, and your Father knows that you need these things. But seek the kingdom of God, and all these things shall be added to you." In this way Jesus emphasized the importance of eternal things over earthly success and prosperity.

Yet, on the other hand, it is not God's plan for the believer to neglect or sacrifice the needs of his family in his service for God, as 1 Timothy 5:8 makes clear when it says that whoever neglects to provide for his household "has denied the faith." We are to value family and others more than personal status and success in the eyes of the world.

I don't think we can ever adequately understand ourselves in isolation. We are relational beings, and come to more fully understand ourselves when in relationship with others, and with God.

Just as righteousness and justice are core characteristics of God, so is love, which includes compassion and mercy. The Bible says that God is a Father to the fatherless and orphans and cares for widows (Deut 10:18; Ps 146:9; 68:6), and His people are commanded to care for them (Deut 24:17, 19–21; 26:12; Zech 7:10).

James 1:27 says, "Pure and undefiled religion before God and the Father is this: to visit orphans and widows in their trouble, and to keep oneself unspotted from the world." (This is a reflection of God's own heart, as the "A father of the fatherless, [and] a defender of widows" [Ps 68:5–6].)

The biblical value of compassion on the poor and the infirm (or needy) is a value that is shared today by many in the secular world also.

Our compassion for others and actions to help the needy are to be an example to others of what it means to be spiritual children of God (Matt 5:16; Phil 2:15). The biblical phrase in Matthew 5:16 "Let your light so shine before men, that they may see your good works and glorify your Father in heaven" reminds us of Jesus' words "I am the light of the world" (see also John 1:9; 3:19; 8:12; 9:5). We are commanded to "walk as children of light" (Ephs 5:8; see also Phil 2:15; 1 Thess 5:5).

Jesus is "the spiritual light of the world." We are to let our light shine also—like the light of the moon. No light originates from the moon; it simply reflects the light of the sun. In like manner we are to live so pure and compassionate a life that it reflects the light of the Son. The more we grow as Christians, the more our light should progress from that of the new moon to that of the full moon. Matthew 5:16 says, "Let your light so shine before men, that they may see your good works and glorify your Father in heaven."

PART 4—How Significant Is This for Our Daily Lives?

The New Testament uses two interesting phrases to explain our relationship as believers to our greater society. We are "in this world" (1 John 4:17), and we are "not of this world" (John 8:23; 18:36). God wants us to be a demonstration of what happens when someone surrenders to God and lives a life of prayer and obedience to His Word. If the Spirit of God abides in our hearts, like the Bible says is true of the believer, then to a degree we will reflect God's nature to others. According to Lee Strobel, his wife's conversion impressed him because he saw her as a more loving and caring and authentic person, and this led him to take the claims of the Bible more seriously. Paul tells us "But we have this treasure in earthen vessels, that the excellence of the power may be of God and not of us" (2 Cor 4:7).

Recently I sent out a simple questionnaire to a number of believers, asking them what living life to the fullest meant to them. I received responses worded in various ways, but there was one common theme. Important to them was a sense that they wanted to honor God by helping others. It seems one common manifestation of the Spirit of God within us is in the desire to help others.

5. It Fosters an Active Style of Life

The Bible presents a very clear worldview that reveals the big picture on how a believer should live and also very specific directions on how he is to speak and act. James wrote "be doers of the word, and not hearers only, deceiving yourselves" (see Jas 1:22–25).

It is a *comprehensive* worldview. It is designed to cover all aspects of life. The biblical worldview is not a *passive* one. The Apostle Peter admonishes us to "gird up the loins of your mind (for action)" (1 Pet 1:13–16). He was referring to how a fisherman in his day would tie up his robes in order to free his limbs for physical labor, and telling us to prepare our minds for a life of action.

It is an *active* worldview. The believer is admonished in the Bible to make a positive contribution to his society. He or she is to help the poor and sick, visit the shut-ins, and show compassion on those less fortunate than oneself. As James 4:17 puts it, "Therefore, to him who knows to do good and does not do it, to him it is sin." First John 3:17 declares that "whoever has this world's goods, and sees his brother in need, and shuts up his heart from him" shows that he does not have the love of God in him.

On the other hand, the biblical worldview is not *activism*. It is not a crusade or a *jihad*. We should not seek to be martyrs, but should be willing to be martyrs—not out of hatred for the enemies of God but out of our love for God. We are to respect—even love—all people, including those who are very different from us and may even hate us. Luke 6:35–36 admonishes believers to "love your enemies, do good, and lend, hoping for nothing in return; and your reward will be great, and you will be sons of the Most High. For He is kind to the unthankful and evil. Therefore be merciful, just as your

Father also is merciful." Luke 6:28 adds, "Bless those who curse you, and pray for those who spitefully use you."

I don't see it as an acting *for* God, as much as it is an acting *with* God. Joseph, in Genesis 39:2, is a good illustration of this: "The Lord was with Joseph, and he was a successful man." Joseph's life was a demonstration of dependence on God, and of God's guidance and provision even in the presence of those who sought to do him harm.

6. Bible Values That Guide Behavior

When we consider how the values the Bible promotes affect our behavior we must always keep in mind that our motives as humans were greatly affected when Eve fell. Genesis 3:1–7 reports that Satan, taking the bodily form of a snake, tempted Eve in two ways. First, when Eve correctly reported that God told her that if they ate the fruit of one certain tree that they would die, Satan responded by saying first that "you will not surely die"—tempting her to doubt the truthfulness of God's Word. Second, Satan told her, "For God knows that in the day you eat of it your eyes will be opened, and you will be like God, knowing good and evil." This temptation "to be like God" was clearly enticement to think, live and act without living a life of dependence on God.

Genesis 3:6 reports the temptation of Eve: "So when the woman saw that the tree was good for food, that it was pleasant to the eyes, and a tree desirable to make one wise, she took of its fruit and ate. She also gave to her husband with her, and he ate."

This one act of disobedience brought about a drastic change in human personality. "Therefore, just as through one man sin entered the world, and death through sin, and thus death spread to all men, because all sinned" (Rom 5:12). It was, in essence, a choosing of self-interest rather than God's will for her. This temptation, in its modern manifestation, is clearly seen in 1 John 2, that discusses basic human motivations after the fall of mankind.

> Do not love the world or the things in the world. If anyone loves the world, the love of the Father is not in him. For all that is in the world-the lust of the flesh, the lust of the eyes, and the (boastful) pride of life-is not of the Father but is of the world. And the world is passing away, and the lust of it; but he who does the will of God abides forever. (1 John 2:15–17)

The "lust of the flesh" refers to the desire to experience. Sigmund Freud taught that the basic human motive is sexual, but this is only one part of this motive. Horie, the young Japanese fellow who crossed the Pacific Ocean in a rowboat some years ago was clearly not motivated by sex. Neither are those who climb difficult mountains and do other challenging physical things. Neither was Sarah Outen, who in September 2013 became the first woman to row alone the 3,750 miles from Japan to Alaska. The desire to experience extends broadly from our desire to eat tasty food to thrill-seeking experiences.

PART 4—How Significant Is This for Our Daily Lives?

The second motive, "the lust of the eyes," refers to our desire to possess things. This is the desire to accumulate wealth and possessions beyond what is necessary for daily living. It may be to show off our wealth or just the joy of having possessions.

The third motive is more properly translated, I believe, as the desire to be something. It is human nature to be tempted to boast in our skills, qualifications, titles, position and power or authority over others. First John contrasts the desire to use these things for our own selfish purposes with the desire to use them in ways that please God. It is striking to read how much Job valued *prestige* as a reward for his godly living (see Job 29:21–25).

Two kinds of motives common to humanity are described in the New Testament. Both come under the broad category of "self interest." The first is described in 1 John 2. The other is illustrated in the reason given for the Jews of Jesus' day to reject Him and side with their rulers who were out to kill Jesus.

Examples of rejecting the truth about Jesus for "fear of the Jews" are John 7:13 and the parents of the man born blind in John 9:20–23. Paul expressed this basic motive in Galatians 1:10, where he says, "For do I now persuade men, or God? Or do I seek to please men? For if I still pleased men, I would not be a bondservant of Christ." Psychologists call this approval seeking. Paul testified that this is a motive unworthy of a true follower of Christ. There is a most fundamental difference here: do you speak and act to be "seen of God" or to "be seen of men"?

It is this basic motive of self-gain that energizes all of our behavior. It is the only motive that we are born with. It is also the motive that leads us to repent of our sins and surrender our lives to God. Hebrews 11:6 declares that "without faith it is impossible to please Him, for he who comes to God must believe that He is, and that He is a rewarder of those who diligently seek Him." Faith in Jesus and in the gospel certainly affects *what* we seek in life, but the motive for *why* we seek it is still self-gain or self-interest.

Moses, as we said earlier, is a prime example of how faith in God and His Word revolutionizes our behavior. He chose "to suffer affliction with the people of God" over "the pleasures of sin" because he looked to the enduring reward from God rather then the passing rewards of man (Heb 11:25–26). Paul likewise changed his purpose of living after meeting the risen Christ.

Do we have selfish motives for all that we do? I only know of one exception, and that is to be motivated by the same love toward others and to God that God has for us. This love is described in 1 Corinthians 13:4, and at it's core is the attitude that love does not seek its own benefit but that of the one loved.

Is naturalism a manifestation of man's desire to live and philosophize independently of God? If so, it is a reflection of Eve's desire to "be like God, knowing good and evil" that tempted her to disobey God in the garden of Eden. The view of naturalists that creation just came into existence on its own and that creatures like we are just products of nature can be likened to a child who is just learning to walk and refuses his parent's hand.

7. The Biblical Worldview Is an Exclusive Worldview

Do All Paths Upward Lead to the Same Place?

Is the biblical one an exclusive worldview? Is it superior to other worldviews? The Japanese have a saying that there are many paths up Mt. Fuji to the summit, but they all end up at the same place. But does this express spiritual reality?

The Bible says that Jesus is *unique* as Savior in every sense of the word, totally different from all other "saviors" that have promoted themselves as such or declared as such by others. Paul, in 1 Corinthians 15, declares Jesus to be "the second Man" and "the last Adam." This refers to Jesus as the One who was foreordained by God from the beginning of mankind to be a new start of the human race after Adam sinned. He was tempted by Satan in the wilderness, like Eve was, when He was thirty. At the heart of Satan's temptation of Jesus, the Son of God, was to get Him to act independently of the will of God His Father. Jesus always answered Satan by quoting Scripture. From this can be seen how Jesus did exactly what Eve failed to do: He believed God's Word and acted in total dependence on God the Father. He famously responded to Satan by saying, "It is written, 'Man shall not live by bread alone, but by every word of God'" (Luke 4:4).

John reports that Jesus said of Himself, "I do nothing of My own initiative" (John 5:30); "I have come down from heaven, not to do My will, but the will of Him Who sent me" (John 6:38); and "I always do the things that are pleasing to Him [the Father]" (John 8:29). Jesus said, in prayer while speaking of His pending death, "Father, if it is Your will, take this cup away from Me; nevertheless not My will, but Yours, be done" (Luke 22:42). This shows how Jesus lived a life of constant, conscious, willing dependence on God the Father as He lived His sinless life on earth. (These verses are from the New American Standard of 1960.)

What does the Bible say about *the uniqueness of Jesus as Savior*? In John 14:6, Jesus said, "I am the way, the truth, and the life. No one comes to the Father except through Me." Paul declared that "there is one God and one Mediator between God and men, the Man Christ Jesus, Who gave Himself a ransom for all, to be testified in due time" (1 Tim 2:5–6). (Acts 4:12 declares, "Nor is there salvation in any other, for there is no other name under heaven given among men by which we must be saved.")

In His life of perfect obedience to his Father, and in His death on the cross for all mankind, Jesus is the unique and only Savior of mankind. Romans 5:10 says, "For if when we were enemies we were reconciled to God through the death of his Son, much more, having been reconciled, we shall be saved by His life." He lived a perfect life on earth as man, and died on the cross for the sins of all mankind to make salvation possible for all who come to God in repentance and in faith.

The Old and New Testaments Present the Same Basic Worldview—But in Very Different Circumstances.

The other day a Christian friend told me that he doesn't believe in the God of the Old Testament; he believes in the God of the New Testament. He described his concept of the God of the Old Testament as a God of vengence, and Jesus in the New Testament as the God of love. But God's love is also seen in the Old Testament (Deut 4:37 and 23:5; 1 Kgs 10:9; 2 Chron 9:8), and His wrath in the New (Heb 10:31). For example, Deuteronomy 4:37 says of the descents of Abraham, "Because He [God] loved your fathers, therefore He chose their descendants after them; and He brought you out of Egypt with His Presence, with His mighty power," and Hebrews 10:31 in the New Testament says, "It is a fearful thing to fall into the hands of the living God."

What my friend said was expressed in a dramatic way, but I wonder how many there are who read the Bible and think that the nature of God as revealed in the New Testament is different from that of God in the Old Testament. It is easy to infer from many Old Testament verses that God was only commanding His people to outward acts of obedience. But a careful study of all of Scripture gives a very different picture. The first person whose life is reported in much detail in the Bible is Noah. Hebrews 11:7 says that "*by faith* Noah, being divinely warned of things not yet seen, moved with godly fear, prepared an ark for the saving of his household, by which he condemned the world and became heir of the righteousness which is according to faith." The next verse says, "*By faith* Abraham obeyed when he was called to go out to the place which he would receive as an inheritance. And he went out, not knowing where he was going." Hebrews 11 lists many prominent individuals in Old Testament days who believed God's Word and obeyed Him because they trusted his word.

Genesis 15:6 says Abraham "believed in the Lord, and He accounted it to him for righteousness." This verse was repeated by Paul in Romans 4:3 and in Galatians 3:6, and by James in James 2:23 to emphasize that God saves and considers righteous only those who believe him and obey him.

When Paul wrote in Romans 1:17 and Galatians 3:11 that "the just shall live by faith," he was quoting from Habakkuk 2:4 in the Old Testament. It is clear that many religious Jews in Old Testament days and during the time Jesus was on earth depended on their own efforts to be accepted by God, but the consistent message of both the Old and New Testaments is that God requires "the obedience of faith."

When Jesus was revealed in glory on the Mount of Transfiguration in Mark 9:4, Moses and Elijah appeared with Him. The main message that God was giving the disciples here is clear. This is that the same God Who was speaking through Moses when He gave Israel the law, and was speaking through Elijah during the days of the prophets, was now speaking through Jesus.

The Old Testament looks forward to "the Lamb of God" Who will atone for sin. The Gospels proclaim, "Behold! The Lamb of God Who takes away the sin of the world!" (John 1:29). The New Testament looks back to "the Lamb of God" and recognizes that Jesus died on the cross to atone for the sins of all men, from the days of Adam until the last person born on earth. In both cases it is faith in God and in His Word that are required for salvation.

The daily sacrifices of animals in the tabernacle in the wilderness and later in the temple were meant to symbolize the coming of a Messiah or Christ Who would die for the sins of all (see Heb 7–10). Faith in the coming of a Redeemer is seen in the faith of Job, Abraham, Moses, David, the Old Testament prophets, and many others whose experiences and prophesies are recorded in the Old Testament.

The law in the Old Testament had as a purpose separating and protecting a people group from the beliefs and practices of the nations and people groups around them. An example of an Old Testament prohibition that is not in effect today is the forbidding of a Jew from marrying someone from another nation. This was one of the most violated laws in ancient days, and its purpose is abundantly illustrated in the life of King Solomon. He "loved many foreign women" and married many wives, and his wives turned his heart away from the exclusive worship of the true God to worshipping idols, and as a result, "God became angry with Solomon" and as a result severely punished him and his family line (see 1 Kgs 11:1–11 and Neh 13:26).

The New Testament, rather than teaching geographical separation of the people of God, teaches them to be distinct from unbelievers in their beliefs and religious behaviors while dwelling in their midst. They are to be "in the world" of humanity (John 17:11; Eph 2:12), but not "of the world" (John 15:19; 17:14–16).

An emphasis of the New Testament is to separate and protect a people group or subculture within dominant cultures and worldviews. But even in the New Testament the believer is admonished against marrying an unbeliever (see 2 Cor 6:14–18).

The ritual rules for cleansing were meant to guard the people from physical sickness, but also to teach or illustrate the need for spiritual cleansing. According to Matthew 15:10–11, "When He [Jesus] had called the multitude to Himself, He said to them, 'Hear and understand: Not what goes into the mouth defiles a man; but what comes out of the mouth, this defiles a man.'"

It can be said that both the Old and New Testaments emphasized loving and obeying God, separation from those of other religious beliefs, compassion on the poor and needy, and raising ones' children in their parents' religious faith.

Because of the dramatic effects of the Holy Spirit entering the hearts of the believers gathered in the upper room on the Day of Pentecost, it is easy to get the impression that the Holy Spirit was not active in the lives and hearts of God's people until after the resurrection of Jesus. But He certainly was active in Old Testament days. The

PART 4—How Significant Is This for Our Daily Lives?

first mention of the Spirit of God in the Old Testament is in Genesis 1:2 and the last in Malachi 2:15.

The New Testament teaches that the Holy Spirit brings about the new birth (John 3), enters the believer when he is born again (Eph 1:13), and indwells every true believer (1 Cor 3:16; 6:19; Jas 4:5; 1 John 3:24).

The centrality of the Holy Spirit in the life of the believer can be illustrated by a trolley car that runs on two rails with an overhead power line. One rail is the regular reading of the Bible and meditating on its application to our lives. The other rail is prayer. But the believer lacks in spiritual power if he relies on these two as "works" that enable him to live as he should. The Holy Spirit is the "energizing power" that enables us to understand and obey God's Word, and He is the One Who gives life to our prayers as we pray in active faith and dependence on Him. Jesus' life on earth is an example of this. Jesus was filled with the Holy Spirit (Luke 4:1); was led by the Spirit (Matt 4:1; Luke 4:14); and prayed in the Spirit (Luke 10:21).

To us moderns, it appears that the status of women in the Old Testament was very low. But it should be remembered that wealth and possessions then were almost always based on owning or at least having the right to use land. A woman almost always married into the extended family of her husband, and her occupation, besides being a wife and mother, was usually farming or tending sheep. As a result, inheritance was from father to son. A prophetess or female ruler was the exception. (In Genesis 31:14–15 we see that the daughters of Laban, Rachel, and Leah, felt that their father had sold them for money to Jacob to be his wives.) But males and females are of equal "value" before God (see Gen 1:27; 5:2; Gal 3:28).

The worldwide church is composed of men, women, and children from all ethnic groups, languages and cultures, scattered throughout the world, who are "all one in Christ Jesus."

Conclusion

The biblical worldview starts with God and ends with God and eternity. Its guide book is the Bible. The believer's priorities are to put God first, and then self and others. It's time orientation is eternity.

The primary attitude it promotes is to think and act with love as your motive. First, love toward God and then toward others. Our loving and serving others should flow from our loving God. Colossians 3:23 says, "Whatever you do, do it heartily, as to the Lord and not to men." First Corinthians 10:31 adds, "Therefore, whether you eat or drink, or whatever you do, do all to the glory of God."

This is the teaching of both the Old and New Testaments. Deuteronomy is an important part of the "law of Moses" which was given to Israel as the basis of their faith-culture. Deuteronomy 10:12–13 says, "And now, Israel, what does the Lord your God require of you, but to fear the Lord your God, to walk in all His ways and to love

Him, to serve the Lord your God with all your heart and with all your soul, and to keep the commandments of the Lord and His statutes which I command you today for your good?"

Micah 6:6–8 declares, "With what shall I come before the Lord, and bow myself before the High God? Shall I come before Him with burnt offerings, with calves a year old? Will the Lord be pleased with thousands of rams, ten thousand rivers of oil? Shall I give my firstborn for my transgression, the fruit of my body for the sin of my soul? He has shown you, O man, what is good; And what does the Lord require of you but to do justly, to love mercy, and to walk humbly with your God?"

How does one develop a biblical worldview? By spending time reading, thinking about, and applying to one's daily life the truths of the Bible. Romans 12:2 tells us, "Do not be conformed to this world, but be transformed by the renewing of your mind, that you may prove what is that good and acceptable and perfect will of God." The renewing of one's mind can come only from thoughtful and extended exposure to the Word of God.

When I asked what "life to the fullest means," one of my sons, a pastor in Japan, quoted Jesus' words in John 10:10, "I have come that they may have life, and that they may have it more abundantly." He wrote, "To me 'living life to the fullest' means to take advantage of every opportunity God gives me to use my spiritual gifts and natural abilities in a way that furthers His purposes. This includes spending time daily in prayer and reading the Bible and taking advantage of opportunities to encourage believers and to witness to unbelievers. 'Living life to the fullest' means having a goal that is worth devoting your life to and making definite progress towards accomplishing that goal. It means a life without fear, trusting God to meet all my needs and to guide me each step of the way."

It seems clear from my son's personal experience, as well as from mine, that one's personal experience with God and the truths of God's Word can give fuller and richer meaning to daily life on this earth. We can also live without fear, not only because God promises to meet our needs on this earth but has also assured us of a far better life in eternity.

To know that the God Who created us loves us and has promised to be with us and to meet all our needs is the most encouraging and comforting thought there is! "Perfect love casts out fear!" (1 John 4:18).

> Blessed is the man who walks not in the counsel of the ungodly, nor stands in the path of sinners, nor sits in the seat of the scornful; but his delight is in the law of the Lord, and *in His law he meditates day and night*. He shall be like a tree planted by the rivers of water, that brings forth its fruit in its season, whose leaf also shall not wither; and whatever he does shall prosper (Ps 1:1–3).

God created the first man and woman, the ancestors of the whole human race, in His image. They were created morally righteous, but when tempted, fell into sin.

PART 4—How Significant Is This for Our Daily Lives?

This changed their moral natures, and all humans born since were born with a sinful nature. This helps to explain the evils in the world today. Jesus, the eternal Son of God, came to earth as a human to display God's character, to redeem God's people, to provide salvation and liberation from the bondage to sin, and He will return again some day to usher in God's eternal kingdom. As a result we can have in our lifetime forgiveness for sins, God's enabling to lead righteous and meaningful lives, and have the assurance of a life beyond the grave that has no pain, suffering, or loss of any kind. This is the biblical worldview in a nutshell.

John 8:32 says: "And you shall know the truth, and the truth shall make you free."

Appendix

Introduction

Several of the authors presented in part 3 gave detailed interpretations of key verses in Genesis when explaining their theories of biblical creation. Some of this material is presented in this appendix.

Here we will look at some of the explanations by Hugh Ross, Whitcomb and Morris, Davis Young, and John Collins. We will begin by looking at a Bible commentary to explore the meaning of key verses in Genesis.

I. Commentary on the Whole Bible

Robert Jameson, A. R. Fausset, and David Brown authored a commentary that has many significant and interesting explanations of the creation events. The following comments are taken from pages 17 and 18 of their *Commentary on the Whole Bible*. They are written primarily to explain the meaning of the Hebrew text.

> Genesis 1, verse 1: "*In the beginning*—a period of remote and unknown antiquity, hid in the depths of eternal ages; and so the phrase is used in Proverbs 8:22, 23."
>
> "*Created*—not formed from any pre-existing materials, but made out of nothing."
>
> "*The heavens and the earth*—the universe."
>
> Verse 2: "*The earth was without form and void*—or in 'confusion and emptiness,' as the words are rendered in Isaiah 34:11. This globe, at some undescribed period, having been convulsed and broken up, was a dark and watery waste for ages."
>
> "*The Spirit of God moved*—literally, continued brooding over it, as a fowl does, when hatching eggs. The immediate agency of the Spirit, by working on the dead and

Appendix

discordant elements, combined, arranged, and ripened them into a state adapted for being the scene of a new creation."

Verse 5: "*First day*—a natural day, as the mention of its two parts clearly determines; and Moses reckons, according to Oriental usage, from sunset to sunset, saying not day and night as we do, but evening and morning." Here they interpret the first day as "a natural day," like Morris and other young earth creationists do, but they do not refer to it as a twenty-four-hour day.

Verse 14: "*Let there be lights in the firmament*—the atmosphere being completely purified, the sun, moon, and stars were for the first time unveiled in all their glory in the cloudless sky."

Verse 16: "*Two great lights* . . . Both these lights may be said to be 'made' on the fourth day—not created, indeed, for a different word is here used, but constituted, appointed to the important and necessary office of serving as luminaries to the world, and regulating by their motions and their influence the progress and divisions of time."

Verse 26: "*God said, Let us make man*—words which show the peculiar importance of the work to be done, the formation of a creature, who was to be God's representative, clothed with authority and rule as visible head and monarch of the world."

"*In our image, after our likeness*—This was a peculiar distinction, the value attached to which appears in the words being twice mentioned. And in what did this image of God consist? Not in the erect form or features of man, nor in his intellect, for the devil and his angels are, in this respect, far superior, . . . but in the moral dispositions of his soul, commonly called *original righteousness* (Eccl. 7:29). As the new creation is only a restoration of this image, the history of one throws light on the other; and we are informed that it is renewed after the image of God in knowledge, righteousness, and true holiness (Col. 3:10; Eph. 4:24)."

This is an interesting argument, that the "image of God" is not in human intellect—but this is assuming that the angels were not also created in God's image in that they have intellect. In any case, it is clear that the "image of God" is not *limited* to mankind's intellect, but includes his moral nature and social nature as well. Yet again, the angels may well have been created with moral and social natures too.

Verse 31: "*Behold it was very good* . . . He saw everything that He had made answering the plan which His eternal wisdom had conceived."

Jameson, Fausset, and Brown remind us that there were no chapter and verse divisions in the original text when they say, "The course of the narrative is improperly broken by the division of the chapter." It seems a more natural division would be after Genesis 2:3, which would put all seven days together. Or, alternatively, the division could come after Genesis 2:6, after which the text focuses on the creation of mankind.

Chapter 2, verse 7: "*Breathed into his nostrils the breath of life*—not that the Creator literally performed this act, but respiration being the medium and sign of life, this phrase is used to show that man's life originated in a different way from his body—being implanted directly by God (Eccles. 12:7), and hence in the new creation of the soul Christ breathed on His disciples (John 20:22)." (John 20:22 refers to Jesus breathing on them to give them the Holy Spirit.)

II. Hugh Ross

Ross is an astrophysicist. He explains the conditions of the early earth based on his knowledge of cosmology on pages 27–57 of his book *The Genesis Question: Scientific Advances and the Accuracy of Genesis*. He infers what the original conditions were and says he finds that what Genesis 1 says about it rings true to recent discoveries in the cosmos.

He says that Genesis 1 reports only a handful of creation events. He adds that God "does not seek to bury us under a mountain of data that might distract us from His main themes: communicating His plan for our redemption and securing forever all that is good against the presence, and even the possibility of evil."[579]

1. Prior to the First Day

> And the earth was formless and void, and darkness was over the surface of the deep: and the Spirit of God was moving over the surface of the waters. (Gen 1:2)

He says, "The 'formless' and 'empty' conditions of earth would be expected, given the initial opaque atmosphere and interplanetary debris. Without light, photosynthesis could not occur . . . We can easily understand why no evidence of life on Earth dates earlier than four billion years ago."[580]

> Genesis 1:2 gives a hint, at least, that God's work of creating life on Earth may have begun before the first events recorded in the following six days of creation.[581]

He says geology supports this interpretation.

> Earth's geology testifies that marine life did indeed arise before all other lifeforms. The oldest fossils found to date show the unicellular, marine-like organisms in clearly identified marine sediments.[582]

Appendix

2. Day One: Light Comes

THE FIRST DAY

> Then God said, "Let there be light"; and there was light. (Gen 1:3)

Ross notes that when God said, "Let there be light," none of the Hebrew words for "create" are used. "God created physical light, that is, electro-magnetic radiation, 'in the beginning,' when He brought the cosmos into existence. The matter and energy of the cosmos included light. Indeed, light is the dominant form of energy for both the primordial and present-day universe."[583]

> Remembering Earth's initial condition and that the frame of reference for the passage is earth's surface, we can comprehend what happened on Day One: light penetrated Earth's dark shroud for the first time. God cleared away some of the debris that had previously kept light from coming through. Earth's atmosphere changed, too, from opaque to translucent—not transparent, yet, but clear enough to permit light's passage.[584]

> The change in Earth's atmosphere presents a serious challenge to our understanding. The rule of thumb in planetary formation is that the greater a planet's surface gravity and the greater a planet's distance from its star, the heavier and thicker its atmosphere. Yet Earth departs from that rule. Theoretically, Earth should have an atmosphere heavier and thicker than that of Venus, but in fact it has a far lighter and much thinner atmosphere.[585]

He thinks the reason for this seems to be in how our moon came into being. Ross describes, based on what is known and assumed of how moons form, how a body at least the size of Mars hit the earth and was absorbed into earth's core. (Mars is nine times the size of our moon) A cloud of debris arose from this collision and eventually coalesced to form our moon. This thinned earth's atmosphere and provided the right chemical composition to permit light to reach earth's surface.

> In summary, this amazing collision, for which we have an abundance of circumstantial evidence, appears to have been perfectly timed and designed to transform Earth from a "formless and empty" place into a site where life could survive and thrive. In fact, the number of conditions that must be fine-tuned—and the degree of fine-tuning needed for each of these conditions—for life to possibly survive that is manifested in a single event argues powerfully on its own for a divine Creator.[586]

In chapter 13 we discussed where Harold Booher presented more background on this theory of how the moon came to exist.

Distinguishing Day and Night

> With sunlight now penetrating the interplanetary medium and Earth's atmosphere, an observer on the planet's surface could detect, for the first time, the cycle of day and night. Not until this time, though Earth had been rotating since its beginning, do "day" and "night" become discernible (Genesis 1:5)...
>
> The much-thinned atmosphere would generate greater temperature modulations so that life would not face a single unrelenting temperature. Air and ground temperatures would now vary smoothly and continuously from daytime highs to nighttime lows.[587]

He says light from the sun would now penetrate to allow for the development of plant life.

3. Day Two: Water Cycle Begins

The Second Day

> Then God said, "Let there be an expanse in the midst of the waters, and let it separate the waters from the waters." (1:6)
>
> And God made the expanse, and separated the waters which were below the expanse from the waters which were above the expanse; and it was so. (1:7)
>
> And God called the expanse heaven. And there was evening and there was morning, a second day. (1:8)

Ross describes a stable water cycle that is needed for advanced life. "For advanced life to be well supported by an abundance of lower life-forms and a diversity of species, the rainfall must vary from one geographical area to another. The range must fall between about 2 inches and 600 inches per year—not just for a few millennia, but for a few billion years."[588]

He explains that the sun burns its fuel at an unusually constant and stable rate, compared to other stars. But it gradually increases in brightness. This increase is suited to development on earth of life-forms that increase from primitive to more complex. "If the sun's luminosity and Earth's biomass and biodiversity fall out of sync by even a slight amount, the result would be either a runaway greenhouse (heating) effect, or a runaway freeze."[589]

He says the Hebrew of Genesis 1:7 implies God designed and built earth's atmosphere. This illustrates that God continued to "fine-tune" earth.

Appendix

4. Day Three

> Then God said, "Let the waters below the heavens be gathered into one place, and let the dry land appear"; and it was so. (1:9)

> And God called the dry land earth, and the gathering of the waters He called seas; and God saw that it was good. (1:10)

> With light coming through the still permanently overcast sky, with day distinguishable from night, with poisonous atmospheric gases almost gone and protective gases building up, and with a gentle water cycle established, the stage is set for the introduction of land life. All Earth needs is a place to put it, and that is what God arranged on Day Three.[590]

Then he notes that

> plate tectonics and volcanic activity cause the wrinkling of a planet's surface. Compared with other planets, Earth experiences an extremely high level of both kinds of activity. These levels not only maintain the conditions necessary for life support but also lead eventually to the emergence of land masses above sea level, masses large enough and high enough to resist Earth's powerful erosion forces and remain above sea level for very long periods of time.[591]

Ross adds that "our calculations show that Earth's ratio of continents to oceans and the placements of the continents allow for the greatest possible biomass of advanced species of life."

LATER ON DAY THREE: PRODUCTION OF PLANTS

> Then God said, "Let the earth sprout vegetation, plants yielding seed, and fruit trees bearing fruit after their kind, with seed in them, on the earth," and it was so. (1:11)

> And the earth brought forth vegetation, plants yielding seed after their kind, and trees bearing fruit, with seed in them, after their kind; and it was good. (1:12)

> And there was evening and there was morning, a third day. (1:13)

He explains that the Hebrew words in Genesis 1:11 can refer to any plant and any means of reproduction. It can refer to the relatively primitive plants that scientists say were the first vegetation on land. It could have been some natural process or a miracle.

> The text does not say that all land vegetation appeared at this time, but emphasizes, rather, that God chose this time for dry land to abound with vegetation

> ... this particular text cannot be used to state definitely to what extent God may or may not have employed natural processes in the development of plant life on Earth This is an issue science is attempting to comprehend.[592]

The Miracle of Life

> Plants appear to have a limited capacity for speciation (production of new species) through natural processes. Botanists have actually observed a few plants develop new "species." However, in these cases the word "species" may be a misnomer. Boundaries between plant species are much less distinct than boundaries between animal species. No plant species radically different from already existing species has arisen under human observation. The rapid rate of which plant species go extinct, both today and during the fossil era, implies that supernatural rather than natural processes are responsible for the major changes in plant species evident throughout plant history.[593]

5. Day Four: Lights in the Sky

> Then God said, "Let there be lights in the expanse of the heavens to separate the day from the night, and let them be for signs, and for seasons, and for days and years; and let them be for lights in the expanse of the heavens to give light on the earth," and it was so. (1:14–15)

> And God made the two great lights, the greater light to govern the day, and the lesser light to govern the night. He made the stars also. (1:16)

> And God placed them in the expanse of the heavens to give light on the earth, and to govern the day and the night, and to separate the light from the darkness; and God saw that it was good. (1:17–18)

> And there was evening and there was morning, a fourth day. (1:19)

Ross explains that a situation similar to heavy overcast would cover the sky for many million years. Then volcanic activity would subside, cooling and a reduction in humidity would lead to the heavy overcast of clouds to weaken. Where clouds disappeared the sky would go from translucent to transparent.

Ross believes that Earth's rotation rate gradually began to slow. This would decrease wind speeds and quiet the oceans, and this would thin the cloud cover.

> The driving force behind Earth's level of plate tectonics and volcanism is the release of heat from slow-decaying radioisotopes in the crust. The spectacular

collision that resulted in the formation of our moon helps explain how Earth acquired some of its enormous abundance of long-lasting heat-releasing isotopes.[594]

He says this heat is strategic for life. The heat released diminishes over time, slowing tectonics and volcanic activity. This slowing favors life.

> Advanced species show the least tolerance for earthquakes and volcanoes; primitive species, the greatest tolerance. We can reasonably surmise that God created primitive life on Earth at the first opportune moment and created human beings near the last (and, thus, the most) opportune moment. He exquisitely designed the sun, Earth, and moon to maximize the proliferation and duration of life on Earth. In so doing, God endowed humans with abundant biological resources (for example, several feet of topsoil, a trillion barrels of oil, hundreds of billions of tons of coal, quadrillions of cubic feet of natural gas, billions of tons of limestone and marble, millions of diverse species of life, and so forth).[595]

He adds that it was on the fourth day of creation that for the first time the sun, the moon, and the stars became distinctly visible from Earth's surface.

He explains that the word "made" [*asa*] in Genesis 1:16 is a verb-form showing an action already completed. (Verse 16 says, "God made two great lights: the greater light to govern the day, and the lesser light to govern the night. He also made the stars.") The verse tells us why God made them but doesn't say when they were made. Ross thinks the Hebrew text suggests the sun was made on the first creation day, and perhaps the moon was also.

Ross explains how the ozone layer enveloping the earth formed. Plants consumed carbon dioxide and expelled oxygen for many thousands and perhaps millions of years. The ozone layer absorbs ultraviolet radiation from the sun and protects advanced life. He adds that, "unless the stratospheric ozone layer is just right, neither too thin nor too thick, Earth's biomass, biodiversity, and biovitality will be impaired."[596]

> The existence and integration of the rare and perfect conditions necessary for the development of our delicate, life-essential ozone shields (nowhere else in the solar system or beyond can we find any) constitutes a set of miracles. (I use the word *miracle* here to mean something in the natural realm manifesting supernatural design and occurring with supernatural timing and placement).[597]

In referring to days 5 and 6, Ross explains that "with light and breathable air, dry land and oceans, an abundance and a diversity of plants, a water cycle and ozone shields to help them flourish, and discernable sky lights to govern biological clocks, Earth is ready for a new wonder: advanced animal life."[598]

Appendix

6. Day Five: Lower Vertebrates

> Then God said, "Let the waters abound with an abundance of living creatures, and let birds fly above the earth across the face of the firmament of the heavens." (1:20)

Ross says that the Hebrew nouns in this case include mollusks, crustaceans, fish, and amphibians.

Ross says some land animal species were introduced on the fifth day. He calls them *nephesh* or "soulish" animals.

> So God created great sea creatures and every living thing that moves, with which the waters abounded, according to their kind, and every winged bird according to its kind. And God saw that it was good. (1:21)

Ross explains what he thinks "nephesh" in the Old Testament means as follows:

> More than physical bodies with nervous, digestive, respiratory, circulatory, and reproductive systems, as miraculous and intricate as those systems may be, these creatures manifest attributes of mind, will, and emotions—what the ancient Hebrews and others would call "soulish" attributes.[599]

> Anyone who has had much contact with birds and mammals realizes that such creatures are uniquely endowed with the capacity to form relationships—with each other and with humans. They have unique ways of expressing their understanding, their choices, and their feelings. Unlike other animals, birds and mammals can be trained to perform tasks that are irrelevant to their survival. They respond to human authority and personality. They show delight and sadness, anger and fear, among other feelings. They form emotional bonds with humans.[600]

Ross notes that in verse 21 the word "create" is used for only the second time. The Hebrew words *bara'* and *'asa* are both used with it here. He sees a suggestion here that God manufactured (*'asa*) some aspect of the *nephesh* from existing resources while at the same time created something that did not previously exist. He thinks *'asa* refers to the construction of the body, and *bara'* refers to the creation of the soul.

> As the only creatures with the capacity to form relationships with humans, birds and mammals can be influenced (in ways that other creatures cannot) by human sin . . . Because the behavior patterns of birds and mammals can be altered by sin, at times they alone, of all Earth's animals, are designated to receive with humans the consequences of God's wrath against profound wickedness.[601]

Appendix

Creation of Sea Mammals

Ross says that some paleontologists "discredit" Genesis 1 because it places the appearance of sea animals before land animals—which appeared on day six. But Ross says that the fifth creation day mentions sea animals generically, but the sixth day mentions only "three specialized kinds of land animals." The text does not say when the other land animals are introduced.

Then Ross says that the fossils of four extinct species of whales have been found and dated as far back as fifty-two million years—which "eliminate any credible challenge to the placement of sea mammals on the fifth creation day." (Then he explains that the oldest whales drank only fresh water. Later their internal organs "mutated" to enable them to drink salt water.)

Are Transitional Forms a Proof of Evolution?

Ross says that naturalists point to the "evolution" of the whale as one proof of evolution. They point to the change from fresh water drinking whales to salt water drinking ones as an illustration. But Ross says that their "best example" of the whale is actually a poor one. Due to their nature whales adapt very slowly. The same is true of the "evolution" of horses. He says that "the many 'transitional' forms of whales and horses suggest that God performed more than just a few creative acts here and there, letting natural evolution fill in the rest. Rather, God was involved and active in creating all the whale and horse species, the first, the last, and the 'transitional' forms."[602]

7. Day Six: Specialized Land Mammals

> Then God said, "Let the earth bring forth the living creature according to its kind: cattle and creeping thing and beast of the earth, each according to its kind"; and it was so. (1:24)

> And God made the beast of the earth according to its kind, cattle according to its kind, and everything that creeps on the earth according to its kind. And God saw that it was good. (1:25)

The Hebrew words are *behema*, *remes*, and *chayya*. Although *remes* ("creeping things") occasionally refers in Hebrew literature of reptiles, Genesis 1:25 makes clear that these are mammals; they are *nephesh*. Ross says insects are not mentioned in the Genesis creation account.

> Then God said, "Let Us make man in Our image, according to Our likeness; let them have dominion over the fish of the sea, over the birds of the air, and over the cattle, over all the earth and over every creeping thing that creeps on the

earth." So God created man in His own image; in the image of God He created him; male and female He created them. (1:26–27)

In creating Adam, God used "create" for the third time. But *'asa* is also used again.

> Just as in the case of the *nephesh*, a second verb, *asa* is used. Again, something about *'adam* is completely new, and something about him is not. Because creatures with body and soul already existed, the spiritual dimension of this one creature set him apart from all others.[603]

I don't think Ross means that Adam was made from pre-existing material, but that some pattern or like object pre-existed. He speaks in his books about human-like creatures called hominids who existed long before the human race. Of course the "new" thing about humans was being created in the image of God.

Some explain this to mean that man is a triunity (body, soul and spirit) and God is a triunity. Ross thinks it means much more. Ross says the human spirit includes the following characteristics:

1. Awareness of a moral code "written" or impressed within a conscience
2. Concern about death and about life after death
3. Propensity to worship and desire to communicate with a higher being
4. Consciousness of self
5. Drive to discover and capacity to recognize truth and absolutes

These traits find expression or conscious repression in every human being, regardless of time and place and intellect. They can even be seen, to some extent, in infants and in people with severe mental or emotional impairment. These qualities help define human uniqueness, and any creature who lacks them cannot be considered *'adam*.

Human History Begins

Ross explains that the Hebrew words for father and son (*'ab* and *ben*) also refer to grandfather and grandson. *'Ab* also can refer to great-grandfather and great-great-grandfather, and so on. *Ben* can refer to great-grandchild, and so on. He gives examples from Scripture. This view is commonly shared by most Old Testament scholars, and even by many proponents of young earth creationism—who commonly state that mankind may have been created as long ago as ten thousand years or so.

Regarding the completeness of the geological records in the Old Testament, Ross remarks,

Appendix

> Comparative analysis of overlapping genealogies throughout the Bible suggest that they may range anywhere from about 90 percent complete at best to about 10 percent complete at worst. Using genealogical data alone, we can place the date for the creation of Adam and Eve very roughly between about seven thousand and about sixty thousand years ago.[604]

Ross declares that "God created the universe for the purpose of conquering all evil once and for all." He infers this from Romans 8 and Revelation 20–22.

Ross refers to Bernard Ramm in this book and seems of similar opinion to Ramm in one area—that to provide humanity with topsoil, coal, oil, ozone, oxygen, and much more "millions of generations of life would need to predate us."

> Because the physical realm changes with respect to time, God apparently created different species at different times to suit the changing environment. For instance, only the most primitive and tiny forms of life could survive the eight-hour-per-day rotation period of early Earth. Because highly advanced life requires a more delicately balanced set of characteristics for survival than primitive life, such life forms are much more vulnerable to extinction. But when such species were extinct, God created new ones, sometimes the same, and more often different (according to environmental and ecological conditions and divine timing) to replace them.[605]

Ross Refers to Man's Assignment

> Then God blessed them, and God said to them, "Be fruitful and multiply; fill the earth and subdue it; have dominion over the fish of the sea, over the birds of the air, and over every living thing that moves on the earth." (1:28)

He says here that "God clearly expects humankind to exercise wise, cautious, ecologically sound management. The Scriptures from Genesis 3 to Revelation tell the story of humanity's failure to meet God's management standards."

III. Morris's and Whitcomb's View

Henry Morris and John Whitcomb gave the following interpretation of Genesis 1, on pages 228–29, 215, and 454–73 of *The Genesis Flood*.

The First Day

> And the earth was formless and void, and darkness was over the surface of the deep: and the Spirit of God was moving over the surface of the waters. Then God said, "Let there be light"; and there was light. (1:2–3)

They say that during the first day the primitive materials of the earth, perhaps mainly the core and mantle, "were molded into physical and chemical forms suitable for habitation and use by man and other forms of life. These reactions were initiated by the introduction of light—the most basic and all-pervasive form of energy—to the surface of the earth. This light, however, was not that of the sun as presently constituted, the 'making' of which occurred only on the fourth day."[606]

They pictured the geologic activities of the first day as a great upheaval, which caused impulses of light energy to reach the surface from the depth of the earth and provide light on the earth. Molten rocks rose to the surface and caused the igneous and metamorphic rocks that are visible today.

They put Genesis 1:2 and 1:3 together and insisted that God did all this in twenty-four hours. It would be foolish to say that God *could not* have done it in twenty-four hours, because God can do *anything* (se Gen 18:14; Jer 32:17, 27; Luke 1:37). But I also think it is foolish to say that God *must* have done it in twenty-four hours. The text does not mention how long the first day was, and does not say that verse 2 was a part of the first day. This is the authors' interpretation. What need did God have to hurry? He could have begun the whole creation process earlier, as He existed from eternity, and when to "begin" was fully under His control. I suggest we acknowledge that with our human reasoning we cannot decide this issue with any finality, but acknowledge that the process may have taken much longer.

The Second Day

> Then God said, "Let there be an expanse in the midst of the waters, and let it separate the waters from the waters." (1:6)

On the second day the waters covering the earth were divided into one great reservoir above the earth (the troposphere or area where the clouds are) and another on earth. Although Morris was a hydraulic engineer, he said he did not know how this was done. However, he thought that the waters on the earth may have been in intense motion, brought about by disturbances in the earth's crust.

Morris says that disturbances on the earth's crust may have continued, which would cause the water to pound intensely, grinding and mixing the loose elements on the surface. They insisted that here also God did all of this in twenty-four hours.

The Third Day

> Then God said, "Let the waters below the heavens be gathered into one place, and let the dry land appear"; and it was so. (1:9)

Appendix

They explain that on the third day dry land appeared for the first time. The lands sank to form oceans and rose in other places to form one or more continents. Heavier materials of greater density were gathered where they forced down the earth's crust and simultaneously a lateral squeeze caused the lighter materials to go outward and upward. Volcanic action may have contributed to the forming of the continents. The result was a balancing of the heavier materials and the water above them with the greater thickness of lighter materials that formed the continents. Morris claimed that this process is basic to both geology and geophysics. He also claimed that God accomplished all this on a third day of twenty-four hours.

The fourth and fifth days of creation are not covered in detail in *The Genesis Flood*. The fourth day is mentioned only briefly on page 215 of *The Genesis Flood*. (It says, "The fourth day witnessed the establishment of the sun and moon in their functions with respect to the earth. Since the sun now provides all the energy received by the earth for its geological processes, this event also has profound geological implications."[607]) The creation of Adam on the sixth day is mentioned, but primarily in relation to the curse pronounced in Genesis 3 when Adam sinned.

IV. Young's View

Because the major contribution of *Creation and the Flood* is Davis Young's attempt to reconcile Genesis One with the geological record as he understood it in 1977, a few of his more significant comments on these verses are reported below. They can be found on pages 117–132 of *Creation and the Flood*.

The Second Day

> Then God said, "Let there be an expanse in the midst of the waters, and let it separate the waters from the waters." (1:6)

> And God made the expanse, and separated the waters which were below the expanse from the waters which were above the expanse; and it was so. (1:7)

> And God called the expanse heaven. And there was evening and there was morning, a second day. (1:8)

Young says that "God is said to make the firmament and to divide the waters. The usage of these words by no means implies the purely miraculous or rules out God's using natural laws and processes that He had already implanted into the structure of creation."[608]

> What then might the events of the second day mean in terms of earth history? Isn't Moses here giving a sweeping, general account of the formation of the

earth's atmosphere as a distinctive entity that encircles the planet, an entity that is now sharply delineated from the surface of the oceans?[609]

Young says that current scientific thought also considers "the development of the atmosphere and oceans as a very early event in the history of the earth." He in this way seeks to demonstrate in detail that scientific thought currently agrees with the *order* of events presented in Genesis One.

The Fifth Day

> Then God said, "Let the waters teem with swarms of living creatures, and let birds fly above the earth, in the open expanse of the heavens." And God created the great sea monsters, and every living creature that moves, with which the waters swarm with their kind, and every winged bird after its kind; and God saw that it was good. (1:20–21)

> The fact that many marine invertebrate animals such as corals and trilobites appear in the fossil record prior to land plants implies a contradiction between Genesis and geology. We must, however, keep in mind the incompleteness of the plant record and our lack of knowledge as to the exact limits of the categories described in verses 20–22. It is important to point out that the major groups in view here, that is, birds, most fish, swimming reptiles such as crocodiles or the extinct mosasaurs, flying reptiles like pterodactyls, seals and whales, do appear later in the fossil record than most land plants.[610]

The Sixth Day

> Then God said, "Let the earth bring forth living creatures after their kind; cattle and creeping things and beasts of the earth after their kind"; and it was so. (1:24)

> And God made the beasts of the earth after their kind, and the cattle after their kind, and everything that creeps on the ground after its kind; and God saw that it was good. (1:25)

Young believes that the term "cattle" refers to animals that would be domesticated by man. He says, of the animals created on the sixth day,

> Again these animals are said to reproduce after their kind and wide varieties are formed. These assertions seem to rule out the possibility of a general theory of evolution, although specific evolution of a type of animal, such as the horse, does not seem to be ruled out. We need not review the difficulties that

confront the Christian paleontologist in applying the facts of Genesis to the fossil record. The Christian paleontologist is always confronted by the problem of knowing exactly *what* was created on the day six and by the problem of not knowing whether the gaps in the fossil record are real or apparent.[611]

Young says day six describes chiefly the formation of higher animals, mainly the mammals. The evidence available from the fossil record likewise indicates that mammals are a very late development. He emphasizes that again there seems to be a general agreement between Genesis and geology.

In this way Young is seeking to reconcile a modern scientific view of historical geology with a Bible account that he considers historical and without factual error. He interprets Genesis One as reporting a *process* of creation of *indeterminate length* that *in its order* is consistent with scientific discoveries which suggest the order that living things appeared on earth.

V. Collin's View: Hints from the Hebrew

> You will find that many writers call Genesis 1:1—2:3 a *cosmogony*, meaning a story about how the universe came to be (*cosmo-* for the cosmos, *-gony* for the origin). I would say that this description is only partly true, and really misses the point. The cosmology part gets taken care of in verse 1 ("in the beginning God created the heavens and the earth"), and then the narrative moves on to its main point, the making and preparing earth as a place for humans to live.[612]

He notes that many readers of the Bible in English have mentioned a problem in that Genesis 1:3 says, "Let there be light," and verse 14—speaking of the fourth day—says, "Let there be lights," or, "Let there be light-bearers." But Collins says this causes no problem in the Hebrew because the term "let there be" is used in phrases like "may the Lord be with you" that do not imply that God wasn't with them before. The Hebrew does not mean the light-bearers didn't exist before, but can mean God "worked out" something that existed before. (This view is somewhat different from that of Hugh Ross, but results in the same general conclusion.)

Collins thinks "let there be light" in 1:3 means God summoned the "dawn" of the first day, and the fourth day "involves God appointing the heavenly lights to mark the set times for worship on man's calendar." In other words, "We don't have to suppose that there was some other source of light than the sun before the first day, when Genesis says nothing of the sort."[613] We don't have to assume a cloud cover that cleared to allow the light of the sun, moon and stars to appear. And we don't have to assume the days are not in sequence—that day four cannot come after day three when plants begin to grow.

He emphasizes all that Genesis 2:5–25 says clearly happened on the sixth day. He translates verses 4 to 7 as follows:

> This is the history of the heavens and the earth when they were created, in the day that the Lord God made the earth and the heavens, (2:4)
>
> before any plant of the field was in the land and before any herb of the field had grown. For the Lord God had not caused it to rain on the *land*, and there was no man to till the ground; (2:5)
>
> but a mist went up from the earth and watered the whole face of the ground. (2:6)
>
> And the Lord God formed man of the dust of the ground, and breathed into his nostrils the breath of life; and man became a living being. (2:7)

Collins explains that the Hebrew in verse 5 he translated as "land" can also mean the earth as a whole, as it is often translated. But it can also mean "some particular land." If so translated, he says we get a picture of some particular land in some particular year, at the time the rainy season began. He says this helps to clarify the time of year when God created Adam. This is because in Palestine it doesn't rain in the summer, and the autumn rains bring a burst of plant growth.

His translation also shows that verses 5 through 25 are a clarification of the chapter 1 account of the sixth day. He points out this means we should not confuse anything in 2:5–6 with what happened on the third day. Also, it implies the cycle of seasons has gone on for some time, which indicates the creation week could not have been an ordinary week.

This long period of time is suggested because of the climate in the Middle East, where long rainless summer days are followed by rains and a burst of plant growth in autumn. He says, "The only way I can make any sense out of this ordinary providence explanation that the Bible itself gives is if I imagine that the cycle of rain, plant growth, and dry season had been going on for some number of years before this point—because the text says nothing about God not yet having made the plants."[614] By "ordinary providence" he means the way we see nature work.

Collins explains in detail why it is clear from Genesis 1 and 2 that at least some of the six days of creation were not "ordinary days," but were much longer periods of time. He agrees with many others that the things Moses listed on happening on the sixth day simply could not have been accomplished in one twenty-four-hour day. Then he points out that the Hebrew of Genesis 2:23 is best read, "And Adam said: 'This is *at last* bone of my bones and flesh of my flesh; she shall be called Woman, because she was taken out of Man.'" This suggests a very long period of time between Adam's creation and that of Eve.

Endnotes

Introduction

1. Schaeffer, *How Should We Then Live*, 145.
2. Strobel, *Case for a Creator*, 16, citing Johnson, *Darwin on Trial*, 126–27.
3. Ibid., 19–20.
4. Ibid., 21.
5. Ibid., 22.
6. Ibid., 23.
7. Ibid.
8. Ibid.
9. Ibid., 24.
10. Ibid.
11. Ibid.
12. Ibid., 25.
13. Ibid.
14. Ibid., 27.
15. Ibid.
16. Ibid., 28.
17. Ibid., 29.
18. Ibid., 28.

Preface to Part 1

19. Meyer, "Demarcation of Science and Religion," 18.
20. Charles Townes, Associated Press report, *Corvallis (Oregon) Gazette-Times*, March 12, 2005.
21. Meyer, "Qualified Agreement," 140–41.
22. Ibid., 145.

Chapter 1

23. Meyer, "Demarcation of Science and Religion," 21.
24. Alfred North Whitehead, quoted in Meyer, "Qualified Agreement," 127.
25. Schaeffer, *How Should We Then Live*, 19.
26. Ibid., 145.
27. Ibid., 132.
28. Whitehead, *Science and the Modern World*, Harvard University Lowell Lectures (1925), quoted in Schaeffer, *How Should We Then Live?*, 132.

29. Meyer, "Demarcation of Science and Religion," 18.

Chapter 2

30. Hummel, *Galileo Connection*, 14.
31. Ibid., 15.
32. Frair and Patterson, "Creationism," 23.
33. Hummel, *Galileo Connection*, 254.
34. Galileo, "Letter to Christina".
35. Van Till in Carlson, *Science and Christianity*, 201.
36. Ibid., 202.
37. Lewontin, "Billions and Billions of Demons," 31.

Chapter 3

38. Collins, *Language of God*, 169.
39. Ross, *Creator and the Cosmos*, 71.
40. Ibid., 70.
41. Thiessen, *Introductory Lectures*, 89.
42. Ibid., 112.
43. Kenneth R. Samples, "Historic Alliance of Christianity and Science," *Reasons to Believe* newsletter, September 1, 1998.
44. Van Till, "Partnership," 198.
45. Schaeffer, *Genesis in Space and Time*, 165.
46. Meyer, "Qualified Agreement," 112.
47. Frair and Patterson, "Creationism," 46.
48. Ibid., 50.

Chapter 4

49. Denton, *Evolution*, 17.
50. Ibid., 34.
51. Ibid., 44.
52. Ibid., 53.
53. Darwin, *Origin of Species*, 243.
54. Ibid., 230.
55. Ibid., 233.
56. Ibid.
57. Ibid., 234.
58. Ibid., 239.
59. Ibid., 241.
60. Ibid., 243.
61. Bowler, "Evolution," 460.
62. Rachels, *Created from Animals*, 39.
63. Darwin, *Origin of Species*, 230.
64. Ibid., 233.
65. Ibid., 152.
66. Ibid., 165.
67. Ibid., 179.
68. Ibid., 237.
69. Ibid., 231.
70. Ibid., 235.
71. Ibid., 165.
72. Ibid., 162.
73. Ibid., 180.

74. Ibid., 242.
75. Ibid., 232.
76. Ibid.
77. Ibid., 238.
78. Darwin, *Descent*, 591
79. Bowler, "Evolution," 461.
80. Darwin, *Descent of Man*, 590.
81. Ibid.
82. Ibid.
83. Ibid., 591.
84. Ibid., 592.
85. Ibid., 593.
86. Ibid., 302.
87. Ibid., 286.
88. Darwin, *Origin of Species*, 87.
89. Ibid., 152.
90. Ibid., 239.

Chapter 5

91. Dawkins, *Selfish Gene*, 12.
92. Rachels, *Created from Animals*, 64–65.
93. Bowler, "Evolution," 464.
94. Dawkins, *Selfish Gene*, 2.
95. Rachels, *Created from Animals*, 70.
96. Ibid., 76.
97. Watson, "Adaptation," *Nature*, August 10, 1929.
98. Denton, *Evolution*, 344.
99. Ibid., 86.
100. Meyer, Qualified Agreement Response, in chap. 2 of Carlson, *Science and Christianity*, 117.
101. Denton, *Evolution*, 345.
102. Collins, *Science and Faith*, 281.
103. Meyer, "Demarcation of Science and Religion," 22.
104. Ibid., 22.

Chapter 6

105. Ross, *Creation as Science*, 85–86.
106. John A. O'Keefe, "The Theological Impact of the New Cosmology," in Robert Jastrow, *God and the Astronomers, 118* (emphasis added), in Strobel, page 191.
107. Ross, *Creation and Time*, 12.
108. Ibid., 11–12.
109. Ross, *Creator and the Cosmos*, 27.
110. Ross, *Creation and Time*, 77.
111. Ibid., 129.
112. Ibid., 132.
113. Ross, *Creator and the Cosmos*, 115.
114. Ross, *Creation and Time*, 137.
115. Ross, *Creator and the Cosmos*, 135.
116. Ross, *Creation as Science*, 93.
117. Ibid., 94.
118. Ross, *Creator and the Cosmos*, 128.
119. Ibid., 137.
120. Ross, *Creation and Time*, 141.

121. Ross, *Genesis Question*, 55.
122. Carroll, "Cosmic Origin of Time's Arrow," *Scientific American*, June 2008, 48.
123. Ibid., 51–52.
124. Clifton and Ferreira, "Does Dark Energy Really Exist?," *Scientific American*, April 2009.
125. Ross, *Creation as Science*, 45–46.
126. Ibid., 47.
127. Ibid.
128. Ibid., 48.
129. Ibid., 41.
130. John Gribbin, "Oscilating Universe Bounces Back," *Nature* 259 (1976) 15–16, cited in Ross, *Creation as Science*, 50.
131. Arno Penzias, in *Cosmos, Bios, Theos*, edited by Morgenau and Varghese (La Salle, IL: Open Court, 1992), 83, cited in Ross, *Creation as Science*, 51.
132. Stephen W. Hawking, *Brief History of Time: From the Big Bang to Black Holes* (New York: Bantam, 1988), 127, cited in Ross, *Creation as Science*, 51.
133. Ross, *Creation as Science*, 51.
134. Ibid.
135. Ibid., 89.
136. Ibid., 94.
137. Ibid., 124.

Chapter 7

138. Behe, *Darwin's Black Box*, 26.
139. Ibid., 183.
140. Denton, *Evolution*, 15.
141. Ibid., 70.
142. Ibid., 86.
143. Ibid., 74–75.
144. Ibid., 194.
145. Ibid., 345.
146. Ibid., 324.
147. Ibid., 327.
148. Ibid., 73.
149. Denton, "Anti-Darwinian Intellectual Journey," 175.
150. Ibid., 176.
151. Gould and Eldredge, in Denton, *Crisis*, 347.
152. Denton, *Crisis*, 358–59.
153. Ibid., 358.

Chapter 8

154. Behe, "Catholic Scientist," 135.
155. Behe, *Darwin's Black Box*, 4.
156. Ibid., 4–5.
157. Ibid., 187.
158. Ibid., 175–76.
159. Ibid., 69.
160. Ibid., 72.
161. Ibid., 139.
162. Behe, *Edge of Evolution*, 201.
163. Ibid., 191.
164. Behe et al., *Science and Evidence*, 142.
165. Ibid., 138.

166. Ibid., 144.
167. Strobel, *Case for a Creator*, 36.
168. Ibid., 54–55.

Chapter 9

169. Oldroyd, "Theories of the Earth," 391.
170. Young, *Creation and the Flood*, 8.
171. Young, *Christianity and the Age of the Earth*, 142.
172. Ibid., 143.
173. Ibid.
174. Klotz, *Genes, Genesis, and Evolution*, 3.
175. Ibid., 12.
176. Ibid., 20.
177. Ibid., 9.
178. Ibid., 16.
179. Ibid., 548.
180. Ibid., 115.
181. Ibid., 389.
182. Darwin, *Origin*, 243.
183. Ibid., 165.
184. Ibid., 152.
185. Strobel, *Case for a Creator*, 64.
186. Ibid., 238.
187. Ibid., 57.

Chapter 10

188. Wilson, "Historiography of Science and Religion," 9.
189. Numbers, "Creationism since 1859," 316.
190. Whitcomb and Morris, *Genesis Flood*, xx.
191. Ibid., 203.
192. Ibid., 89.
193. Ibid., 97.
194. Morris, *Biblical Cosmology*, 106.
195. Ibid., 110.
196. Morris and Morris, *Science, Scripture*, 35.
197. Ibid., 40.
198. Ibid., 51.
199. Parker, *Creation*, 138.
200. Ibid., 141.
201. Ibid., 143.
202. Ibid., 144.
203. Ibid., 169.
204. Ibid., 169–70.
205. Ibid., 172.
206. Denton, *Evolution*, 347.
207. *Time*, April 10, 2006.
208. Parker, *Creation*, 175.
209. Ibid., 176–77.
210. Ibid., 180–81.
211. Ibid., 185.
212. Ibid., 187.
213. Ibid., 191.

214. Ibid., 197–98.
215. Ibid., 205–7.
216. Morris and Morris, *Science, Scripture*, 35.
217. Parker, *Creation*, 202.
218. Ibid., 91.
219. Ibid., 108.
220. Ibid., 114.
221. Ibid.
222. Ross, *Creation and Time*, 96–97.
223. Young, *Christianity and the Age of the Earth*, 91.

Chapter 11

224. Darwin, *Descent of Man*, 392.
225. Young, *Creation and the Flood*, 144.
226. Strobel, *Case for a Creator*, 25.
227. Darwin, *Descent of Man*, 591.
228. Rachels, *Created from Animals*, 173.
229. Zimmer, "Great Mysteries of Human Evolution," *Discover*, September 2003, 34–43.
230. Ibid., 37.
231. Ibid., 40.
232. Ibid., 42.
233. Ibid.
234. Ibid., 34.
235. Pollard, "What Makes Us Human?," *Scientific American*, May 2009, 44.
236. Wells, in Strobel, *Case for a Creator*, 54–55.
237. Klotz, *Genes, Genesis, and Evolution*, 388.
238. Ibid., 389.
239. Ibid., 384.
240. Young, *Creation and the Flood*, 154.
241. Strobel, 62. Here Strobel, in his interview with Jonathan Wells, is quoting him as quoting Henry Gee in *In Search of Deep Time, Beyond the Fossil Record to a New History of Life* (New York: Free Press, 1999), 23.
242. Pinker, *Blank Slate*, 372.
243. Ibid., 372–73.
244. Ibid., 25.
245. Rachels, *Created from Animals*, 4.
246. Ibid., 1.
247. Ibid., 174.
248. Davis and Collins, "Scientific Naturalism," 206.
249. Collins, *Language of God*, 268.

Chapter 12

250. Dembski, *Design Revolution*, 65.
251. Johnson, "Intelligent Design in Biology," *Think* (Royal Institute of Philosophy), February 19, 2007.
252. Behe, *Darwin's Black Box*, 196.
253. Rachels, *Created from Animals*, 10.
254. Darwin, *Origin of Species*, 87.
255. Behe, *Darwin's Black Box*, 213.
256. Denton, *Evolution*, 340.
257. Ibid., 335.
258. Ibid., 339.

259. Ibid., 341.
260. Behe, *Darwin's Black Box*, 202.
261. Ibid., 203.
262. Ibid., 222.
263. Ibid., 223.
264. Ibid., 228.
265. Miller, *Finding Darwin's God*, 164.
266. Behe, *Edge of Evolution*, 204–5.
267. Ibid., 214.
268. Ibid., 216.
269. Ibid.
270. Ibid., 217.
271. Ibid., 233.
272. Ibid., 235.
273. Ibid., 81.
274. Ibid., 74.
275. Denton, *Evolution*, 358.
276. Ibid., 74.
277. Ibid., 357.
278. Behe, *Darwin's Black Box*, 192.
279. Ibid., 193.
280. Ibid., 238.
281. Ibid., 242.
282. This can be found on Ibid., 243–45.
283. Ibid., 232–33.
284. Ibid., 185–86.
285. Ibid., 186.
286. Johnson, *Wedge of Truth*, 167.
287. Behe, *Darwin's Black Box*, 252.
288. Behe, *Edge of Evolution*, 235.
289. Dawkins, *Greatest Show on Earth*, 9.
290. Meyer et al., "Cambrian Explosion," 353.
291. Dawkins, *Greatest Show on Earth*, 147.
292. Meyer, *Signature in the Cell*, 85.
293. These can be found on ibid., 324–26.
294. Ibid., 154.
295. Ibid., 109.
296. Ibid., 330.
297. Ibid., 341.
298. Ibid., 343.
299. Ibid., 347.
300. Ibid., 435.
301. Ibid., 410.
302. The above 6 points are from Meyer, *Signature in the Cell*, 402–15.

Chapter 13

303. Booher, *Origins, Icons, and Illusions*, xi.
304. Ibid., xiv.
305. Ibid., xiii.
306. Ibid., xv.
307. Ibid., xiii.
308. Ibid., xv–xvi.
309. Ibid., xii.

310. Ibid., 190.
311. Ibid., 205.
312. Ibid., 207.
313. Ibid., 211.
314. Ibid., 212.
315. Ibid., 222.
316. Ibid., 21.
317. Ibid., 22.
318. Ibid., 41.
319. Ibid., 171.
320. Ibid., 173.
321. Ibid., 189.
322. Ibid., 138.
323. Ibid., 143.
324. Ibid., 141.
325. Ibid., 144.
326. Ibid., 151.
327. Ross, *Creation as Science*, 85–86.
328. Booher, *Origins, Icons, and Illusions*, 154.
329. Ibid., 166–167.
330. Ibid., 281.
331. Ibid., 271.
332. Ibid., 263.
333. Ibid.
334. Ibid., 266.
335. Ibid., 269.
336. Ibid., 57.
337. Ibid., 59.
338. Ibid., 63.
339. Ibid., 65.
340. Ibid., 112–13.
341. Ibid., 119.
342. Ibid., 127.
343. Ibid., 74.
344. Ibid., 85.
345. Ibid., 224.
346. Ibid., 226.
347. Ibid., 293.
348. Ibid., 294.
349. Ibid., 294–295.
350. Ibid., 297.
351. Ibid., 298.
352. Ibid., 299.
353. Ibid., 303.
354. Ibid., 387.
355. Ibid., xi.

Chapter 14

356. Hawking, *Teacher's Guide to Stephen Hawking's Universe*, 9.
357. Rumsfeld, quoted in Wimbush, *Theorizing Scriptures*, 62.
358. Darwin, *Origin of Species*, 87.
359. Behe, *Edge of Evolution*, 7.
360. Darwin, *Origin of Species*, 165.

361. Meyer et al., "Cambrian Explosion," 353.
362. Darwin, *Descent of Man*, 591.
363. Ibid., 592.
364. Ibid., 597.
365. Kornfield, "Early-Date Genesis Man," *Christianity Today*, June 8, 1973, 42.
366. Strobel, *Case for a Creator*, 54–55.

Preface to Part 3

367. Ross, *Creation and Time*, 11–12.
368. Ramm, *Christian View of Science and Scripture*, 173, quoting W. F. Albright, "Old Testament and Archaeology," in *Old Testament Commentary*, edited by Alleman and Flack, 135.

Chapter 15

369. Ramm, *Christian View of Science and Scripture*, 17–18.
370. Ibid., 22.
371. Ibid.
372. Ibid., 136.
373. Ibid.
374. These 3 points are found on pp. 104–5.
375. Ibid., 108. Ramm is quoting *Aquinas in Summa*, I.8,3.
376. Ibid., 65.
377. Ibid., 66–67.
378. Ibid., 69.
379. Ibid., 70. Ramm is quoting W. B. Dawson, *Bible Confirmed by Science*, 32–33.
380. Ibid., 70.
381. Ibid., 77.
382. Ibid., 82.
383. Ibid., 84.
384. Ibid.
385. Ibid., 85.
386. Ibid.
387. Ibid., 86.
388. Ibid., 92.
389. Ibid., 95.
390. Ibid., 349–50.

Chapter 16

391. Rimmer, *Modern Science and the Genesis Record*, 32–34.
392. *Scofield Reference Bible*, 3.
393. Ibid.
394. Rimmer, *Modern Science and the Genesis Record*, 12.
395. Ibid., 13.
396. Ibid., 11–12.
397. Ibid., 102.
398. Ibid., 105.
399. Ibid., 198.
400. Ibid., 25.
401. Ibid., 353.

Chapter 17

402. Ramm, *Christian View of Science and Scripture*, 227.
403. Ibid., 347–48.

404. Ibid., 117.
405. Ibid., 115–16.
406. Ibid., 116.
407. Ibid., 215.
408. Ibid., 227.
409. Ibid., 227–28.
410. Ibid., 272.
411. Ibid., 219.
412. Ibid., 220.
413. Ibid., 223.
414. Ibid., 226.
415. Ibid.
416. Ibid., 209.

Chapter 18

417. Whitcomb and Morris, *Genesis Flood*, 440.
418. Ibid., 118.
419. Morris, *Biblical Cosmology*, 33.
420. Whitcomb and Morris, *Genesis Flood*, 212.
421. Ibid., 214.
422. Ibid.
423. Ibid., 215.
424. Ibid.
425. Ibid., 216.
426. Ibid., 219.
427. Ibid., 233.
428. Ibid., 489.
429. Ibid., 393.
430. Ibid., 215.
431. Ibid., 455.
432. Ibid., 454.
433. Ibid., 457.
434. Whitcomb and Morris, *Genesis Flood*, 459.
435. Morris, *Biblical Cosmology*, 26.
436. Whitcomb and Morris, *Genesis Flood*, 489.
437. Ibid., 393.

Chapter 19

438. Young, *Christianity and the Age of the Earth*, 10.
439. Young, *Creation and the Flood*, 14–15.
440. Ibid., 15.
441. Ibid., 21.
442. Ibid., 87.
443. Ibid., 82.
444. Ibid., 113.
445. Ibid.
446. Ibid., 115.
447. Ibid., 101.
448. Ibid., 116.
449. Ibid., 132.
450. Ibid., 80.
451. Ibid., 114.

Chapter 20

452. Ross, *Genesis Question*, 19.
453. Ibid.
454. Ross, *Creation and Time*, 154.
455. Ross, *Genesis Question*, 20.
456. Ross, *Creation and Time*, 149.
457. Ibid., 149–50.
458. Ibid. The above 10 points are from *Creation and Time*, 152–53.
459. Ibid., 151.
460. Ibid., 153.
461. Ibid., 45.
462. Ibid., 49.
463. Ibid., 154.
464. Ibid., 141.
465. Ibid.
466. Ibid.
467. Ibid., 154.
468. Ibid., 101.
469. Ibid., 102.

Chapter 21

470. Carlson, *Science and Christianity*, 13.
471. Ibid., 17.
472. Frair and Patterson, "Creationism," 20.
473. Ibid., 46–47.
474. Ibid., 23.
475. Ibid., 31.
476. Ibid., 41.
477. Ibid., 48.
478. Ibid., 46.
479. Pond, "Independence," 81.
480. Ibid., 103.
481. Van Till, "Partnership," 233.
482. Behe, *Edge of Evolution*, 229.
483. Meyer, "Demarcation of Science and Religion," 18.
484. Meyer, in Carlson, *Science and Christianity*, 57.
485. Meyer, "Qualified Agreement," 140–41.
486. Ibid., 144.
487. Ibid., 153.
488. Ibid., 158.
489. Ibid., 171–72.
490. Ibid., 174.
491. Waltke, "Literary Genre," 2.
492. Ibid., 9.
493. Frair and Patterson, 39.

Chapter 22

494. Collins, *Science and Faith*, 339.
495. Ibid., 38.
496. Ibid., 334–335.
497. Ibid., 40.
498. Ibid., 42.

499. Ibid., 47.
500. Ibid., 230.
501. Ibid., 242.
502. Ibid., 248.
503. Ibid.
504. Ibid., 264.
505. Ibid., 265.
506. Ibid., 267.
507. Haldane, *Possible Worlds*, 28, in Collins, *Science and Faith*, 281.
508. Collins, *Science and Faith*, 105.
509. Ibid., 64.
510. Ibid., 65.
511. Ibid., 70.
512. Ibid., 71.
513. Ibid., 72.
514. Ibid., 282.

Chapter 23

515. Booher, *Origins, Icons, and Illusions*, 303.
516. Ibid.
517. Ibid.
518. Ibid., 304.
519. Ibid., 305.
520. Ibid.
521. Ibid., 306.
522. Ibid.
523. Ibid., 307.
524. Ibid.
525. Ibid., 441.
526. Ibid., 308.
527. Ibid., 309.
528. Ibid.
529. Ibid., 343.
530. Ibid., 344.
531. Ibid.
532. Ibid., 347.
533. Ibid.
534. Ibid.
535. Ibid., 349.
536. Ibid., 354.
537. Ibid.
538. Ibid., 355.
539. Ibid.
540. Ibid., 361.
541. Ibid.
542. Ibid., 362.
543. Ibid.
544. Ibid., 363.

Chapter 24

545. Whitcomb and Morris, *Genesis Flood*, 457.
546. Ross, *Creation and Time*, 154.

547. Schaeffer, *Genesis in Space and Time*, 35.

Chapter 25

548. Collins, *Language of God*, 23.
549. Here Collins is quoting Penzias, quoted in M. Browne, "Clues to the Universe's Origin Expected," *New York Times*, March 12, 1978.
550. Collins, *Language of God*, 25.
551. Ibid., 27
552. Ibid., 80–81.
553. Ibid., 3.
554. Ibid., 30.
555. Ibid., 123.
556. Ibid., 187.
557. Behe, *Edge of Evolution*, 7.
558. Collins, *Language of God*, 189.
559. Ibid., 193.
560. Ibid., 95.
561. Ibid., 94–95.
562. Ibid., 94.
563. Ibid., 133–34.
564. Ibid., 138.
565. Ibid., 141.
566. Ibid., 126.
567. Ibid., 207.
568. Ibid., 209.
569. Ibid., 136–37.
570. Strobel, *Case for a Creator*, 54.
571. Collins, *Language of God*, 140.
572. Sy Garte, "Stochastic Grace," *BioLogos Forum*, July 1, 2013, http://biologos.org/blog/stochastic-grace.
573. Collins, *Language of God*, 210.
574. Ibid., 207–8.

Chapter 26

575. Schaeffer, *How Should We Then Live*, 145.

Chapter 27

576. Booher, *Origins, Icons, and Illusions*, 303.
577. Thiessen, *Introductory Lectures*, 89.

Chapter 30

578. Collins, *Language of God*, 30.

Appendix

579. Ross, *Genesis Question*, 30.
580. Ibid., 27.
581. Ibid., 29.
582. Ibid.
583. Ibid., 31.
584. Ibid.
585. Ibid.

586. Ibid., 32–33.
587. Ibid., 33.
588. Ibid., 34.
589. Ibid., 35.
590. Ibid., 37.
591. Ibid.
592. Ibid., 39.
593. Ibid., 42.
594. Ibid., 43.
595. Ibid., 43–44.
596. Ibid., 45.
597. Ibid., 46.
598. Ibid., 47.
599. Ibid., 49.
600. Ibid.
601. Ibid., 50.
602. Ibid., 51–52.
603. Ibid., 53.
604. Ibid., 54.
605. Ibid., 56.
606. Whitcomb and Morris, *Genesis Flood*, 228.
607. Ibid., 454–73.
608. Young, *Creation and the Flood*, 121.
609. Ibid., 124.
610. Ibid., 130.
611. Ibid., 131–32.
612. Collins, *Science and Faith*, 63.
613. Ibid., 90–91.
614. Ibid., 88.

Bibliography

Behe, Michael J. "A Catholic Scientist Looks at Darwinism." Chapter 8 of *Uncommon Dissent: Intellectuals Who Find Darwinism Unconvincing*, edited by William A. Dembski. Wilmington, DE: ISI, 2004.

———. *Darwin's Black Box: The Biochemical Challenge to Evolution*. New York: Free Press, 1996.

———. *The Edge of Evolution: The Search for the Limits of Darwinism*. New York: Free Press, 2007.

Behe, Michael J., et al. *Science and Evidence for Design in the Universe*. Papers presented at a conference sponsored by the Wethersfield Institute, New York City, September 25, 1999. San Francisco: Ignatius, 2002.

Booher, Harold R. *Origins, Icons, and Illusions: Exploring the Science and Psychology of Creation and Evolution*. St. Louis: Green, 1998.

Bowler, Peter J. "Evolution." Chapter 85 of *The History of Science and Religion in the Western Tradition*, edited by Gary B. Ferngren. 458–465. New York: Garland, 2000.

Carlson, Richard F., ed. *Science and Christianity: Four Views*. Downers Grove: InterVarsity, 2000.

Carroll, Sean M. "Cosmic Origin of Time's Arrow." *Scientific American*, June 2008.

Clifton, Timothy, and Pedro Ferreira. "Does Dark Energy Really Exist?" *Scientific American*, April 2009.

Collins, C. John. *Science and Faith: Friends or Foes?* Wheaton, IL: Crossway, 2003.

Collins, Francis S. *The Language of God: A Scientist Presents Evidence for Belief*. New York: Free Press, 2006.

Darwin, Charles. *The Descent of Man and Selection in Relation to Sex*. In *Great Books of the Western World*, vol. 49, *Darwin*, edited by Robert M. Hutchins and Mortimer J. Adler. 1–243. Chicago: Encyclopaedia Britannica, 1952.

———. *The Origin of Species by Means of Natural Selection*. In *Great Books of the Western World*, vol. 49, *Darwin*, edited by Robert M. Hutchins and Mortimer J. Adler. 253–597. Chicago: Encyclopaedia Britannica, 1952.

Davis, Edward B., and Robin Collins. "Scientific Naturalism." Chapter 39 of *The History of Science and Religion in the Western Tradition*, edited by Gary B. Ferngren. 201–207. New York: Garland, 2000.

Bibliography

Dawkins, Richard. *The Greatest Show on Earth: The Evidence for Evolution*. New York: Free Press, 2009.

———. *The Selfish Gene*. New York: Oxford University Press, 1989.

Dembski, William A. *The Design Revolution: Answering the Toughest Questions about Intelligent Design*. Downers Grove: InterVarsity, 2004.

———. "The Design Argument", Chapter 10 of *The History of Science and Religion in the Western Tradition*, edited by Gary B. Ferngren. 63–67, New York: Garland, 2000.

———. *Intelligent Design: The Bridge between Science and Theology*. Downers Grove: InterVarsity, 1999.

———, ed. *Uncommon Dissent: Intellectuals Who Find Darwinism Unconvincing*. Wilmington, DE: ISI, 2004.

Denton, Michael J. "An Anti-Darwinian Intellectual Journey." Chapter 9 of *Uncommon Dissent*, edited by William A. Dembski. 153–176. Wilmington, DE: ISI, 2004.

———. *Evolution: A Theory in Crisis*. Bethesda, MD: Adler & Adler, 1986.

———. *Nature's Destiny: How the Laws of Biology Reveal Purpose in the Universe*. New York: Free Press, 1998.

Ferngren, Gary B., ed. *The History of Science and Religion in the Western Tradition: An Encyclopedia*. New York: Garland, 2000.

Frair, Wayne, and Gary D. Patterson. "Creationism: An Inerrant Bible and Effective Science." Chapter 1 of *Science and Christianity: Four Views*, edited by Richard F. Carlson. 19–51. Downers Grove: InterVarsity, 2000.

Galileo Galilei. Letter to Grand Duchess Christina of Tuscany. 1615. In Hummel, *The Galileo Connection*.

Hawking, Stephen. *A Brief History of Time: From the Big Bang to Black Holes*, New York: Bantam Press, 1988.

Hawking, Stephen. *Teacher's Guide to Stephen Hawking's Universe*.

Hummel, Charles E. *The Galileo Connection: Resolving Conflicts between Science and the Bible*. Downers Grove: InterVarsity, 1986.

Johnson, Phillip E. "Intelligent Design in Biology." *Think* (Royal Institute of Philosophy), February 19, 2007.

———. *The Wedge of Truth: Splitting the Foundations of Naturalism*. Downers Grove: InterVarsity, 2000.

Klotz, John W. *Genes, Genesis, and Evolution*. 2nd ed. St. Louis: Concordia, 1959.

Kornfield, William. "Early-Date Genesis Man." *Christianity Today*, June 8, 1973.

Lewontin, Richard. "Billions and Billions of Demons." Review of *The Demon-Haunted World*, by Carl Sagan. *New York Review of Books*, January 9, 1997.

Mayr, Ernst. "The Growth of Biological Thought", 438–439. Cambridge, Mass: Harvard University Press, 1982.

Meyer, Stephen C. "The Demarcation of Science and Religion." Chapter 3 of *The History of Science and Religion in the Western Tradition*, edited by Gary B. Ferngren. 17–23. New York: Garland, 2000.

———. "Qualified Agreement." Chapter 3 of *Science and Christianity: Four Views*, edited by Richard F. Carlson. 127–174. Downers Grove: InterVarsity, 2000.

———. *Signature in the Cell: DNA and the Evidence for Intelligent Design*. New York: HarperOne, 2009.

Meyer, Stephen C., et al. "The Cambrian Explosion: Biology's Big Ban." In *Darwinism, Design, and Public Education*, edited by John A. Campbell and Stephen C. Meyer, 323–402. East Lansing: Michigan State University Press, 2003.

Miller, Kenneth R. *Finding Darwin's God: A Scientist's Search for Common Ground between God and Evolution*. New York: HarperCollins, 1999.

Morris, Henry M. *Biblical Cosmology and Modern Science*. Nutley, NJ: Craig, 1970.

Morris, Henry M., and John D. Morris. *Science, Scripture, and the Young Earth: An Answer to Current Arguments against the Biblical Doctrine of Recent Creation*. El Cajon, CA: Institute for Creation Research, 1989.

Numbers, Ronald L. "Creationism since 1859." Chapter 56 of *The History of Science and Religion in the Western Tradition*, edited by Gary B. Ferngren. 313–319. New York: Garland, 2000.

Oldroyd, David R. "Theories of the Earth and Its Age before Darwin." Chapter 73 of *The History of Science and Religion in the Western Tradition*, edited by Gary B. Ferngren. 391–396. New York: Garland, 2000.

Parker, Gary. *Creation: Facts of Life*. Green Forest, AR: Master, 1994.

Pinker, Stephen. *The Blank Slate: The Modern Denial of Human Nature*. New York: Viking, 2002.

Pollard, Katherine S. "What Makes Us Human?" *Scientific American*, May 2009.

Pond, Jean. "Independence: Mutual Humility in the Relationship between Science and Christian Theology." Chapter 2 in *Science and Christianity: Four Views*, edited by Richard F. Carlson. 67–104. Downers Grove: InterVarsity, 2000.

Rachels, James. *Created from Animals: The Moral Implications of Darwinism*. New York: Oxford University Press, 1990.

Ramm, Bernard. *The Christian View of Science and Scripture*. Grand Rapids: Eerdmans, 1956.

Rimmer, Harry. *Modern Science and the Genesis Record*. 7th ed. Berne, IN: Berne Witness, 1939.

Ross, Hugh. *Creation and Time: A Biblical and Scientific Perspective on the Creation-Date Controversy*. Colorado Springs: NavPress, 1994.

———. *Creation as Science: A Testable Model Approach to End the Creation/Science Wars*. Colorado Springs: NavPress, 2006.

———. *The Creator and the Cosmos: How the Greatest Scientific Discoveries of the Century Reveal God*. Colorado Springs: NavPress, 1994.

———. *The Genesis Question: Scientific Advances and the Accuracy of Genesis*. Colorado Springs: NavPress, 1998.

———. *A Matter of Days: Resolving a Creation Controversy*. Colorado Springs: NavPress, 2004.

———. *More than a Theory: Revealing a Testable Model for Creation*. Grand Rapids: Baker, 2009.

———. *Why the Universe Is the Way It Is*. Grand Rapids: Baker, 2008.

Schaeffer, Francis A. *Genesis in Space and Time: The Flow of Biblical History*. Downers Grove: InterVarsity, 1972.

———. *How Should We Then Live: The Rise and Decline of Western Thought and Culture*. Old Tappan, NJ: Revell, 1976.

The Scofield Reference Bible. Edited by C. I. Scofield. New York: Oxford University Press, 1917.

Bibliography

Strobel, Lee. *The Case for a Creator: A Journalist Investigates Scientific Evidence That Points toward God.* Grand Rapids: Zondervan, 2004.

Thiessen, Henry C. *Introductory Lectures in Systematic Theology.* Grand Rapids: Eerdmans, 1952.

Van Till, Howard J. "Partnership." Chapter 4 of *Science and Christianity: Four Views*, edited by Richard F. Carlson. 195–234. Downers Grove: InterVarsity, 2000.

Watson, D.M.S., "Adaptation," 231–234. Nature Magazine, August 10, 1929.

Waltke, Bruce. "The Literary Genre of Genesis, Chapter One." *Crux Vol. XXVII, No. 4* (Regent College) 27 (1991) 2–10.

Whitcomb, John C., Jr., and Henry M. Morris. *The Genesis Flood: The Biblical Record and Its Scientific Implications.* Grand Rapids: Baker, 1961.

Wilson, David B. "The Historiography of Science and Religion." Chapter 1 of *The History of Science and Religion in the Western Tradition*, edited by Gary B. Ferngren. 3–11. New York: Garland, 2000.

Wimbush, Vincent L. *Theorizing Scriptures: New Critical Orientations to a Cultural Phenomenon.* New Brunswick, NJ: Rutgers University Press, 2008.

Young, Davis A. *The Biblical Flood: A Case Study of the Church's Response to Extrabiblical Evidence.* Grand Rapids: Eerdmans, 1995.

———. *Christianity and the Age of the Earth.* Grand Rapids: Zondervan, 1982.

———. *Creation and the Flood: An Alternative to Flood Geology and Theistic Evolution.* Grand Rapids: Baker, 1977.

Zimmer, Carl. "Great Mysteries of Human Evolution." *Discover*, September 2003.

Seeking the Truth by Scientific Discovery:

I. Natural Science:

A. Biology:

Age of the Human Race: 27, 64, 99, 101, 133, 135–136, 138–140, 143, 234, 242, 243, 285, 288–289, 308–309, 312, 380, 383, 389
Common ancestor/Common descent: 14, 40, 54, 55, 85, 88, 90–91, 104, 116, 117, 124, 125, 134, 136, 143, 153, 158, 171, 209–210, 293, 325–330
Darwin's Theory: 15, 20, 38–43, 45, 50–52, 55, 59, 66, 69, 71–86, 88–93, 118, 123, 125–126, 142, 156–157, 162, 171, 175, 181, 192–194, 201, 205, 208, 250, 256, 322–323, 326, 387.
Descent of Man: ix, 23, 36, 38–47, 54, 85, 88, 90, 132–133, 141, 171, 209–210, 268
Devolution of Humans: 98, 139–140, 232 .
Effects of Mutations: 42, 54, 42, 64, 85, 87–90, 94, 104, 124, 125, 128, 131, 142, 151, 153, 154, 155, 171, 172, 180, 185, 196, 199, 202, 205, 206, 284, 291, 314, 319, 323, 331
Genetics: 50, 52, 72, 73, 95, 101, 118, 125, 141, 156, 170, 176, 185, 206, 325.
Human DNA: 300, 321–323, 325–330, 333, 347
Human Evolution: 133–138, 329, 333.
Homology: 90, 210.
Human Genome: 136, 320–330, 332, 333, 347
Hominids: 64, 97–98, 133–134, 39,143, 193, 260, 309, 329–330, 383.
Icons of Evolution: xi-xii, 75, 88–90, 91, 100–101, 205. 324.
Intelligent Design: 18, 83, 88, 118, 146–164, 166–175, 195–197, 201, 206–207, 209–210, 212, 263, 264, 270–271, 275–277, 290, 294, 304, 313, 322–324, 331, 342, 355.
Irreducible complexity: 84, 86–87, 161–162, 170, 322, 324,
Micro-evolution: ix, 71–79, 81, 85, 88, 90–91, 97–98, 101–102, 119, 124, 140, 161, 164, 171, 180–181,196, 199, 207, 211, 262,292, 294, 314, 379
Natural selection: xiii, 37, 39–44, 46, 47–55, 77, 81–88, 91, 99, 101, 104, 107, 109,116–117, 123- 125, 131, 135, 142, 149, 151, 153–155, 162, 170- 171, 176, 180, 202, 205–206, 211, 284, 291, 299–302, 314–315, 319, 323, 331, 355.
Nature and Development of the Human Race: vii, viii, xiii, 3, 7–8, 9, 13, 23, 31, 46–48, 51, 55, 64, 67, 75, 85, 90, 98, 104, 113, 125, 131–145, 147, 148, 150, 152, 157, 161–163, 192–193, 197, 209, 211, 227, 232, 240, 242, 243, 244, 254, 258–260, 263, 269, 274, 275–277, 280, 282, 286, 289, 291–293, 296, 302, 306, 309, 310, 311, 312, 315, 317, 320–322, 325–332, 343–344, 345, 346–350, 358, 360–362, 365–366, 369, 371–372, 374, 380. 383, 384, 388.

Seeking the Truth by Scientific Discovery:

Neo-Darwinism: 49–51, 55, 88, 90, 131, 136, 147, 169, 180, 185, 284, 289
Origin of the Human Race (See also "Common ancestor"): 3, 7, 13, 45, 47, 64, 98, 116, 131–144, 161, 162, 192–193, 200, 204, 209, 232, 234, 235, 282, 284, 317, 322, 326, 329, 374, 375.
Origin of intelligence in the cell: 160, 166–169, 178, 312
The Origin of Species: ix, xi, xiii, 38–45, 49, 54, 57, 71, 76–79, 8, 92, 103–104, 118, 163, 174, 208
Pathway: 88, 91, 162, 171, 206, 269, 323
Presuppositions of evolutionism: 20 21, , 51–52
Problem of Imperfections in nature: 151–152, 197, 226, 237, 277, 296, 297, 302, 308.
Random mutations: 73, 81, 85, 87, 88, 91, 125, 132, 153, 155, 171, 196, 206, 284, 314, 319, 331
Spontaneous generation: xi, 36, 40, 48, 55, 74. 87. 89, 146, 176–178, 205, 278, 342.
Theistic evolution: 192, 208, 232, 235, 249, 250, 260, 294, 295, 298, 322, 331, 332, 408
Were There Prehumans?: ix, xi, 14, 64, 69, 309.
    Cro-Magnon: 139–140
    Java Man: 140–141.
    Lucy: 134, 138, 193, 200.
    Neanderthal: 64, 134, 135, 137–138, 139–140, 144
    Piltdown Man: 193

B. Cosmology:

Age of the Earth: 63, 69, 92–37, 108, 111, 113, 126, 212, 230, 246, 251, 282–285, 306–307
Age of the Universe (Cosmos): 28, 61–63, 262–264, 272, 283, 315,
Anthropic Principle: 59, 61, 62, 282, 320
Big Bang: 4, 27, 57, 59, 60, 65, 66, 68, 69, 70, 99, 100, 127, 159, 181, 182, 183, 184, 197, 198, 205, 215, 216, 256, 265, 268, 269, 271, 272, 273, 275, 277, 282, 289, 310, 311, 314, 315, 320, 321, 393, 406.
Cosmological Time: 12, 16, 23, 52, 58, 60, 61, 62–63, 65–66, 69, 127, 184–185, 194, 200, 203, 205, 212, 226, 234, 235, 255, 256, 272, 285, 312, 314, 359
Cosmology: 4, 27, 56–70, 79, 104, 108, 127, 130, 181, 183–184, 198–199, 205, 216, 245, 256, 264, 265, 271–272, 282, 283, 299, 302, 310, 312, 314, 375, 388.
Creationism: 4, 7, 19, 22, 25–27, 30–31, 35, 45, 47, 58, 64, 69–70, 76–78, 92, 93–94, 96, 100, 107, 109, 113, 128, 131, 140, 142, 143, 148, 152, 161, 162, 184, 216–217, 218, 223, 225, 227–228, 229–230, 233, 236–238, 239–247, 248–254, 255–260, 265, 267, 274–22, 279–289, 291, 292–293, 297, 302, 310, 311, 312, 313, 315–317, 326–333, 345–356, 360, 362, 373–389.
Einstein, Albert: vii, 4, 16, 21, 57, 65–66, 69, 130, 158, 194, 256, 272, 303
Entropy: 59, 194, 200, 244–245, 292, 308–309
Geological Time: 12, 16, 23, 37, 52, 54, 58, 60, 61, 62–63, 65–66, 69, 76, 94, 98, 102, 103,107, 108, 113, 119, 120–121, 122–123, 126, 128, 129, 141, 191–192, 195, 199, 208, 231, 243, 246, 250, 252, 259, 262, 287–289, 307–308, 312, 389.
Just-right Universe: 58–59, 60–64. 166.
Laws of Thermodynamics: 21, 130, 194, 200, 292, 293
Natural Science: vii, ix, 15, 36, 49, 53, 55, 58, 95, 103, 161, 226, 263, 273–274, 276–277, 279, 312
Origin of the Human Race: 3, 7, 13, 45, 47, 64, 98, 116, 131–144, 161, 162, 192–193, 200, 204, 209, 232, 234, 235, 282, 284, 317, 322, 326, 329, 374, 375.
Origin of the Universe: 7, 12–13, 62, 68, 96, 113, 128, 173, 198, 229–230, 241–242, 276, 282, 294, 300, 342

C. Historical Geology:

Anthropology: 93, 138, 209, 222, 233, 260, 282.
Cambrian fossils: 41, 43, 98 - 102, 115, 116, 129, 157, 167, 189, 207, 324, 331
Catastrophism: 107–113, 187–189, 199, 244, 250
Dating methods: 101, 128, 192, 208, 283

Flood Geology: 104, 105, 107, 113, 126, 128, 408
Stephen Gould, who openly struggled with the fossil evidence against evolution: ix, x, 20, 51, 89, 100, 102, 117, 119, 124, 164–166, 171, 185, 186, 190, 200, 204, 207–208, 267, 275, 299, 302, 314, 324–325
Grand Canyon: 104, 114, 120, 122–123, 126, 129, 188, 191, 194, 199, 283
Mount St. Helens: 112, 114, 121–123, 129, 188, 199
Old Earth Creationism 248–254, 255–262, 265–267, 269–270, 272–273, 274–277, 279–289, 307, 310–311, 315, 319–333, 375–384, 386–389.
Origin of life on earth: 3, 12–13, 28, 30–31, 36, 38–45, 63, 72, 100, 123, 125, 175, 176, 272–274
Uniformitarianism: 95, 106, 108–113, 126, 189, 248, 283
Young Earth creationism: 261, 292, 308–310, 384–386.

D. Philosophical (or Other) Issues in Science:

Absolutes: x, xv, 3, 11, 18, 20, 21, 30, 39, 48, 78, 83, 96, 101, 118, 142, 191–192, 195, 206, 211, 244, 282, 285, 312, 339, 340, 355, 358, 383.
Faith commitments of science: 3, 4, 15, 17,19, 27
Limits of Science: 17, 85, 129,130, 141, 145, 148, 155, 158, 162, 163, 166, 195, 197–198, 200, 203, 204, 216, 217, 223, 242, 275–276, 281, 288, 298, 301, 303, 310–13, 315, 322, 340.
Naturalism: 15, 16, 18, 19, 20, 21, 56, 68, 70, 72, 132, 141, 144, 163, 174, 195, 196, 198, 215, 240, 268, 271, 273, 277, 281, 301, 302, 303, 304, 313, 321, 324, 340, 366, 395, 405
Presuppositions of Modern Science:
xii-xiv, 3, 7, 8, 11–12, 14–21, 27–29, 46, 51, 52–55, 58, 76, 82. 85, 93–96, 146, 158–163, 167, 172, 175, 179, 186, 190, 193–196, 201, 208, 211–212, : 217, 221, 298, 288, 300.

II. Social Science (Anthropology):

Conscience: 252, 297, 300, 301, 320, 347, 383.
Free Will: x, 14, 293, 296, 331, 355.
Heredity and environment on personality 141–142
Human soul: xiii, 18, 46, 142, 260, 293, 310, 326, 329, 330, 331, 332, 333, 347, 350, 351, 352, 360, 362, 371, 374, 375, 382–383.
Man, nature of : xii, xiii, 8, 13, 14, 29, 31, 45–48, 51, 64. 67. 75–76, 90, 98, 104, 113, 124–125, 131–144, 147–150, 152, 154, 161–163, 178, 192–194, 196–197, 200, 203, 204, 208–211, 215—216, 222, 226, 227, 229, 230, 232, 234–240, 243–244, 249, 253, 255–256, 258–260, 263, 269, 275–276, 280, 286–288, 293, 296, 307–310, 315, 317, 321, 326–329, 332, 343, 345–352, 360–366, 372, 374, 382, 388–389.
Moral nature: 142; 161, 317, 320–321, 332, 333, 347–348, 372, 374, (Updated 12–18)
Origin of the Human Race: 3, 7, 13, 45, 47, 64, 98, 116, 131–144, 161, 162, 192–193, 200, 204, 209, 232, 234, 235, 282, 284, 317, 322, 326, 329, 374, 375.
Personality -- 141–142, 209, 317, 347, 362, 365.
Social science. ix, 9, 10, 15, 35, 46, 47, 50, 51, 55, 73. 132, 141, 152, 161, 173, 174–175, 176, 204, 209, 300, 317, 347, 349, 362, 374.

III. Characteristics of the Scientific Method:

Definition of science: 16, 96–97, 154, 195
Limits of science: 17–18, 57, 63, 85–87, 129–130, 155, 158–159, 195, 211, 203, 340.
Philosophies of science: 51, 54, 66, 68,95, 96, 103, 110, 130, 132, 139, 148,154, 157, 159–160, 166–168, 174–177, 195–196, 198, 212, 217–218, 221–223, 227, 239, 240, 264, 268, 270–271,281, 290, 298–300,303–304,306. 313, 340..
Presuppositions of science: 9 -13, 17, 20, 21, , 51, 126.
Testing scientific theories: 4, 12, 65, 68, 78, 158, 198, 340.

Seeking the Truth by Revelation:

IV. Beyond Science—God and the Supernatural:

A. The Act of God in Creation: 20, 23, 51, 67, 68, 128, 217, 237, 249, 250, 261, 295, 298, 299, 302, 340, 357

Bible teaching about creation: 4, 7, 19, 22, 25–27, 30–31, 35, 45, 47, 58, 64, 69–70, 76–78, 92, 93–94, 96, 100, 107, 109, 113, 128, 131, 140, 142, 143, 148, 152, 161, 162, 184, 216–217, 218, 223, 225, 227–228, 229–230, 233, 236–238, 239–247, 248–254, 255–260, 265, 267, 274-22, 279–289, 291, 292–293, 297, 302, 310, 311, 312, 313, 315–317, 326–333, 345–356, 360, 362, 373–389.

B. The Bible as Truth: xv, 3–5, 7–8.11–12, 23–25, 26–31,35, 58–60, 68–69, 72, 96–97, 101, 105, 128, 215–216, 231, 237–240, 247, 248, 249, 250, 252, 254, 261, 264–266, 268, 274–276, 278, 279–291, 298, 304, 305, 309, 311, 312–313, 315, 321, 326–329, 332, 337, 339–346, 353, 355–356, 359–361, 365, 367, 371–372.

Biblical Interpretation: 15, 25–26, 30, 31, 58, 105, 215, 217, 223–226, 257, 258, 262, 265, 267, 275, 313, 316–318.
Claims the Bible Makes About Itself:
    It Claims to Be Authoritative: 24–25
    It Claims to Be a Revelation from God: 20, 23–24, 25, 26, 30, 69, 224, 239, 312, 341, 359
    It Claims to Be Without Error: 96, 332
Internal consistency of the Bible: 24, 152, 243, 312, 316, 327, 343, 362, 368–370.

C. The Nature of God:

Essential Nature of God: 7–8, 12, 18, 19, 23, 24, 59,60, 62, 174, 201, 205 255, ,259, 264,273, 293, 303, 304, 307, 319, 321, 345, 350–351, 356, 362, 370, 371, 383, 385.
Non-moral attributes: (omnipresence, omniscience, omnipotence, immutabilty. personal): 3,12, 23, 25, 26,30, 49, 96, 131, 163, 194, 215–216, 226–227, 235 280, 293, 294,296, 297, 332,342–343, 345–347, 349, 353, 355–356, 360, 362, 385,
Moral attributes: (holiness, righteousness and justice, goodness): 132, 161, 215, 280, 295–296, 306, 317, 321, 325, 327, 345–349, 351, 360–363, 368, 374.

Seeking the Truth by Revelation:

Relationship to Humanity: 8, 19, 25, 96, 131, 260, 261, 268, 274, 293, 306, 327, 331, 343, 345-346, 347, 348-350, 351, 352, 356, 360, 352, 363.

God is Trustworthy: 26, 215, 280, 343, 353.371.

D. The Nature of Man:

Fall of mankind: 104, 209, 215, 226, 229, 243-244, 246, 292, 296, 297, 302, 307-308, 327-328, 347, 348, 360, 361, 365-266, 371-372.

Man in the Image of God: xiii, 8, 64, 131, 142, 145, 152, 215, 249, 260, 280, 284, 293, 306, 310, 317, 321, 333, 346, 347, 349-350, 355, 360-362, 371, 374, 382-383.

V. How This Knowledge can Improve Our Lives.

A. Worldview:

1. Biblical Worldview: iii, 4, 12, 15, 23, 25-28, 30, 31, 37, 64, 76-77, 103-130, 131, 138, 142, 152, 181, 184, 216, 222 223-227, 234-333, especially 337-372, 373-389.
2. Examples of Some Who Changed Their Worldviews:
    The author, vi, vii, vii, viii, ix, 356-357.
    Collins, Francis, 320-322.
    Darwin, 35, 37.
    Moses, 356
    Paul, 356-357
    Ross, 58, 356
    Strobel, x-xiv,355, 357
    Strobel's Wife Leslie xiii, xiv. 351
3. Worldview in General: x-xiv, 9-13, 14-15, 17, 19, 22. 27, 39, 130, 132, 159, 173-174, 176, 179, 186, 191, 215, 221, 264, 270-274, 276, 300, 302-303, 312, 316, 321, 332-333, 340-341, 355-372.

B. Life's Questions:

Life's Questions: viii, ix, xv, 8, 9, 280, 316, 340, 345-352, 358.

C. More Effective Living: viii, 5, 304, 337, 345-352.

Living Life to the Fullest: viii, 29, 219, 337, 355-372.

www.ingramcontent.com/pod-product-compliance
Lightning Source LLC
Chambersburg PA
CBHW081146290426
44108CB00018B/2460